VOLUME FIVE HUNDRED AND THIRTY SEVEN

METHODS IN ENZYMOLOGY

Methods of Adipose Tissue Biology, Part A

METHODS IN ENZYMOLOGY

Editors-in-Chief

JOHN N. ABELSON and MELVIN I. SIMON
Division of Biology
California Institute of Technology
Pasadena, California

ANNA MARIE PYLE
Departments of Molecular, Cellular and Developmental
Biology and Department of Chemistry
Investigator, Howard Hughes Medical Institute
Yale University

GREGORY L. VERDINE
Department of Chemistry and Chemical Biology
Harvard University

Founding Editors

SIDNEY P. COLOWICK and NATHAN O. KAPLAN

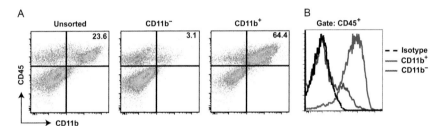

Kae Won Cho et al., Figure 16.2 Enrichment of CD11b$^+$ SVCs using magnetic beads. SVCs were isolated as described in Section 3.1 and CD11b$^+$ cells were positively selected using Miltenyi microbeads as described in Section 3.5. Unsorted and sorted cells were analyzed by flow cytometry. (A) Scatterplots showing CD45 versus CD11b expression on unsorted (left), CD11b-depleted (CD11b$^-$; middle), and CD11b-enriched (CD11b$^+$; right) SVCs. The percentage of total cells for the CD45$^+$CD11b$^+$ population is indicated. (B) Histogram showing enrichment efficiency of CD11b$^+$ SVCs following positive selection.

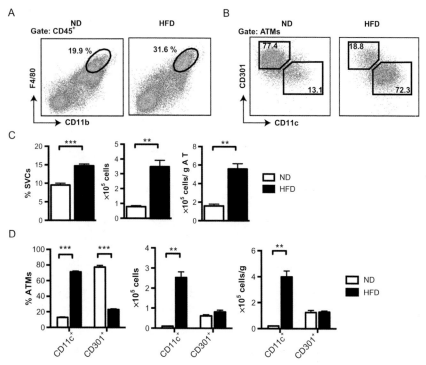

Kae Won Cho *et al.*, **Figure 16.3** High-fat diet increases ATM content and alters the distribution of M1 and M2 ATMs in visceral fat. C57BL/6J mice were fed a normal diet (ND) or high-fat diet (HFD; 60 kcal from fat) for 12 weeks before stromal vascular cells (SVCs) were isolated from epididymal adipose tissue and examined by flow cytometry as described in these protocols. (A) Representative scatterplots of SVCs from ND (left) and HFD (right) mice stained with CD11b and F4/80. Cells are gated on viable $CD45^+$ SVCs as shown in Fig. 16.1A. (B) Representative scatterplots showing gates for M1 ($CD11c^+CD301^-$) and M2 ($CD11c^-CD301^+$) ATMs. (C) ATM content in epididymal adipose tissue shown as percentage of total SVCs, absolute number per fat pad, and normalized to visceral adipose tissue mass. (D) Distribution of ATM subsets in visceral fat from lean and obese mice shown as percentage of total ATMs, absolute number per fat pad, and normalized to adipose tissue mass. $^{**}P<0.01$, $^{***}<0.001$ versus ND.

VOLUME FIVE HUNDRED AND THIRTY SEVEN

METHODS IN ENZYMOLOGY

Methods of Adipose Tissue Biology, Part A

Edited by

ORMOND A. MACDOUGALD

Department of Molecular and Integrative Physiology; Division of Metabolism, Endocrinology and Diabetes, Department of Internal Medicine, School of Medicine, University of Michigan, Ann Arbor, Michigan, USA

AMSTERDAM • BOSTON • HEIDELBERG • LONDON
NEW YORK • OXFORD • PARIS • SAN DIEGO
SAN FRANCISCO • SINGAPORE • SYDNEY • TOKYO

Academic Press is an imprint of Elsevier

Academic Press is an imprint of Elsevier
525 B Street, Suite 1800, San Diego, CA 92101-4495, USA
225 Wyman Street, Waltham, MA 02451, USA
Radarweg 29, PO Box 211, 1000 AE Amsterdam, The Netherlands
The Boulevard, Langford Lane, Kidlington, Oxford, OX5 1GB, UK
32 Jamestown Road, London NW1 7BY, UK

First edition 2014

Copyright © 2014, Elsevier Inc. All Rights Reserved.

No part of this publication may be reproduced, stored in a retrieval system or transmitted in any form or by any means electronic, mechanical, photocopying, recording or otherwise without the prior written permission of the publisher

Permissions may be sought directly from Elsevier's Science & Technology Rights Department in Oxford, UK: phone (+44) (0) 1865 843830; fax (+44) (0) 1865 853333; email: permissions@elsevier.com. Alternatively you can submit your request online by visiting the Elsevier web site at http://elsevier.com/locate/permissions, and selecting *Obtaining permission to use Elsevier material*

Notice
No responsibility is assumed by the publisher for any injury and/or damage to persons or property as a matter of products liability, negligence or otherwise, or from any use or operation of any methods, products, instructions or ideas contained in the material herein. Because of rapid advances in the medical sciences, in particular, independent verification of diagnoses and drug dosages should be made

> For information on all Academic Press publications
> visit our website at store.elsevier.com

ISBN: 978-0-12-411619-1
ISSN: 0076-6879

Printed and bound in United States of America
14 15 16 17   11 10 9 8 7 6 5 4 3 2 1

CONTENTS

Contributors   *xi*
Preface   *xvii*

1. Adipocyte-Specific Transgenic and Knockout Models   1
Sona Kang, Xingxing Kong, and Evan D. Rosen

  1. Introduction   1
  2. Loss-of-Function Strategies in Adipocytes *In Vivo*   3
  3. Gain-of-Function Strategies in Adipocytes *In Vivo*   9
  References   13

2. Imaging White Adipose Tissue with Confocal Microscopy   17
Gabriel Martinez-Santibañez, Kae Won Cho, and Carey N. Lumeng

  1. Introduction   18
  2. Materials for Imaging Adipose Tissue   24
  3. Methods   25
  4. Summary   28
  Acknowledgments   28
  References   28

3. Isolation and Study of Adipocyte Precursors   31
Christopher D. Church, Ryan Berry, and Matthew S. Rodeheffer

  1. Introduction   32
  2. Adipose Tissue Depots and Cell Populations   33
  3. Digestion of Whole WAT for Isolation of SVF   33
  4. Flow Cytometry and FACS   36
  5. Primary Adipocyte Precursor Cell Culture and Differentiation   41
  6. Digestion of Whole WAT for Isolation of Mature Adipocytes   43
  Acknowledgments   45
  References   45

4. Imaging of Adipose Tissue   47
Ryan Berry, Christopher D. Church, Martin T. Gericke, Elise Jeffery, Laura Colman, and Matthew S. Rodeheffer

  1. Introduction   48
  2. Imaging of Whole Mounted Adipose Tissue   48

3.	Imaging of Sectioned Adipose Tissue	53
	Acknowledgments	72
	References	72

5. Adipose Tissue Angiogenesis Assay — 75

Raziel Rojas-Rodriguez, Olga Gealekman, Maxwell E. Kruse, Brittany Rosenthal, Kishore Rao, SoYun Min, Karl D. Bellve, Lawrence M. Lifshitz, and Silvia Corvera

1.	Introduction	76
2.	Materials	77
3.	Methods	79
4.	Method Limitations	90
	Acknowledgment	90
	References	91

6. Quantifying Size and Number of Adipocytes in Adipose Tissue — 93

Sebastian D. Parlee, Stephen I. Lentz, Hiroyuki Mori, and Ormond A. MacDougald

1.	Introduction	94
2.	Technical Aspects	96
3.	Future Challenges and Conclusions	114
	Acknowledgments	119
	References	119

7. Use of Osmium Tetroxide Staining with Microcomputerized Tomography to Visualize and Quantify Bone Marrow Adipose Tissue *In Vivo* — 123

Erica L. Scheller, Nancy Troiano, Joshua N. VanHoutan, Mary A. Bouxsein, Jackie A. Fretz, Yougen Xi, Tracy Nelson, Griffin Katz, Ryan Berry, Christopher D. Church, Casey R. Doucette, Matthew S. Rodeheffer, Ormond A. MacDougald, Clifford J. Rosen, and Mark C. Horowitz

1.	Introduction	124
2.	Materials	129
3.	Methods	131
4.	Summary	137
	Acknowledgments	137
	References	137

8. **Brown Adipose Tissue in Humans: Detection and Functional Analysis Using PET (Positron Emission Tomography), MRI (Magnetic Resonance Imaging), and DECT (Dual Energy Computed Tomography)**   141

Magnus Borga, Kirsi A. Virtanen, Thobias Romu, Olof Dahlqvist Leinhard, Anders Persson, Pirjo Nuutila, and Sven Enerbäck

    1. Introduction   142
    2. Positron Emission Tomography   143
    3. Magnetic Resonance Imaging   147
    4. Dual Energy Computed Tomography   150
    References   156

9. **Analyzing the Functions and Structure of the Human Lipodystrophy Protein Seipin**   161

M.F. Michelle Sim, Md Mesbah Uddin Talukder, Rowena J. Dennis, J. Michael Edwardson, and Justin J. Rochford

    1. Introduction   162
    2. Technical Aspects   163
    3. Discussion   173
    References   174

10. **Differentiation of Human Pluripotent Stem Cells into Highly Functional Classical Brown Adipocytes**   177

Miwako Nishio and Kumiko Saeki

    1. Introduction   178
    2. Experimental Components and Considerations   179
    3. Notes with Troubleshooting   195
    References   197

11. **Analysis and Measurement of the Sympathetic and Sensory Innervation of White and Brown Adipose Tissue**   199

Cheryl H. Vaughan, Eleen Zarebidaki, J. Christopher Ehlen, and Timothy J. Bartness

    1. Introduction   200
    2. Surgical Denervation of WAT and BAT Nerves   202
    3. Chemical Denervation of Adipose Tissue Using 6-Hydroxy-Dopamine   207
    4. Local Sensory Denervation of Adipose Tissue Using Capsaicin   209
    5. Assessment of Sympathetic Denervation Using NETO or Content as Measured by HPLC-EC   211

6.	Assessment of Sensory Denervation Using CGRP ELIA	218
7.	Expression of Data	220
	Acknowledgments	220
	References	221

12. Measurement and Manipulation of Human Adipose Tissue Blood Flow Using Xenon Washout Technique and Adipose Tissue Microinfusion — 227

Richard Sotornik and Jean-Luc Ardilouze

1.	Introduction	228
2.	AT ^{133}Xenon Washout—Principle of the Method	229
3.	AT Microinfusion—Principle of the Method	230
4.	Materials	231
5.	Procedure	232
6.	Calculations	236
7.	Other Techniques Used in ATBF Measurement	237
8.	Conclusion	239
	Acknowledgments	239
	References	240

13. Isolation and Quantitation of Adiponectin Higher Order Complexes — 243

Joseph M. Rutkowski and Philipp E. Scherer

1.	Introduction	244
2.	Sample Collection and Preparation	246
3.	Gel Fractionation by FPLC	247
4.	Western Blot Analysis	251
5.	Complex Distribution Quantitation and Presentation	254
6.	Comparison to Other Techniques	256
7.	Concluding Remarks	258
	Acknowledgments	258
	References	258

14. Genome-Wide Profiling of Transcription Factor Binding and Epigenetic Marks in Adipocytes by ChIP-seq — 261

Ronni Nielsen and Susanne Mandrup

1.	Introduction	261
2.	Technical Aspects	262
3.	Future Challenges	277

Acknowledgments	277
References	277

15. Analysis and Isolation of Adipocytes by Flow Cytometry   281

Susan M. Majka, Heidi L. Miller, Karen M. Helm, Alistaire S. Acosta, Christine R. Childs, Raymond Kong, and Dwight J. Klemm

1.	Introduction	282
2.	Preparation of Adipocytes by Collagenase Digestion of Adipose Tissue	286
3.	Staining of Single Cell Suspensions from Adipose Tissue for Flow Cytometry Analysis	287
4.	Analysis and Sorting of Adipocytes by Flow Cytometry	291
5.	Summary	294
	Acknowledgment	295
	References	295

16. Flow Cytometry Analyses of Adipose Tissue Macrophages   297

Kae Won Cho, David L. Morris, and Carey N. Lumeng

1.	Introduction	298
2.	Materials	299
3.	Methods	302
4.	Discussion	310
	Acknowledgments	313
	References	313

Author Index	*315*
Subject Index	*335*

CONTRIBUTORS

Alistaire S. Acosta
Cancer Center Flow Cytometry Core, University of Colorado Anschutz Medical Campus, Aurora, Colorado, USA

Jean-Luc Ardilouze
Division of Endocrinology, Department of Medicine, Université de Sherbrooke, Quebec, Canada

Timothy J. Bartness
Department of Biology, Neuroscience Institute and Center for Obesity Reversal, Georgia State University, Atlanta, Georgia, USA

Karl D. Bellve
Program in Molecular Medicine, University of Massachusetts Medical School, Worcester, Massachusetts, USA

Ryan Berry
Section of Comparative Medicine; Department of Molecular, Cell, and Developmental Biology, and Yale Stem Cell Center, Yale University School of Medicine, New Haven, Connecticut, USA

Magnus Borga
Center for Medical Image Science and Visualization, and Department of Biomedical Engineering, Linköping University, Linköping, Sweden

Mary A. Bouxsein
Center for Advanced Orthopedic Studies, Beth Israel Deaconess Medical Center and Harvard Medical School, Boston, Massachusetts, USA

Christine R. Childs
Cancer Center Flow Cytometry Core, University of Colorado Anschutz Medical Campus, Aurora, Colorado, USA

Kae Won Cho
Department of Pediatrics and Communicable Diseases, University of Michigan Medical School, Ann Arbor, Michigan, USA

Christopher D. Church
Section of Comparative Medicine, and Yale Stem Cell Center, Yale University School of Medicine, New Haven, Connecticut, USA

Laura Colman
Section of Comparative Medicine, Yale University School of Medicine, New Haven, Connecticut, USA

Silvia Corvera
Program in Molecular Medicine, University of Massachusetts Medical School, Worcester, Massachusetts, USA

Rowena J. Dennis
Institute of Metabolic Science, Addenbrooke's Hospital, University of Cambridge Metabolic Research Laboratories, Cambridge, United Kingdom

Casey R. Doucette
Maine Medical Center Research Institute, Scarborough, Maine, USA

J. Michael Edwardson
Department of Pharmacology, University of Cambridge, Cambridge, United Kingdom

J. Christopher Ehlen
Department of Biology, Neuroscience Institute and Center for Obesity Reversal, Georgia State University, and Department of Neurobiology, Neuroscience Institute, Morehouse School of Medicine, Atlanta, Georgia, USA

Sven Enerbäck
Department of Clinical and Medical Genetics, Institute of Biomedicine, University of Gothenburg, Gothenburg, Sweden

Jackie A. Fretz
Department of Orthopaedics and Rehabilitation, Yale University School of Medicine, New Haven, Connecticut, USA

Olga Gealekman
Program in Molecular Medicine, University of Massachusetts Medical School, Worcester, Massachusetts, USA

Martin T. Gericke
Institute of Anatomy, University of Leipzig, Leipzig, Germany

Karen M. Helm
Cancer Center Flow Cytometry Core, University of Colorado Anschutz Medical Campus, Aurora, Colorado, USA

Mark C. Horowitz
Department of Orthopaedics and Rehabilitation, Yale University School of Medicine, New Haven, Connecticut, USA

Elise Jeffery
Department of Cell Biology, Yale University School of Medicine, New Haven, Connecticut, USA

Sona Kang
Division of Endocrinology, Beth Israel Deaconess Medical Center, Boston, Massachusetts, USA

Griffin Katz
Department of Orthopaedics and Rehabilitation, Yale University School of Medicine, New Haven, Connecticut, USA

Dwight J. Klemm
Department of Medicine, University of Colorado Anschutz Medical Campus, Aurora, Colorado, USA

Raymond Kong
Amnis Corporation, Seattle, Washington, USA

Xingxing Kong
Division of Endocrinology, Beth Israel Deaconess Medical Center, Boston, Massachusetts, USA

Maxwell E. Kruse
Program in Molecular Medicine, University of Massachusetts Medical School, Worcester, Massachusetts, USA

Olof Dahlqvist Leinhard
Center for Medical Image Science and Visualization, and Department of Medical and Health Sciences, Linköping University, Linköping, Sweden

Stephen I. Lentz
Division of Metabolism, Endocrinology and Diabetes, Department of Internal Medicine, School of Medicine, University of Michigan, Ann Arbor, Michigan, USA

Lawrence M. Lifshitz
Program in Molecular Medicine, University of Massachusetts Medical School, Worcester, Massachusetts, USA

Carey N. Lumeng
Department of Pediatrics and Communicable Diseases, University of Michigan Medical School, Ann Arbor, Michigan, USA

Ormond A. MacDougald
Department of Molecular and Integrative Physiology, and Division of Metabolism, Endocrinology and Diabetes, Department of Internal Medicine, School of Medicine, University of Michigan, Ann Arbor, Michigan, USA

Susan M. Majka
Department of Medicine, Vanderbilt University, Nashville, Tennessee, USA

Susanne Mandrup
Department of Biochemistry and Molecular Biology, University of Southern Denmark, Odense, Denmark

Gabriel Martinez-Santibañez
Cellular and Molecular Biology Program, University of Michigan, Ann Arbor, Michigan, USA

Heidi L. Miller
Department of Medicine, University of Colorado Anschutz Medical Campus, Aurora, Colorado, USA

SoYun Min
Program in Molecular Medicine, University of Massachusetts Medical School, Worcester, Massachusetts, USA

Hiroyuki Mori
Department of Molecular and Integrative Physiology, School of Medicine, University of Michigan, Ann Arbor, Michigan, USA

David L. Morris
Department of Medicine, Indiana University School of Medicine, Indianapolis, Indiana, USA

Tracy Nelson
Department of Orthopaedics and Rehabilitation, Yale University School of Medicine, New Haven, Connecticut, USA

Ronni Nielsen
Department of Biochemistry and Molecular Biology, University of Southern Denmark, Odense, Denmark

Miwako Nishio
Department of Disease Control, Research Institute, National Center for Global Health and Medicine, Tokyo, Japan

Pirjo Nuutila
Turku PET Centre, University of Turku, and Department of Endocrinology, Turku University Hospital, Turku, Finland

Sebastian D. Parlee
Department of Molecular and Integrative Physiology, School of Medicine, University of Michigan, Ann Arbor, Michigan, USA

Anders Persson
Center for Medical Image Science and Visualization; Department of Medical and Health Sciences, Linköping University, and County Council of Östergötland, Linköping, Sweden

Kishore Rao
Program in Molecular Medicine, University of Massachusetts Medical School, Worcester, Massachusetts, USA

Justin J. Rochford
Rowett Institute of Nutrition and Health, Institute of Medical Sciences, University of Aberdeen, Aberdeen, United Kingdom

Matthew S. Rodeheffer
Section of Comparative Medicine; Department of Molecular, Cell, and Developmental Biology, Yale Stem Cell Center, Yale University School of Medicine, New Haven, Connecticut, USA

Raziel Rojas-Rodriguez
Program in Molecular Medicine, University of Massachusetts Medical School, Worcester, Massachusetts, USA

Thobias Romu
Center for Medical Image Science and Visualization, and Department of Biomedical Engineering, Linköping University, Linköping, Sweden

Clifford J. Rosen
Maine Medical Center Research Institute, Scarborough, Maine, USA

Evan D. Rosen
Division of Endocrinology, Beth Israel Deaconess Medical Center, and Harvard Medical School, Boston, Massachusetts, USA

Brittany Rosenthal
Program in Molecular Medicine, University of Massachusetts Medical School, Worcester, Massachusetts, USA

Joseph M. Rutkowski
Touchstone Diabetes Center, Department of Internal Medicine, UT Southwestern Medical Center, Dallas, Texas, USA

Kumiko Saeki
Department of Disease Control, Research Institute, National Center for Global Health and Medicine, Tokyo, Japan

Erica L. Scheller
Department of Molecular & Integrative Physiology, University of Michigan, Ann Arbor, Michigan, USA

Philipp E. Scherer
Touchstone Diabetes Center, Department of Internal Medicine, and Department of Cell Biology, UT Southwestern Medical Center, Dallas, Texas, USA

M.F. Michelle Sim
Institute of Metabolic Science, Addenbrooke's Hospital, University of Cambridge Metabolic Research Laboratories, Cambridge, United Kingdom

Richard Sotornik
Division of Endocrinology, Department of Medicine, Université de Sherbrooke, Quebec, Canada

Md Mesbah Uddin Talukder
Department of Pharmacology, University of Cambridge, Cambridge, United Kingdom

Nancy Troiano
Department of Orthopaedics and Rehabilitation, Yale University School of Medicine, New Haven, Connecticut, USA

Joshua N. VanHoutan
Department of Internal Medicine, Endocrinology, Yale University School of Medicine, New Haven, Connecticut, USA

Cheryl H. Vaughan
Department of Biology, Neuroscience Institute and Center for Obesity Reversal, Georgia State University, Atlanta, Georgia, USA

Kirsi A. Virtanen
Turku PET Centre, University of Turku, Turku, Finland

Yougen Xi
Department of Orthopaedics and Rehabilitation, Yale University School of Medicine, New Haven, Connecticut, USA

Eleen Zarebidaki
Department of Biology, Neuroscience Institute and Center for Obesity Reversal, Georgia State University, Atlanta, Georgia, USA

PREFACE

This book will be informative to those interested in obesity, stem cells, or the development and physiology of adipose tissues. Although white adipose tissue is often maligned due to societal pressures and the diseases associated with obesity, in reality these dynamic tissues have important but incompletely understood roles in regulation of whole-body metabolism and maintenance of health. Thus, this volume presents a wide array of state-of-the-art methods to facilitate further study of development, physiology, and pathophysiology of adipocytes in cultured cells, animal models, and humans. In addition, research on energy-consuming brown adipocytes has exploded over the past few years because of the potential for activation or expansion of brown adipose tissues in humans to help alleviate incidence of obesity. Consequently, a number of methods for visualization and investigation of brown adipose tissue are also detailed. Finally, white adipose depots are a source of readily available stem cells, whose multipotency and other properties have considerable potential to treat human diseases. Accordingly, methods for purification and study of these important precursors are also described.

The study of adipose tissues goes hand in hand with global efforts to understand and reverse the epidemic of obesity and associated medical complications. Thus, tremendous progress has recently been made in our ability to investigate aspects of white and brown adipose tissue biology. Contributors include those researchers that have made substantive contributions to our ability to explore adipose tissue biology at the biochemical, cellular, tissue, and/or organismal levels. Authors were recruited not only based on their contributions to the field but also on their ability to communicate their cutting-edge methodological advances in a cogent and unambiguous style. These investigators have documented their "lab protocol," including small but critical details for which there is often not space within a standard journal article. Where possible, they have also included general suggestions on how to optimize or modify protocols for the specific application of the reader.

The editor wants to express appreciation to the contributors for providing their contributions in a timely fashion; to the senior editors for guidance; and to the staff at Elsevier for helpful input.

Ormond A. MacDougald
Cambridge, UK
October, 2013

CHAPTER ONE

Adipocyte-Specific Transgenic and Knockout Models

Sona Kang*, Xingxing Kong*, Evan D. Rosen*,†,1
*Division of Endocrinology, Beth Israel Deaconess Medical Center, Boston, Massachusetts, USA
†Harvard Medical School, Boston, Massachusetts, USA
[1]Corresponding author: e-mail address: erosen@bidmc.harvard.edu

Contents

1. Introduction	1
1.1 Adipose tissue: Brown, white, and beige	1
1.2 Cellular heterogeneity of adipose tissue	2
1.3 Developmental stages of adipocytes	3
2. Loss-of-Function Strategies in Adipocytes *In Vivo*	3
2.1 Overall considerations	3
2.2 Knocking out genes in all fat depots	4
2.3 Knocking out genes in brown and beige fat	9
3. Gain-of-Function Strategies in Adipocytes *In Vivo*	9
3.1 Methods	11
References	13

Abstract

Adipose tissue plays a major role in metabolic homeostasis, which it coordinates through a number of local and systemic effectors. The burgeoning epidemic of metabolic disease, especially obesity and type 2 diabetes, has focused attention on the adipocyte. In this chapter, we review strategies for genetic overexpression and knockout of specific genes in adipose tissue. We also discuss these strategies in the context of different types of adipocytes, including brown, beige, and white fat cells.

1. INTRODUCTION

1.1. Adipose tissue: Brown, white, and beige

Adipose tissues play key roles in energy balance, glucose and lipid homeostasis, in addition to a variety of other functions ranging from immunity to hemostasis to angiogenesis (Rosen & Spiegelman, 2006). Because adipose tissue affects so many physiological processes, and is in turn affected by the systemic environment, there has been great interest in establishing systems for determining the molecules and pathways that operate within the adipocyte itself. One cannot assume, for example, that an animal (or person) exhibiting a disorder of adiposity, such as obesity or resistance to weight gain, has a primary defect in the adipocyte.

Traditionally, adipocytes have been divided into two types: white and brown. White adipocytes make up the bulk of mammalian adipose tissue; these cells store lipid and expand during overnutrition. Brown adipocytes, conversely, are rich in mitochondria and burn energy through activity of uncoupling protein-1 (UCP-1), which dissipates the mitochondrial proton gradient with allowing ATP synthesis, thus generating heat (Seale, Kajimura, & Spiegelman, 2009). Recently, brown adipocytes have been further divided into two groups: "classic" or interscapular brown fat, which are derived from $Pax7^+/Myf5^+$ precursor cells, and so-called "beige" or "brite" adipocytes, which are induced within white adipose depots in response to cold, sympathetic stimulation, and other stimuli, and which derive from $Pax7^-/Myf5^-$ cells (Wu, Cohen, & Spiegelman, 2013). The expression profiles of white, brown and beige adipocytes overlap extensively, which have made the identification of transgenic drivers that target one cell type over another extremely difficult (discussed in more detail below).

An additional consideration is the existence of multiple white adipose depots. These are traditionally broken down into "visceral" and "subcutaneous" fat pads, although each group is made up of many smaller depots, each of which may have its own biological properties. The reader is referred to recent reviews on adipose depot heterogeneity for more details (Lee, Wu, & Fried, 2013). Importantly for the purposes of this discussion, however, there are no molecular markers which absolutely distinguish between white depots, and thus one cannot make a true "subcutaneous fat-only" or "visceral fat-only" transgenic or knockout mouse.

1.2. Cellular heterogeneity of adipose tissue

Like most organs, adipose tissue is comprised of several different cell types, including mature adipocytes, preadipocytes, fibroblasts, endothelial cells, and a wide variety of immune cells. These cells interact with one another in very significant ways with profound consequences for local and systemic metabolic function. Many of these cell types express specific markers that can be used to evaluate their contributions to overall metabolic function (e.g., *Tie2* for endothelial cells and *Lyzs* for myeloid-derived cells), but there are no markers as yet that can be used to specifically study the populations of these cells that reside within the adipose depot. Thus, while one can knock out a gene in all endothelial cells and then study the effects on an isolated fat pad, one cannot knock out a gene *only* in intra-adipose endothelial cells. For the purposes of this chapter, we focus our attention on mature adipocytes, which comprise approximately 50% of the cellular content of a fat pad.

1.3. Developmental stages of adipocytes

Adipocytes develop from precursor cells in a process that can be split into two stages: lineage commitment and terminal differentiation. Because most of the heavily studied cellular models of adipogenesis represent already committed preadipocytes, we know much more about terminal differentiation. There has been a recent flurry of work that has defined pericyte-like cells that represent much earlier stages in adipogenesis, and this has spurred the identification of transcription factors that both promote and repress lineage commitment. Furthermore, the study of multipotent bone marrow stromal cells has also enabled the elucidation of pathways involved in the "bone-fat switch." However, from the perspective of this chapter, it is unclear how to use this information to direct transgenic gene expression at specific developmental stages. For example, *Pdgfra* is expressed in early adipose progenitor cells, but it is also associated with vascular cells not destined to become adipocytes (Lee, Petkova, Mottillo, & Granneman, 2012). Other markers of early adipose precursor cells such as CD24 are also expressed in other stem cells (Elghetany & Patel, 2002). This is also true for markers of committed preadipocytes, such as Pref-1, which is "specific" for preadipocytes versus adipocytes, but which is expressed in several other tissues during embryonic development, including pancreatic β-cells (Carlsson et al., 1997).

2. LOSS-OF-FUNCTION STRATEGIES IN ADIPOCYTES IN VIVO

2.1. Overall considerations

In addition to the Cre-loxP-dependent strategies described below, other methods have been employed that should be mentioned, although they will not be described in detail here. First, one can study animals that are global knockouts of the gene of interest. The caveat here of course is that an observed phenotype in adipose tissue might be due to an indirect effect of the gene in another cell type. This can be mitigated somewhat by studying primary adipocytes from such animals (Chiang et al., 2009; Oh et al., 2005;

Shaughnessy, Smith, Kodukula, Storch, & Fried, 2000), or by harvesting preadipocytes, or embryonic fibroblasts for differentiation and subsequent experimentation *ex vivo* (Eguchi et al., 2011; Fisher et al., 2012). Another approach has been to use antisense oligonucleotides or adenoviral-mediated knock-down (Inoue et al., 2008; Jarver et al., 2012; Levine, Jensen, Eberhardt & O'Brien, 1998; Rondinone et al., 2002; Samuel et al., 2006; Yu et al., 2008). When applied via intraperitoneal injection, one can get knock-down in visceral fat pads, but the effect is not seen in all depots nor is it exclusive to fat; liver may be the dominant tissue affected in these studies. Adenovirus can also be directly injected into the fat pad, but again the effect may be seen systemically, it may affect multiple cell types within the pad, and reproducibility between injections can be a major issue. Finally, one can knock-down genes in cultured preadipocytes (e.g., 3T3-F442A cells), implant those cells under the skin of a nude mouse, and then study the fat pad that develops there (Kang et al., 2012; Mandrup, Loftus, MacDougald, Kuhajda, & Lane, 1997). This can be a convenient technique to study gene expression or whether a gene affects adipogenesis, but the amount of fat that grows is typically insufficient to affect the metabolism of the recipient animal.

2.2. Knocking out genes in all fat depots

Pan-adipose gene knockout can be theoretically achieved using any fat-selective marker to drive Cre recombinase. The first and thus most widely used model is the aP2-Cre mouse; aP2 is a lipid-binding protein that is highly expressed in adipocytes, and is encoded by the *Fabp4* gene. Spiegelman's group identified several upstream *cis*-regulatory elements that drive *Fabp4* expression in fat (Graves, Tontonoz, Platt, Ross, & Spiegelman, 1992; Graves, Tontonoz, & Spiegelman, 1992; Ross et al., 1990), and this enabled three different groups to generate mice expressing Cre driven by a 5.4 kb piece of the *Fabp4* flanking sequence (Abel et al., 2001; Barlow et al., 1997; He et al., 2003). Two of these lines (aP2-Cre^{Salk} and aP2-Cre^{BI}) have been used with great frequency to achieve adipose-specific gene deletion. There are general concerns about specificity when using aP2-Cre lines, as aP2 is also expressed in activated macrophages (Fu, Luo, Lopes-Virella, & Garvey, 2002; Makowski et al., 2001). This issue is of more than theoretical concern, as macrophage infiltration into adipose tissue, especially under conditions of overnutrition, has significant effects on local and systemic metabolism (Chawla, Nguyen, & Goh, 2011; Osborn & Olefsky, 2012).

Other issues related to the adipose specificity of available aP2-Cre lines are discussed below.

Other adipose-specific Cre lines have also been developed. Most notably, two lines of mice that express Cre driven by control elements from the adiponectin (encoded by *Adipoq*) locus were generated by the Scherer lab (Adipoq-Cre^S) (Wang, Deng, Wang, Sun, & Scherer, 2010) and the Rosen lab (Adipoq-Cre^R) (Eguchi et al., 2011). The former is a traditional promoter-driven construct using 5.4 kb of the upstream *Adipoq* flanking sequence. Adipoq-Cre^R, in contrast, is a BAC transgenic, with Cre inserted into the translational start site of the *Adipoq* gene in the context of >150 kb of flanking sequence. This means that the majority of the *Adipoq* regulatory elements are present in the Adipoq-Cre^R mouse; a potential caveat with this animal is that there are passenger genes on the large BAC that could also affect the resulting phenotype. Yet another mouse was generated by the Lazar lab using a 33 kb fragment of the resistin (*Retn*) gene (Mullican et al., 2013). Resistin was chosen as a driver because its expression was believed to be restricted to white adipose tissue (Steppan et al., 2001), although the Retn-Cre mouse does not adhere to this expectation (Mullican et al., 2013).

2.2.1 Temporal control of adipose-specific recombination

The transgenes described above are all expressed preferentially in mature adipocytes, but the timing of their expression is not well characterized *in vivo*. Studies of adipogenesis *in vitro* suggest that aP2 is induced by Day 2 of differentiation, while adiponectin appears to be turned on around Day 4. We have no idea, however, how this translates into the biology of a living fat pad, in which cells turn over at a low but constant rate (Arner & Spalding, 2010; Spalding et al., 2008; Tchoukalova et al., 2012). One concern is that as these transgenes express at some point during differentiation, the gene knockout effect may alter the subsequent development of the cells. This means that an observed phenotype would represent both developmental and physiological consequences of gene loss, which can be difficult to deconvolute. To circumvent this, one can consider adding a layer of temporal control, most commonly achieved by fusing the Cre transgene to a mutated form of the estrogen receptor that responds to tamoxifen in preference to native estrogens (Danielian, Muccino, Rowitch, Michael, & McMahon, 1998). Such a mouse has been developed by the Chambon/Metzger labs, using the 5.4 kb aP2 promoter to generate a tamoxifen-inducible aP2-CreERT2 line (Imai, Jiang, Chambon, & Metzger, 2001; Imai et al., 2004).

2.2.2 Tissue specificity

Despite the theoretical concerns about macrophage expression of aP2 that could confound metabolic phenotypes, early studies with both aP2-Cre^{BI} and aP2-Cre^{Salk} lines suggested that both were fairly adipose-specific. However, later studies reported widespread expression during early embryogenesis (Urs, Harrington, Liaw, & Small, 2006), as well as expression in several other tissues including adult lymphatic tissue (Ferrell, Kimak, Lawrence, & Finegold, 2008) and the peripheral and central nervous system (Martens, 2010). This suggests that Cre expression may have become increasingly "leaky" over time in these animals.

Two groups have recently performed a systemic evaluation of several adipose-specific Cre lines by crossing them to reporter mice (R26R-lacZ) in which Cre-mediated recombination can be visualized and quantified by lacZ staining (Lee, Russell, et al., 2013; Mullican et al., 2013; Soriano, 1999). All Cre lines were able to induce recombination in adipose tissue when assessed by X-gal staining or expression of Cre mRNA or protein. However, aP2-Cre^{BI}, aP2-Cre^{Salk}, aP2-CreERT2, and Retn-Cre lines exhibited significant recombination in other tissues including brain and skeletal muscle (see Table 1.1). Such non-adipose recombination can cause unexpected lethality; animals in which *Dicer* or *Hdac3* were knocked out using aP2-Cre^{Salk} die shortly after birth while knockout animals generated with Retn-Cre and/or Adipoq-Cre^S survive without gross abnormalities. Of note, aP2-Cre^{BI} mice showed recombination in the developing spermatogonia of ~2% of the seminiferous tubules, which suggest occasional clonal germ-line recombination during spermatogenesis. Indeed, Cre-mediated excision of a target gene was observed in tail DNA when insulin receptor or *Tfam*^{flox} mice were crossed to aP2-Cre^{BI} (Lee, Russell, et al., 2013). Such germ-line recombination (also called the "delta" effect) can potentially cause lethality (Dubois, Hofmann, Kaloulis, Bishop, & Trumpp, 2006) and may result in whole body heterozygous knockout in the next generation if these litters are not excluded from breeding pairs. Unlike aP2-Cre mice, non-adipose tissue recombination was not observed with Adipoq-Cre lines (Eguchi et al., 2011; Lee, Russell, et al., 2013; Mullican et al., 2013; Wang et al., 2010). It is worth mentioning, however, that adiponectin is also expressed in osteoblasts (Erwin Wagner, personal communication), and this

Table 1.1 Comparison of available adipose-specific care lines

	aP2-Cre^{Salk}	aP2-Cre^{BI}	Retn-Cre	Adipo-Cre^S	Adipo-Cre^R	aP2-CreERT2	UCP1-Cre	UCP1-CreER
Developed by	Evans RM	Kahn BB	Lazar MA	Scherer P	Rosen ED	Chambon/Metzger Laboratory	Benito M	Wolfrum C
Driver	5.4 kb piece of aP2/Fabp4 promoter	5.4 kb piece of aP2/Fabp4 promoter	BAC transgenic containing 23 kb up- and 10 kb downstream of Retn	5.4 kb piece of adiponectin promoter/5′-end-coding region	BAC transgenic containing ~150 kb flanking sequence of Adipoq	5.4 kb piece of aP2/Fabp-4 promoter	8.4 kb piece of promoter and 4.3 kb coding region of Ucp1	BAC transgenic containing up and downstream of Ucp1
Tissue expression	iBAT WAT Liver Skeletal muscle Brain	iBAT WAT Lung Liver Heart Testis Skeletal muscle Brain Clonal germline recombination (~2%) Early embryonic development	iBAT WAT Brain	iBAT WAT Osteoblasts (not tested but expected)	iBAT WAT Osteoblasts	iBAT WAT Lung Heart Skeletal muscle Salivary gland	iBAT Beige/brite fat (not tested but expected)	iBAT Beige/brite fat

needs to be considered when interpreting a metabolic phenotype (Karsenty & Ferron, 2012).

2.2.3 Efficiency of recombination

Differences in recombination efficiency might be due to copy number variation and/or positional effects caused by the location of the Cre transgene. This may explain lower recombination efficacy of aP2-CreERT2 compared to aP2-Cre mice, which were both generated with the same 5.4 kb fragment of *Fabp4* promoter (Lee, Russell, et al., 2013). Adipoq-Cre^R appears to be more efficient at promoting recombination in fat than either aP2-Cre line, despite a lower level of Cre expression (Lee, Russell, et al., 2013). This might be because Adipo-Cre^R directs Cre expression to a more complete population of adipocytes, but this is unclear.

Recombination efficiency can be influenced by the genomic location and distance between the loxP sites of the target locus (Feil, Valtcheva, & Feil, 2009). Adipose-specific Cre lines showed great variation depending on targeted alleles and depots, which is well summarized by Lee, Russell, et al. (2013). It has been shown that knockout efficiency with the use of aP2-Cre^{BI} greatly vary in the adipose tissue depending on the floxed alleles. Multiple genes including *Pparg*, *Hif1a*, *Hif1b*, *Slc2a4*, and *IR* were efficiently ablated but genes like *Tfam* and *Ptp1b* were poorly recombined with the aP2-Cre^{BI}. Similarly, allele-specific differences in recombination efficiency have been shown with aP2-CreERT2. Ablation of *Rxra* and *Pparg* was efficient in this model, while other target alleles such as *Insr* and *Dicer* were poorly recombined. Adipoq-Cre^R has shown efficient targeting for multiple alleles including *Tfam*, *Ptp1b*, and *Dicer*, but may prove to be less efficient for other alleles.

There are also depot-dependent differences in recombination efficiency seen with these models. For example, aP2-Cre^{BI} and Adipoq-Cre mice show greater recombination efficiency in BAT than in WAT, aP2-CreERT2 shows modest recombination in subcutaneous WAT with poor recombination in the BAT and perigonadal fat, and Retn-Cre shows high recombination in WAT and a mosaic pattern of recombination in BAT (Lee, Russell, et al., 2013; Mullican et al., 2013). This is in accordance with previous observations from mouse models using these Cre lines. For example, crossing *Hif1b^{flox}*, *Shox2^{flox}*, or *Pparg^{flox}* mice to aP2-Cre mice resulted in more efficient ablation of target genes in BAT than in WAT. When *Shox2^{flox}* mice were crossed with aP2-Cre^{BI} mice, *Shox2* mRNA was reduced by 48%, 58%, and 81% in perigonadal, subcutaneous, and brown fat, respectively. When $Tfam^{flox}$ mice were crossed with aP2-Cre, a 54% reduction of *Tfam* mRNA was observed in isolated subcutaneous adipocytes, but no difference in Tfam expression in the isolated perigonadal adipocytes, indicating a major depot-specific difference in recombination.

Further, an age-dependent increase in recombination was observed in the fat-specific knockout of *Insr* (FIRKO) mice (Bluher et al., 2002). This might be more apparent than real, as there is a relative reduction in the number of preadipocytes as an animal ages (Alt et al., 2012). Another possibility is that age causes epigenetic modifications to the *Insr* locus that might affect recombination efficiency (Serrano et al., 2005).

Finally, one consideration that is often overlooked is that it is difficult to segregate Cre and floxed alleles when they are present on the same chromosome. This does not affect recombination efficiency at the locus *per se*, but it does make it more difficult to generate large cohorts of floxed, Cre only, and knockout mice.

2.3. Knocking out genes in brown and beige fat

While the field awaits a WAT-specific Cre line, there are options for those who wish to knock out genes in brown fat only. A mouse line that drives Cre from the *Ucp1* promoter was developed by the Benito group in 2001 (Guerra et al., 2001). This line has been used to drive BAT-specific recombination but has not been tested to see if it also expresses in beige cells. Furthermore, there have been concerns about both recombination efficiency and specificity of this line. Additional Ucp1-Cre lines are currently under development. Interestingly, at the time of this writing, a new Ucp1-CreER line has been published which should allow temporal control of recombination in brown and beige fat (Rosenwald, Perdikari, Rulicke, & Wolfrum, 2013).

At present, there is no Cre line that allows for beige-specific recombination without also knocking genes out in interscapular BAT. However, Wu and Spiegelman have identified markers that are selective for beige cells, such as *Slc27a1*, *CD40*, *CD137*, and *TMEM26* (Wu et al., 2012), making such a tool at least theoretically possible.

3. GAIN-OF-FUNCTION STRATEGIES IN ADIPOCYTES *IN VIVO*

In general, strategies to overexpress genes in adipose tissue mirror those used to knock them out, in that both paradigms involve finding an adipose-specific marker to drive expression of a transgene, be it Cre recombinase or another gene of interest. Accordingly, most adipose-specific transgenic studies have used the 5.4 kb aP2 promoter sequence, as it is the best known. As mentioned above, caveats about non-adipose expression, particularly in macrophages, must be considered. Furthermore, the usual issues applying to any transgenic model can confound the results obtained with these models, such as variation in transgene copy number and position effects. Because of this, it has been generally recommended that more than one transgenic line be studied, to be sure that the observed phenotype is truly due to overexpression of the gene of interest.

These concerns can be largely circumvented using a simple knock-in strategy for transgenic expression involving the ROSA26 locus, identified in 1991 by Friedrich and Soriano (1991). The ubiquitous expression of ROSA26 in embryonic and adult tissues, together with the high frequency of gene-targeting events observed at this locus in murine ES cells has enabled a large number of knock-in lines to be generated. The strategy used by our lab and others is to introduce the transgene of interest into the ROSA26 locus downstream of a "STOP" cassette consisting of multiple polyadenylation signals flanked by loxP sites (Soriano, 1999). Upon temporal and cell-type-specific induction of Cre, transcription of the transgene from the ROSA26 promoter (or from an exogenous promoter inserted into the ROSA26 locus) is induced as a result of the deletion of the loxP-flanked STOP cassette. To trace expression of the transgene *in vivo*, it can be useful to introduce a downstream or upstream reporter gene (usually GFP or lacZ) separated from the transgene by a viral 2A peptide (Fig. 1.1A and B). The 2A peptide cleaves autocatalytically (Szymczak et al., 2004) and allows efficient expression of both transgene and reporter after translation. An internal ribosomal entry site (IRES) can be used instead of the 2A peptide, but the large size and difference in expression levels between genes before and after the IRES can make this a less desirable strategy. The Cre/loxP system controls transgene expression in a time- and cell-type-specific fashion. However, once induced, the transgene can no longer be silenced. To overcome this limitation, the tetracycline (Tet)-controlled system can be superimposed to generate inducible ROSA26 transgenes (Beard, Hochedlinger, Plath, Wutz, & Jaenisch, 2006).

Because this strategy is a gene-targeting approach in embryonic stem cells, it involves homologous recombination and selection of positive clones, which adds to the burden of work early in transgenic generation. However, incorporation directly into the ROSA26 locus eliminates copy number and position effects, which means that there is no longer a need to generate

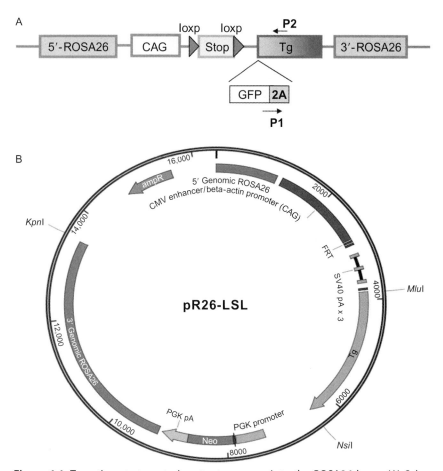

Figure 1.1 Targeting strategy to insert a transgene into the ROSA26 locus. (A) Schematic depicting an example of a targeting construct. ROSA26 targeting arms surround a CAG promoter, a Stop cassette flanked by loxP sites, and a transgene of interest which may be attached to a reporter by a self-cleaving 2A peptide, (B) another view of pR26-LSL showing the entire plasmid. (See the color plate.)

multiple transgenic lines. Most importantly, the transgene of interest can be expressed in any cell type for which an appropriate Cre line is available, without the need for creating a new construct and founder line.

3.1. Methods

1. Amplify your transgene from cDNA and clone into pR26-LSL using *Mlu*I and *Nsi*I.
2. Linearize the targeting cassette with *Kpn*I (ensure that there is no *Kpn*I site in your transgene) and purify the DNA using gel electrophoresis or

Elutip-D. The Elutip-D method yields the purest DNA for microinjection, whereas the gel purification method is a quick and easy and yields adequately clean DNA for microinjection.
3. Measure the DNA concentration using a spectrophotometer.
4. Microinjection of DNA per the specifications of your local facility. ES cells from 129Sv mice are most commonly used.
5. ES cell selection with neomycin.
6. Identification of correctly targeted ES cell clones. Collect DNA from each neo+ clone using digestion buffer (10 mM Tris–HCl, pH 7.6–8.0; 25 mM EDTA; 100 mM NaCl; 0.5% SDS; 0.25 mg/ml Proteinase K) and probe for the correct insertion of the transgene into the ROSA26 locus by Southern blot or long-range PCR. We currently employ long-range PCR to screen ES cells, using the following primers to screen the 5′ insertion site: 5′-GCCAAGTGGGCAGTTTACCG-3′ (outside of the 5′-arm) and 5′-TAGGTAGGGGATCGGGACTCT-3′ (in the CAG). For the 3′ insertion site, the primers are: 5′-GCCAGCTCATTCCTCCCACTC-3′ and 5′-GGCATGGCAATG TTCAAGCAG-3′ (outside of 3′-arm).
   6.1. Expand ES cell till confluent in 24-well plate
   6.2. Aspirate medium; Add 400 µl of digestion buffer
   6.3. Incubate 60 °C for >3 h
   6.4. Transfer the solution into 1.5 ml tubes
   6.5. Add equal volume of phenol/chloroform, vortex well
   6.6. Extract and precipitate with 100% EtOH
   6.7. Wash twice with 70% EtOH
   6.8. Resuspend in 30–40 µl of TE depending on the size of pellet
   6.9. Measure the concentration of DNA
   6.10. Store the samples at −20 °C or do PCR
   6.11. PCR program:

95 °C	4 min
95 °C	45 s
61 °C	30 s
72 °C	7 min
Go to step 2	34 cycles
72 °C	10 min
4 °C	

7. Generation of chimeric mice.
8. Identification of germ-line transmission. Tail biopsies (5 mm) are collected from 2-week-old mice. Genomic DNA is extracted and tested for the presence of ROSA26 transgene DNA with transgene-specific PCR. We design one pair of primers unique to our gene as shown in Fig. 1.1A. The other pair of primers is: forward 5′-GGCATTAAAGC AGCGTATCC-3′ and reverse 5′-CTGTTCCTGTACGGCATGG-3′ as wild-type.

REFERENCES

Abel, E. D., Peroni, O., Kim, J. K., Kim, Y. B., Boss, O., Hadro, E., et al. (2001). Adipose-selective targeting of the GLUT4 gene impairs insulin action in muscle and liver. *Nature*, *409*(6821), 729–733.

Alt, E. U., Senst, C., Murthy, S. N., Slakey, D. P., Dupin, C. L., Chaffin, A. E., et al. (2012). Aging alters tissue resident mesenchymal stem cell properties. *Stem Cell Research*, *8*(2), 215–225.

Arner, P., & Spalding, K. L. (2010). Fat cell turnover in humans. *Biochemical and Biophysical Research Communications*, *396*(1), 101–104.

Barlow, C., Schroeder, M., Lekstrom-Himes, J., Kylefjord, H., Deng, C. X., Wynshaw-Boris, A., et al. (1997). Targeted expression of Cre recombinase to adipose tissue of transgenic mice directs adipose-specific excision of loxP-flanked gene segments. *Nucleic Acids Research*, *25*(12), 2543–2545.

Beard, C., Hochedlinger, K., Plath, K., Wutz, A., & Jaenisch, R. (2006). Efficient method to generate single-copy transgenic mice by site-specific integration in embryonic stem cells. *Genesis*, *44*(1), 23–28.

Bluher, M., Michael, M. D., Peroni, O. D., Ueki, K., Carter, N., Kahn, B. B., et al. (2002). Adipose tissue selective insulin receptor knockout protects against obesity and obesity-related glucose intolerance. *Developmental Cell*, *3*(1), 25–38.

Carlsson, C., Tornehave, D., Lindberg, K., Galante, P., Billestrup, N., Michelsen, B., et al. (1997). Growth hormone and prolactin stimulate the expression of rat preadipocyte factor-1/delta-like protein in pancreatic islets: Molecular cloning and expression pattern during development and growth of the endocrine pancreas. *Endocrinology*, *138*(9), 3940–3948.

Chawla, A., Nguyen, K. D., & Goh, Y. P. (2011). Macrophage-mediated inflammation in metabolic disease. *Nature Reviews Immunology*, *11*(11), 738–749.

Chiang, S. H., Bazuine, M., Lumeng, C. N., Geletka, L. M., Mowers, J., White, N. M., et al. (2009). The protein kinase IKKepsilon regulates energy balance in obese mice. *Cell*, *138*(5), 961–975.

Danielian, P. S., Muccino, D., Rowitch, D. H., Michael, S. K., & McMahon, A. P. (1998). Modification of gene activity in mouse embryos in utero by a tamoxifen-inducible form of Cre recombinase. *Current Biology*, *8*(24), 1323–1326.

Dubois, N. C., Hofmann, D., Kaloulis, K., Bishop, J. M., & Trumpp, A. (2006). Nestin-Cre transgenic mouse line Nes-Cre1 mediates highly efficient Cre/loxP mediated recombination in the nervous system, kidney, and somite-derived tissues. *Genesis*, *44*(8), 355–360.

Eguchi, J., Wang, X., Yu, S., Kershaw, E. E., Chiu, P. C., Dushay, J., et al. (2011). Transcriptional control of adipose lipid handling by IRF4. *Cell Metabolism*, *13*(3), 249–259.

Elghetany, M. T., & Patel, J. (2002). Assessment of CD24 expression on bone marrow neutrophilic granulocytes: CD24 is a marker for the myelocytic stage of development. *American Journal of Hematology*, *71*(4), 348–349.

Feil, S., Valtcheva, N., & Feil, R. (2009). Inducible Cre mice. *Methods in Molecular Biology*, *530*, 343–363.

Ferrell, R. E., Kimak, M. A., Lawrence, E. C., & Finegold, D. N. (2008). Candidate gene analysis in primary lymphedema. *Lymphatic Research and Biology*, *6*(2), 69–76.

Fisher, F. M., Kleiner, S., Douris, N., Fox, E. C., Mepani, R. J., Verdeguer, F., et al. (2012). FGF21 regulates PGC-1alpha and browning of white adipose tissues in adaptive thermogenesis. *Genes & Development*, *26*(3), 271–281.

Friedrich, G., & Soriano, P. (1991). Promoter traps in embryonic stem cells: A genetic screen to identify and mutate developmental genes in mice. *Genes & Development*, *5*(9), 1513–1523.

Fu, Y., Luo, N., Lopes-Virella, M. F., & Garvey, W. T. (2002). The adipocyte lipid binding protein (ALBP/aP2) gene facilitates foam cell formation in human THP-1 macrophages. *Atherosclerosis*, *165*(2), 259–269.

Graves, R. A., Tontonoz, P., Platt, K. A., Ross, S. R., & Spiegelman, B. M. (1992). Identification of a fat cell enhancer: Analysis of requirements for adipose tissue-specific gene expression. *Journal of Cellular Biochemistry*, *49*(3), 219–224.

Graves, R. A., Tontonoz, P., & Spiegelman, B. M. (1992). Analysis of a tissue-specific enhancer: ARF6 regulates adipogenic gene expression. *Molecular and Cellular Biology*, *12*(7), 3313.

Guerra, C., Navarro, P., Valverde, A. M., Arribas, M., Bruning, J., Kozak, L. P., et al. (2001). Brown adipose tissue-specific insulin receptor knockout shows diabetic phenotype without insulin resistance. *The Journal of Clinical Investigation*, *108*(8), 1205–1213.

He, W., Barak, Y., Hevener, A., Olson, P., Liao, D., Le, J., et al. (2003). Adipose-specific peroxisome proliferator-activated receptor gamma knockout causes insulin resistance in fat and liver but not in muscle. *Proceedings of the National Academy of Sciences of the United States of America*, *100*(26), 15712–15717.

Imai, T., Jiang, M., Chambon, P., & Metzger, D. (2001). Impaired adipogenesis and lipolysis in the mouse upon selective ablation of the retinoid X receptor alpha mediated by a tamoxifen-inducible chimeric Cre recombinase (Cre-ERT2) in adipocytes. *Proceedings of the National Academy of Sciences of the United States of America*, *98*(1), 224–228.

Imai, T., Takakuwa, R., Marchand, S., Dentz, E., Bornert, J. M., Messaddeq, N., et al. (2004). Peroxisome proliferator-activated receptor gamma is required in mature white and brown adipocytes for their survival in the mouse. *Proceedings of the National Academy of Sciences of the United States of America*, *101*(13), 4543–4547.

Inoue, N., Yahagi, N., Yamamoto, T., Ishikawa, M., Watanabe, K., Matsuzaka, T., et al. (2008). Cyclin-dependent kinase inhibitor, p21WAF1/CIP1, is involved in adipocyte differentiation and hypertrophy, linking to obesity, and insulin resistance. *The Journal of Biological Chemistry*, *283*(30), 21220–21229.

Jarver, P., Coursindel, T., Andaloussi, S. E., Godfrey, C., Wood, M. J., & Gait, M. J. (2012). Peptide-mediated cell and in vivo delivery of antisense oligonucleotides and siRNA. *Molecular Therapy—Nucleic Acids*, *1*, e27.

Kang, S., Akerblad, P., Kiviranta, R., Gupta, R. K., Kajimura, S., Griffin, M. J., et al. (2012). Regulation of early adipose commitment by Zfp521. *PLoS Biology*, *10*(11), e1001433.

Karsenty, G., & Ferron, M. (2012). The contribution of bone to whole-organism physiology. *Nature*, *481*(7381), 314–320.

Lee, Y. H., Petkova, A. P., Mottillo, E. P., & Granneman, J. G. (2012). In vivo identification of bipotential adipocyte progenitors recruited by beta3-adrenoceptor activation and high-fat feeding. *Cell Metabolism*, *15*(4), 480–491.

Lee, K. Y., Russell, S. J., Ussar, S., Boucher, J., Vernochet, C., Mori, M. A., et al. (2013). Lessons on conditional gene targeting in mouse adipose tissue. *Diabetes*, *62*(3), 864–874.

Lee, M. J., Wu, Y., & Fried, S. K. (2013). Adipose tissue heterogeneity: Implication of depot differences in adipose tissue for obesity complications. *Molecular Aspects of Medicine*, *34*(1), 1–11.

Levine, J. A., Jensen, M. D., Eberhardt, N. L., & O'Brien, T. (1998). Adipocyte macrophage colony-stimulating factor is a mediator of adipose tissue growth. *The Journal of Clinical Investigation*, *101*(8), 1557–1564.

Makowski, L., Boord, J. B., Maeda, K., Babaev, V. R., Uysal, K. T., Morgan, M. A., et al. (2001). Lack of macrophage fatty-acid-binding protein aP2 protects mice deficient in apolipoprotein E against atherosclerosis. *Nature Medicine*, *7*(6), 699–705.

Mandrup, S., Loftus, T. M., MacDougald, O. A., Kuhajda, F. P., & Lane, M. D. (1997). Obese gene expression at in vivo levels by fat pads derived from s.c. implanted 3T3-F442A preadipocytes. *Proceedings of the National Academy of Sciences of the United States of America*, *94*(9), 4300–4305.

Mullican, S. E., Tomaru, T., Gaddis, C. A., Peed, L. C., Sundaram, A., & Lazar, M. A. (2013). A novel adipose-specific gene deletion model demonstrates potential pitfalls of existing methods. *Molecular Endocrinology*, *27*(1), 127–134.

Oh, W., Abu-Elheiga, L., Kordari, P., Gu, Z., Shaikenov, T., Chirala, S. S., et al. (2005). Glucose and fat metabolism in adipose tissue of acetyl-CoA carboxylase 2 knockout mice. *Proceedings of the National Academy of Sciences of the United States of America*, *102*(5), 1384–1389.

Osborn, O., & Olefsky, J. M. (2012). The cellular and signaling networks linking the immune system and metabolism in disease. *Nature Medicine*, *18*(3), 363–374.

Rondinone, C. M., Trevillyan, J. M., Clampit, J., Gum, R. J., Berg, C., Kroeger, P., et al. (2002). Protein tyrosine phosphatase 1B reduction regulates adiposity and expression of genes involved in lipogenesis. *Diabetes*, *51*(8), 2405–2411.

Rosen, E. D., & Spiegelman, B. M. (2006). Adipocytes as regulators of energy balance and glucose homeostasis. *Nature*, *444*(7121), 847–853.

Rosenwald, M., Perdikari, A., Rulicke, T., & Wolfrum, C. (2013). Bidirectional interconversion of brite and white adipocytes. *Nature Cell Biology*, *15*, 659–667, advance online publication.

Ross, S. R., Graves, R. A., Greenstein, A., Platt, K. A., Shyu, H. L., Mellovitz, B., et al. (1990). A fat-specific enhancer is the primary determinant of gene expression for adipocyte P2 in vivo. *Proceedings of the National Academy of Sciences of the United States of America*, *87*(24), 9590–9594.

Samuel, V. T., Choi, C. S., Phillips, T. G., Romanelli, A. J., Geisler, J. G., Bhanot, S., et al. (2006). Targeting foxo1 in mice using antisense oligonucleotide improves hepatic and peripheral insulin action. *Diabetes*, *55*(7), 2042–2050.

Seale, P., Kajimura, S., & Spiegelman, B. M. (2009). Transcriptional control of brown adipocyte development and physiological function of mice and men. *Genes & Development*, *23*(7), 788–797.

Serrano, R., Villar, M., Martinez, C., Carrascosa, J. M., Gallardo, N., & Andres, A. (2005). Differential gene expression of insulin receptor isoforms A and B and insulin receptor substrates 1, 2 and 3 in rat tissues: Modulation by aging and differentiation in rat adipose tissue. *Journal of Molecular Endocrinology*, *34*(1), 153–161.

Shaughnessy, S., Smith, E. R., Kodukula, S., Storch, J., & Fried, S. K. (2000). Adipocyte metabolism in adipocyte fatty acid binding protein knockout mice (aP2-/-) after short-term high-fat feeding: Functional compensation by the keratinocyte [correction of keritinocyte] fatty acid binding protein. *Diabetes*, *49*(6), 904–911.

Soriano, P. (1999). Generalized lacZ expression with the ROSA26 Cre reporter strain. *Nature Genetics*, *21*(1), 70–71.

Spalding, K. L., Arner, E., Westermark, P. O., Bernard, S., Buchholz, B. A., Bergmann, O., et al. (2008). Dynamics of fat cell turnover in humans. *Nature, 453*(7196), 783–787.

Steppan, C. M., Bailey, S. T., Bhat, S., Brown, E. J., Banerjee, R. R., Wright, C. M., et al. (2001). The hormone resistin links obesity to diabetes. *Nature, 409*(6818), 307–312.

Szymczak, A. L., Workman, C. J., Wang, Y., Vignali, K. M., Dilioglou, S., Vanin, E. F., et al. (2004). Correction of multi-gene deficiency in vivo using a single 'self-cleaving' 2A peptide based retroviral vector. *Nature Biotechnology, 22*(5), 589–594.

Tchoukalova, Y. D., Fitch, M., Rogers, P. M., Covington, J. D., Henagan, T. M., Ye, J., et al. (2012). In vivo adipogenesis in rats measured by cell kinetics in adipocytes and plastic-adherent stroma-vascular cells in response to high-fat diet and thiazolidinedione. *Diabetes, 61*(1), 137–144.

Urs, S., Harrington, A., Liaw, L., & Small, D. (2006). Selective expression of an aP2/fatty acid binding protein 4-Cre transgene in non-adipogenic tissues during embryonic development. *Transgenic Research, 15*(5), 647–653.

Wang, Z. V., Deng, Y., Wang, Q. A., Sun, K., & Scherer, P. E. (2010). Identification and characterization of a promoter cassette conferring adipocyte-specific gene expression. *Endocrinology, 151*(6), 2933–2939.

Wu, J., Bostrom, P., Sparks, L. M., Ye, L., Choi, J. H., Giang, A. H., et al. (2012). Beige adipocytes are a distinct type of thermogenic fat cell in mouse and human. *Cell, 150*(2), 366–376.

Wu, J., Cohen, P., & Spiegelman, B. M. (2013). Adaptive thermogenesis in adipocytes: Is beige the new brown? *Genes & Development, 27*(3), 234–250.

Yu, X. X., Murray, S. F., Watts, L., Booten, S. L., Tokorcheck, J., Monia, B. P., et al. (2008). Reduction of JNK1 expression with antisense oligonucleotide improves adiposity in obese mice. *American Journal of Physiology, Endocrinology and Metabolism, 295*(2), E436–E445.

CHAPTER TWO

Imaging White Adipose Tissue with Confocal Microscopy

Gabriel Martinez-Santibañez[*], Kae Won Cho[†], Carey N. Lumeng[†,1]

[*]Cellular and Molecular Biology Program, University of Michigan, Ann Arbor, Michigan, USA
[†]Department of Pediatrics and Communicable Diseases, University of Michigan Medical School, Ann Arbor, Michigan, USA
[1]Corresponding author: e-mail address: clumeng@umich.edu

Contents

1. Introduction — 18
   1.1 Adipocyte morphology — 19
   1.2 Adipose tissue macrophages and crown-like structures — 20
   1.3 Milky spots and fat-associated lymphoid clusters — 21
   1.4 Capillaries and blood vessels — 22
   1.5 Advantages of confocal imaging — 23
2. Materials for Imaging Adipose Tissue — 24
   2.1 Reagents and buffers — 24
   2.2 Antibodies and suggested concentrations — 25
   2.3 Nuclear and lipid stains — 25
   2.4 Slides and equipment — 25
3. Methods — 25
   3.1 Perfusion, collection, and fixation of white adipose tissues — 25
   3.2 Blocking — 26
   3.3 Primary antibody incubation — 26
   3.4 Secondary antibody incubation — 27
   3.5 Antibody-independent staining of nuclei and lipids — 27
   3.6 Imaging and 3D reconstructions — 28
4. Summary — 28
Acknowledgments — 28
References — 28

Abstract

Adipose tissue is composed of a variety of cell types that include mature adipocytes, endothelial cells, fibroblasts, adipocyte progenitors, and a range of inflammatory leukocytes. These cells work in concert to promote nutrient storage in adipose tissue depots and vary widely based on location. In addition, overnutrition and obesity impart significant changes in the architecture of adipose tissue that are strongly associated with metabolic dysfunction. Recent studies have called attention to the importance of adipose tissue microenvironments in regulating adipocyte function and therefore require techniques that preserve cellular interactions and permit detailed analysis of three-dimensional structures in fat. This chapter summarizes our experience with the use of laser scanning confocal microscopy for imaging adipose tissue in rodents.

1. INTRODUCTION

Adipose tissue is comprised of a vast range of cellular and noncellular components. By volume, adipocytes are the most prominent cell within a given white fat depot. However, by cell number, it is likely that mature adipocytes are in the minority due to the presence of a large network of supporting cells (Granneman, 2008). These include cells that comprise an extensive vascular system in fat, including fibroblasts, preadipocytes, and cells with mesenchymal and hematopoietic stem cell capacity (Crandall, Hausman, & Kral, 1997; Nishimura et al., 2007; Zuk et al., 2002). Obesity research has called significant attention to the presence of a wide range of inflammatory leukocytes and lymphocytes in fat that change with obesity. These include myeloid cells (macrophages, neutrophils, etc.), lymphocytes (T cells, B cells), eosinophils, mast cells, and NK cells (Lee, Goldfine, Benoist, Shoelson, & Mathis, 2009; Nagasaki, Eto, Yamashita, Ohsugi, & Otsu, 2009; Nishimura et al., 2009; Ohmura et al., 2010). Quantitation and localization of these cells is a critical part of understanding the link between obesity and inflammation. A significant need to understand this association has spurred the development of techniques that permit a detailed examination of the diverse components and architecture of adipose tissue.

Traditional histologic techniques for adipose tissue analysis have included electron microscopy (transmission and scanning) and freeze fracturing. Many laboratories analyze adipose tissue with light microscopy (LM) techniques such as immunohistochemistry and *in situ* hybridization (Cinti, Zingaretti, Cancello, Ceresi, & Ferrara, 2001). However, because of the high lipid content in fat, sectioning of frozen or paraffin-embedded samples is often inconsistent and can distort adipose tissue architecture. This can lead to biased assessments of adipocyte size. More importantly, this limits our capacity to appreciate the diversity of nonadipocyte cell types in fat, and limits our ability to observe their cell–cell interactions. For those reasons, we and others have developed techniques that permit the imaging of whole-mount tissue samples in a way that maintains native architecture

(Cho et al., 2007; Lumeng, DelProposto, Westcott, & Saltiel, 2008). Here, we present a detailed description of the adipose tissue structures that can be imaged with confocal microscopy in rodents, along with detailed protocols.

1.1. Adipocyte morphology

The mature white adipocyte is primarily composed of a single large lipid droplet that is ~100 μm in diameter in mice (Suzuki, Shinohara, Ohsaki, & Fujimoto, 2011). Nuclear and other subcellular components are localized within a very thin cytoplasmic layer that lines the lipid droplet and forms the ghost-like remnant of the adipocyte seen in traditional paraffin-embedded sections. Immature adipocytes contain multiple small lipid droplets and are described as having a "multilocular" appearance. As the adipocyte matures, these lipid droplets fuse and form the round "unilocular" droplet. The fluorescent stains BODIPY and Nile Red are lipid-soluble compounds that help visualize lipid aggregation (Table 2.1).

The adipocyte plasma membrane contains numerous receptors (e.g., insulin receptors) involved in cell signaling that can regulate lipid uptake and fatty acid trafficking. Of these, Caveolin-1 is enriched in the plasma membrane and is commonly found in lipid rafts (Jasmin, Frank, & Lisanti, 2012). Because Caveolin-1 is abundant on the cell surface, it provides an excellent target for staining and imaging the plasma membrane of

Table 2.1 Adipocyte physiology and vascular structures

Target	Antibody or stain	Dilution	With saponin
Capillary and vasculature	Isolectin GS IB$_4$ (Invitrogen; Cat. # varies by conj. fluorochrome)	1:500	–
Lipid	BODIPY 493/503 (Invitrogen)	0.25 μg/mL	–
	Nile Red (Sigma N3013)	0.26 mg/mL	
Nuclei	DAPI	1 μg/mL	–
Cell membranes	Caveolin-1 (BD Biosciences 610060)	1 μg/mL	+
Lipid droplet surface	Perilipin (abcam ab61682)	2 μg/mL	+

Note: A "+" sign denotes that 0.1% saponin is required for membrane permeabilization due to intracellular localization of antigen.

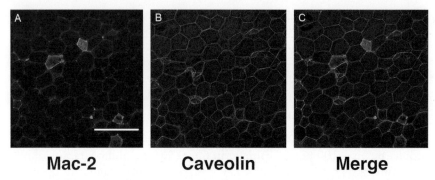

Figure 2.1 Crown-like structures in white adipose tissue. Gonadal fat pads from a high fat diet fed C57Bl/6 mouse were fixed, isolated, and stained as above. Macrophages stain Mac-2 in red (A), Caveolin-1 plasma membrane in green (B), and images were merged (C). Scale bar = 350 μm. (See the color plate.)

adipocytes. Lipid droplets are surrounded by PAT proteins (i.e., perilipin, ADRP, TIP47), which regulate both storage and release of lipids. Perilipin is a useful marker of lipid droplet structures in white fat. Stimulation by adrenergic agonists changes the conformation of perilipin, which allows access of lipases, like hormone-sensitive lipase, to the lipid droplet. This results in the mobilization of triglycerides (Greenberg et al., 1991). Perilipin is also useful for identifying dead or dying adipocytes where loss of perilipin staining is noted (Feng et al., 2011). For reagents useful in visualizing these structures, refer to Table 2.1.

1.2. Adipose tissue macrophages and crown-like structures

The death of adipocytes results in marked remodeling of the adipose tissue microenvironment. H&E sections and immunohistochemistry studies have revealed that areas with adipocyte death create regions called crown-like structures (CLSs) that are described as accumulations of proinflammatory macrophages and extracellular matrix material (Cinti, 2005; Spencer et al., 2010) (Fig. 2.1). Dying adipocytes leave behind Perilipin-negative lipid droplets that also lack Caveolin-1 staining (Feng et al., 2011; Lumeng et al., 2008; Lumeng, Deyoung, Bodzin, & Saltiel, 2007). CLSs are a hallmark of adipose tissue inflammation and fibrosis in human and rodent adipose tissue.

A major cellular component of adipose tissue and CLSs is a population of adipose tissue macrophages (ATMs). Total ATMs can be detected in adipose tissue using a variety of macrophage-specific surface stains such as Mac-2 and F4/80 (Table 2.2). An example of how CLSs can be visualized is by using a

Table 2.2 Adipose tissue macrophage and crown-like structure stains

Target	Antibody	Dilution
Total ATMs	F4/80 (abcam ab6640)	1:300
	Galectin-3 (Mac-2) (eBioscience 14-5301-85)	1:200
M1 ATMs	CD11c (abD Serotech MCA1369)	1:200
M2 ATMs	MGL-1 (abcam ab15635)	1:300
Capillary and vasculature	Isolectin GS IB$_4$ (Invitrogen; cat. no. varies by conj. fluorochrome)	1:500

combination of macrophage stain and perilipin stain, where Mac-2 and/or F4/80 will reveal a circular organization of macrophages that is void of perilipin stain. Resident CD11c$^-$/MGL-1$^+$ M2 ATMs are seen in interstitial spaces between adipocytes and have morphologic characteristics that are distinct from CD11c$^+$ ATMs (Lumeng, Bodzin, & Saltiel, 2007; Xu et al., 2003). In contrast, CD11c$^+$ "classically activated" M1 ATMs are rare in lean mice, but are abundant in obese mice. These are enriched in CLS and are frequently found to contain triglyceride-laden lipid droplets. Resident ATMs lack lipid accumulation and are enriched for markers of M2 polarization such as CD206 and CD301/MGL-1. In addition to ATMs, CLS have been shown to be sites of accumulation of numerous other lymphocytes and leukocytes that include T cells (adipose tissue T cells, or ATTs), B cells, mast cells, and eosinophils (Nishimura et al., 2009; Ohmura et al., 2010; Tsui, Wu, Davidson, Alonso, & Leong, 2011). The trafficking of these cells to fat, and the mechanism by which they are enriched in CLSs, is still unclear.

1.3. Milky spots and fat-associated lymphoid clusters

Hypercellular clusters are known to reside on the surface of numerous adipose tissue depots. Milky spots (MSs) are found primarily in omental adipose tissue depots and are composed of macrophages and B and T lymphocytes (Fig. 2.2A–C). They have been shown to participate in the clearing of debris from the peritoneum and may play a role in adaptive immunity (Rangel-Moreno et al., 2009). Fat-associated lymphoid clusters (FALCs) have recently been described in mesenteric fat, as well as in gonadal fat depots in mice (Moro et al., 2010; Morris et al., 2013) (Fig. 2.2D–G). These contain a unique population of Lin$^-$Kit$^+$Sca1$^+$ innate lymphocytes. If FALC are identical or related to milky spots is not clear, but they both appear to participate in phagocytosis and immune surveillance in several contexts. While

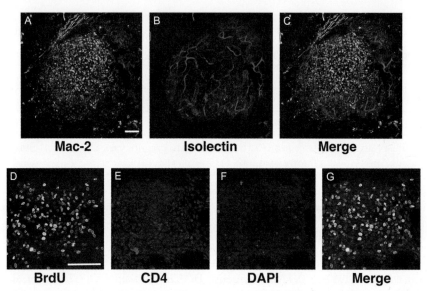

Figure 2.2 Milky spots and FALCS in adipose tissue. Omental fat pads from a C57Bl/6 mouse was fixed and stained and surface imaged to identify milky spots on the fat pad surface. Macrophages (Mac-2) are shown in green (A) and vasculature (isolectin) shown in red (B) with merged image (C). Gonadal fat pads were stained and imaged for surface collections of monocytes to identify FALCS. Proliferating cells are stained for BrdU in green (D), $CD4^+$ T cells are shown in red (E), nuclei are stained with DAPI in blue (F), and all three channels shown merged (G). Scale bar=100 μm. (See the color plate.)

they appear to resemble lymph nodes, there is little evidence of associated lymphatic vessels.

The sizes of these uncapsulated structures range between 100 and 500 μm in diameter and are in direct contact with adipocytes (Rangel-Moreno et al., 2009). They appear to expand in concert with obesity, adipose tissue inflammation, and also in response to aging (Lumeng et al., 2011). The localization of such structures is a challenge in tissue sections as they are relatively rare on the surface of fat. However, whole-mount techniques facilitate the localization and characterization of FALCs and MSs. To image these structures, combination stains for nuclei (DAPI), T cells (CD4), and macrophages (F4/80) can be implemented (Table 2.2).

1.4. Capillaries and blood vessels

Adipose tissue contains an extensive vascular network that participates in the transport of nutrients and leukocytes in and out of fat. Many of the vascular structures are tightly associated with CLS and FALCSs and are believed to facilitate cellular trafficking of leukocytes and lymphocytes (Nishimura et al., 2008). The formation of the primitive fat organ is dependent on the development of an extensive vascular bed (Crandall et al., 1997). Groups have found that the expansion of vascular networks occurs in concert, and even precedes adipogenesis (Han et al., 2011; Hausman & Richardson, 1983; Kimura, Ozeki, Inamoto, & Tabata, 2002). Adipose tissue growth, referring to both the expansion of number (hyperplasia) and size (hypertrophy), is tightly linked with angiogenesis. It has been demonstrated that limiting angiogenesis can also block adipogenesis (Brakenhielm et al., 2004; Liu & Meydani, 2003). In parallel with this idea, proangiogenic therapies can promote adipogenesis (Tabata et al., 2000). A general stain for adipose tissue vasculature is isolectin, which binds tightly to the surface of vascular endothelial cells (Cho et al., 2007; Nishimura et al., 2008) (Fig. 2.3).

1.5. Advantages of confocal imaging

The features in adipose tissue mentioned are complex, three-dimensional structures, thus the limitations of standard LM techniques are apparent. Furthermore, limiting sampling of fat to cross-sections will underrepresent structures such as FALCs and will not fully capture the three-dimensional tortuous route of many adipose tissue blood vessels. In addition, due to the high lipid content of fat, LM often results in significant auto-fluorescence, which is further promoted by light diffraction. The amount of auto-fluorescence can be determined simply by viewing a specimen that is unstained. Often the resolution of images taken with traditional microscopes is compromised because of the fluorescence signals that may arise

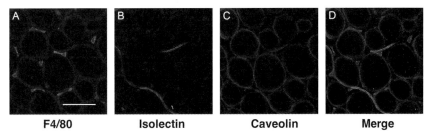

Figure 2.3 Adipose tissue vasculature. Gonadal fat pads from C57Bl/6 mice fed a chow diet. F4/80$^+$ macrophages are stained in green (A), vasculature (isolectin) in red (B), Caveolin-1 denotes the adipocyte plasma membrane in blue (C), images merged in (D). Scale bar = 100 μm. (See the color plate.)

from other optical layers that are not within the plane of focus. Groups have attempted to alleviate this issue by cryosectioning methods and by flattening tissue fragments (Paddock & Eliceiri, 2014). However, such methods can significantly disturb the unique architecture of adipose tissue and may lead to highly variable results.

Laser scanning confocal imaging is an imaging technique that can address many of the limitations of traditional fluorescence microscopy. Because confocal imaging allows for visualizing a very narrow plane of focus, much of the interference that results from auto-fluorescence and out-of-focus blur can be removed. With proper fixation techniques and appropriate staining procedures whole-mount samples of adipose tissue can be imaged with confocal microscopy much more rapidly than with conventional LM techniques. Other advantages of confocal microscopy and its sophisticated optics include the use of specific excitation wavelengths, as well as the ability to employ detectors that exclude auto-fluorescence from other emission spectra. This feature comes into play when there is cross-fluorochrome excitation in neighboring light channels. In addition, z-stack series of images can be easily combined to assemble three-dimensional reconstruction of many of the unique structures within adipose tissue.

2. MATERIALS FOR IMAGING ADIPOSE TISSUE

2.1. Reagents and buffers

16% paraformaldehyde (PFA) 16% EM Grade (Electron Microscopy Sciences, Hatfield, PA; Cat. # 15710)

Phosphate-buffered saline (PBS), pH 7.4 (GIBCO Invitrogen, Carlsbad, CA; Cat. # 10010-023)

Bovine serum albumin (BSA) (Sigma-Aldrich, St. Louis, MO; Cat. # A7030)

Saponin (Sigma-Aldrich, St. Louis, MO; Cat. # 47036)

Glycerol (Sigma-Aldrich, St. Louis, MO; Cat. # G5516)

Tris base (Tris (Hydroxymethyl)Aminomethane) (EMD, Darmstadt, Germany; Cat. # 9230)

Fixing buffer: 1% PFA in PBS, pH 7.4 (v/v)

Blocking buffer: 5% BSA in PBS, pH 7.4 (w/v)

Intracellular stain buffer: 5% BSA in PBS, pH 7.4 (w/v) + 0.1% saponin

Intracellular stain wash buffer: PBS, pH 7.4 + 0.1% saponin

Buffered glycerol solution (optional)—for use as imaging stabilization agent 90% glycerol in 0.1 M Tris–HCl pH 9.0

2.2. Antibodies and suggested concentrations

See Tables 2.1-2.3 for antibody suppliers and suggested working concentrations.

2.3. Nuclear and lipid stains

BODIPY (4,4-difluoro-1,3,5,7,8-pentamethyl-4-bora-3a,4a-diaza-s-indacene) (Molecular Probes, Invitrogen, Carlsbad, CA; Cat. # D3922)
DAPI (40,6-diamidino-2-phenylindole dihydrochloride; Sigma-Aldrich, St. Louis, MO; Cat. # D9542)
Nile Red (Sigma-Aldrich, St. Louis, MO; Cat. # N3013)

2.4. Slides and equipment

Lab-Tek II Chambered #1.5 German borosilicate cover glass system (Nunc, Rochester, NY; Cat. # 155360)
2 mL cylindrical microcentrifuge tubes
Rotating/rocking platform
Inverted laser scanning confocal microscope—Olympus FluoView 500 is routinely employed in our lab but other brands and models are appropriate

3. METHODS

In order to best preserve structure and cellular components within adipose tissue and eliminate background fluorescence, perfusion fixation is recommended. Lower concentrations of fixative are typically employed to minimize auto-fluorescence. In addition, a gentle postfixation is beneficial immediately after tissue is removed. All buffers should be used at room temperature unless stated otherwise.

This protocol is routinely applied for confocal imaging of mouse adipose tissue from different depots. Similar protocols are possible in rats or other experimental animal models with alterations in the staining antibodies. The use of similar techniques in human adipose tissue samples may also be possible with some adaptation, but this has not been extensively tested.

3.1. Perfusion, collection, and fixation of white adipose tissues

1. Euthanize animals via cervical dislocation or CO_2 asphyxiation and perfuse via slow intracardiac injection with 10 mL of fresh fixing buffer (1% PFA in PBS, pH 7.4 (v/v), see Section 2.1) over 2–3 min. Injection can be via manual or pump through the left ventricle. Removal of the right atrium will facilitate removal of blood and clearance of fixative.

2. Excise white adipose tissue depots (i.e., epididymal, peri-renal and retroperitoneal, inguinal, omental). Tissues may be subdivided into 0.5–1 cm^3 sized pieces or left intact.
3. Incubate in 5–10 mL of fixing buffer for 30 min at room temperature with gentle rocking on a rotating platform. Make sure the tissue is completely submerged in fixation buffer. Longer fixation may increase auto-fluorescence.
4. Rinse away the fixing buffer with three, 10-min washes with PBS, pH 7.4, under gentle rocking at room temperature. Tissues may be stored at 4 °C in PBS, pH 7.4, for up to 2 weeks.

3.2. Blocking

To prevent fluorochrome-conjugated antibodies from nonspecific binding, a blocking step is suggested. The addition of a detergent during the blocking step is required for the detection of intracellular antigens (see Section 2.1).

5. Cut a piece from the fixed adipose tissue for staining. Typically this is 0.5–0.75 cm^3. Subsequent incubations are performed in 2 mL cylindrical microcentrifuge tubes.
6. Perform a 30-min block with 1 mL of blocking buffer (5% BSA in PBS, pH 7.4 (w/v)) with gentle rocking at room temperature. For staining intracellular antigens, perform this step with intracellular stain buffer.

3.3. Primary antibody incubation

7. Using blocking buffer (or intracellular stain blocking buffer if appropriate), prepare the primary antibody cocktail. Using a 2 mL microcentrifuge tube, 300 µL volume is typically sufficient to fully submerge one piece of tissue in antibody solution and allow for efficient mixing. (See Tables 2.1–2.3 for suggested staining targets and staining conditions.)

Table 2.3 Milky spot and FALC stains

Target	Antibody or stain	Dilution
CD4 T cells	CD4 (AbBiotech 250592)	1:200
Total ATMs	F4/80 (abcam ab6640)	1:200
Capillary and vasculature	Isolectin GS IB$_4$	1:500
Nuclei	DAPI	1 µg/mL

Note: Useful controls to include are: (1) unstained tissue to observe autofluorescence and (2) tissue stained with secondary antibody only to evaluate for any nonspecific staining.

8. Incubate tissues in primary antibody cocktail for 1 h at room temperature (or overnight at 4 °C) with gentle rocking. Antibody cocktail can be saved and reused at least two more times if stored at 4 °C for up to 4 weeks.
9. Perform three washes with gentle rocking, 10 min per wash, with PBS, pH 70.4 (or intracellular wash buffer if appropriate).

3.4. Secondary antibody incubation

10. Prepare fresh secondary antibody cocktail in 300 µL of antibody staining buffer (or use intracellular stain buffer) and incubate with gentle rocking for 1 h, covered, at room temperature.

Note: Titration may be necessary depending on the antibody source. Our laboratory typically uses AlexaFluor conjugated secondary antibodies (Molecular Probes by Life Technologies) at a dilution of 1:250 for a final concentration of 8 µg/mL.

If imaging vasculature structures, add the anti-isolectin antibody to the secondary antibody cocktail, refer to Table 2.1 for recommended concentrations.

11. Repeat step 9.
12. Tissues are ready for imaging. They can be stored, shielded from light, at 4 °C in PBS, pH 7.4, for 1–2 weeks. Continue to next step to stain lipid and nuclei.

3.5. Antibody-independent staining of nuclei and lipids

BODIPY (green channel), Nile Red (red channel), and DAPI (Ultraviolet or blue channel) stains can be performed independent of antibody-based stains. It is suggested that staining of lipids and nuclei be performed after antibody-based staining to prevent saturation of chemical-based fluorescence. Short incubations are recommended.

13. Prepare staining reagents in 5% BSA in PBS using suggested concentrations in Table 2.1.
14. Incubate tissues in lipid/nuclei staining cocktail for 20 min.
15. Perform three washes with gentle rocking, 10 min per wash, with PBS.
16. Tissues are ready for imaging or can be stored, shielded from light, in PBS at 4 °C in PBS for about 1–2 weeks.

3.6. Imaging and 3D reconstructions

Imaging of the stained tissue will take place on an inverted laser scanning confocal microscope. The tissue is placed atop confocal imaging optimized #1.5 borosilicate glass chamber slide. A small drop of PBS, pH 7.4 should be placed atop the piece of tissue to prevent drying. Alternatively, a glycerol-based stabilizing agent can be used (buffered glycerol solution) although storage of the fat samples after imaging can be difficult in this media.

Imaging capture will vary between experiment and system. We have been successful staining appropriately fixed samples with up to four laser lines (405, 488, 568, and 647 nm). Emission must be collected through the appropriate narrow band-pass filters by the confocal microscope. Adjustments for PMT, gain, and black levels vary and should be optimized per staining condition. If auto-fluorescence is an issue, it is possible to digitally remove it via image subtraction (Stockert, Villanueva, Cristóbal, & Cañete, 2009). Finally, assembly of three-dimensional reconstructions is accomplished by taking z-stack images using software-determined levels along the vertical axis.

4. SUMMARY

The protocols and images captured above provide a new depth of insight into the architecture and cell–cell interactions in adipose tissue with better resolution than LM techniques. While limited to static measurements, it is possible to use similar imaging modalities to evaluate dynamic leukocyte trafficking events into adipose tissue (Nishimura et al., 2008). We hope that these protocols provide a starting point for many researchers to explore and identify novel markers for use in adipose tissue imaging that can expand our knowledge of this dynamic organ.

ACKNOWLEDGMENTS

This work was supported by NIH Grants DK-090262 and DK-092873. This work used services at the University of Michigan NORC supported by NIH Grant DK-089503 and the Michigan Diabetes Research and Training Center funded by P60-DK-020572 from the National Institute of Diabetes and Digestive and Kidney Diseases. Special thanks to the Morphology and Image Analysis Core for training and equipment use.

REFERENCES

Brakenhielm, E., Cao, R., Gao, B., Angelin, B., Cannon, B., Parini, P., et al. (2004). Angiogenesis inhibitor, TNP-470, prevents diet-induced and genetic obesity in mice. *Circulation Research*, *94*(12), 1579–1588.

Cho, C. H., Jun Koh, Y., Han, J., Sung, H. K., Jong Lee, H., Morisada, T., et al. (2007). Angiogenic role of LYVE-1-positive macrophages in adipose tissue. *Circulation Research*, *100*(4), e47–e57.

Cinti, S. (2005). Adipocyte death defines macrophage localization and function in adipose tissue of obese mice and humans. *Journal of Lipid Research*, *46*(11), 2347–2355.

Cinti, S., Zingaretti, M. C., Cancello, R., Ceresi, E., & Ferrara, P. (2001). Morphologic techniques for the study of brown adipose tissue and white adipose tissue. *Methods in Molecular Biology (Clifton, NJ)*, *155*, 21–51.

Crandall, D. L., Hausman, G. J., & Kral, J. G. (1997). A review of the microcirculation of adipose tissue: Anatomic, metabolic, and angiogenic perspectives. *Microcirculation*, *4*(2), 211–232.

Feng, D. D., Tang, Y. Y., Kwon, H. H., Zong, H. H., Hawkins, M. M., Kitsis, R. N. R., et al. (2011). High-fat diet-induced adipocyte cell death occurs through a cyclophilin D intrinsic signaling pathway independent of adipose tissue inflammation. *Diabetes*, *60*(8), 2134–2143.

Granneman, J. G. (2008). Delivery of DNA into adipocytes within adipose tissue. *Methods in Molecular Biology (Clifton, NJ)*, *423*, 191–195.

Greenberg, A. S., Egan, J. J., Wek, S. A., Garty, N. B., Blanchette-Mackie, E. J., & Londos, C. (1991). Perilipin, a major hormonally regulated adipocyte-specific phosphoprotein associated with the periphery of lipid storage droplets. *Journal of Biological Chemistry*, *266*(17), 11341–11346.

Han, J., Lee, J.-E., Jin, J., Lim, J. S., Oh, N., Kim, K., et al. (2011). The spatiotemporal development of adipose tissue. *Development*, *138*(22), 5027–5037.

Hausman, G. J. G., & Richardson, R. L. R. (1983). Cellular and vascular development in immature rat adipose tissue. *Journal of Lipid Research*, *24*(5), 522–532.

Jasmin, J.-F., Frank, P. G., & Lisanti, M. P. (2012). *Caveolins and caveolae*. New York, NY: Springer.

Kimura, Y. Y., Ozeki, M. M., Inamoto, T. T., & Tabata, Y. Y. (2002). Time course of de novo adipogenesis in matrigel by gelatin microspheres incorporating basic fibroblast growth factor. *Tissue Engineering (United States)*, *8*(4), 603–613.

Lee, J., Goldfine, A. B., Benoist, C., Shoelson, S., & Mathis, D. (2009). Lean, but not obese, fat is enriched for a unique population of regulatory T cells that affect metabolic parameters. *Nature Medicine*, *15*(8), 930–939.

Liu, L., & Meydani, M. (2003). Angiogenesis inhibitors may regulate adiposity. *Nutrition Reviews*, *61*(11), 384–387.

Lumeng, C. N., Bodzin, J. L., & Saltiel, A. R. (2007). Obesity induces a phenotypic switch in adipose tissue macrophage polarization. *Journal of Clinical Investigation*, *117*, 175–184.

Lumeng, C. N., DelProposto, J. B., Westcott, D. J., & Saltiel, A. R. (2008). Phenotypic switching of adipose tissue macrophages with obesity is generated by spatiotemporal differences in macrophage subtypes. *Diabetes*, *57*(12), 3239–3246.

Lumeng, C. N., Deyoung, S. M., Bodzin, J. L., & Saltiel, A. R. (2007). Increased inflammatory properties of adipose tissue macrophages recruited during diet-induced obesity. *Diabetes*, *56*(1), 16–23.

Lumeng, C. N., Liu, J., Geletka, L., Delaney, C., Delproposto, J., Desai, A., et al. (2011). Aging is associated with an increase in T cells and inflammatory macrophages in visceral adipose tissue. *The Journal of Immunology*, *187*(12), 6208–6216.

Moro, K., Yamada, T., Tanabe, M., Takeuchi, T., Ikawa, T., Kawamoto, H., et al. (2010). Innate production of T(H)2 cytokines by adipose tissue-associated c-Kit(+)Sca-1(+) lymphoid cells. *Nature*, *463*(7280), 540–544.

Morris, D. L., Cho, K. W., Delproposto, J. L., Oatmen, K. E., Geletka, L. M., Martinez-Santibanez, G., et al. (2013). Adipose tissue macrophages function as antigen presenting cells and regulate adipose tissue CD4+ T cells in mice. *Diabetes*, *62*(8), 2762–2772.

Nagasaki, M., Eto, K., Yamashita, H., Ohsugi, M., & Otsu, M. (2009). CD8+ effector T cells contribute to macrophage recruitment and adipose tissue inflammation in obesity. *Nature Medicine*, *15*, 914–920.

Nishimura, S., Manabe, I., Nagasaki, M., Eto, K., Yamashita, H., Ohsugi, M., et al. (2009). CD8+ effector T cells contribute to macrophage recruitment and adipose tissue inflammation in obesity. *Nature Medicine*, *15*(8), 914–920.

Nishimura, S., Manabe, I., Nagasaki, M., Hosoya, Y., Yamashita, H., Fujita, H., et al. (2007). Adipogenesis in obesity requires close interplay between differentiating adipocytes, stromal cells, and blood vessels. *Diabetes*, *56*(6), 1517–1526.

Nishimura, S. S., Manabe, I. I., Nagasaki, M. M., Seo, K. K., Yamashita, H. H., Hosoya, Y. Y., et al. (2008). In vivo imaging in mice reveals local cell dynamics and inflammation in obese adipose tissue. *Journal of Clinical Investigation*, *118*(2), 710–721.

Ohmura, K., Ishimori, N., Ohmura, Y., Tokuhara, S., Nozawa, A., Horii, S., et al. (2010). Natural killer T cells are involved in adipose tissues inflammation and glucose intolerance in diet-induced obese mice. *Arteriosclerosis, Thrombosis, and Vascular Biology*, *30*(2), 193–199.

Paddock, S. W., & Eliceiri, K. W. (2014). Laser scanning confocal microscopy: History, applications, and related optical sectioning techniques. *Methods in Molecular Biology*, *1075*, 9–47.

Rangel-Moreno, J., Moyron-Quiroz, J. E., Carragher, D. M., Kusser, K., Hartson, L., Moquin, A., et al. (2009). Omental milky spots develop in the absence of lymphoid tissue-inducer cells and support B and T cell responses to peritoneal antigens. *Immunity*, *30*(5), 731–743.

Spencer, M., Yao-Borengasser, A., Unal, R., Rasouli, N., Gurley, C. M., Zhu, B., et al. (2010). Adipose tissue macrophages in insulin-resistant subjects are associated with collagen VI and fibrosis and demonstrate alternative activation. *American Journal of Physiology: Endocrinology and Metabolism*, *299*(6), E1016–E1027.

Stockert, J. C., Villanueva, A., Cristóbal, J., & Cañete, M. (2009). Improving images of fluorescent cell labeling by background signal subtraction. *Biotechnic and Histochemistry*, *84*(2), 63–68.

Suzuki, M., Shinohara, Y., Ohsaki, Y., & Fujimoto, T. (2011). Lipid droplets: Size matters. *Journal of Electron Microscopy*, *60*(Suppl. 1), S101–S116.

Tabata, Y., Miyao, M., Inamoto, T., Ishii, T., Hirano, Y., Yamaoki, Y., et al. (2000). De novo formation of adipose tissue by controlled release of basic fibroblast growth factor. *Tissue Engineering (United States)*, *6*(3), 279–289.

Tsui, H., Wu, P., Davidson, M. G., Alonso, M. N., & Leong, H. X. (2011). B cells promote insulin resistance through modulation of T cells and production of pathogenic IgG antibodies. *Nature Medicine*, *17*(5), 610–617.

Xu, H. H., Barnes, G. T. G., Yang, Q. Q., Tan, G. G., Yang, D. D., Chou, C. J. C., et al. (2003). Chronic inflammation in fat plays a crucial role in the development of obesity-related insulin resistance. *Journal of Clinical Investigation*, *112*(12), 1821–1830.

Zuk, P. A. P., Zhu, M. M., Ashjian, P. P., De Ugarte, D. A. D., Huang, J. I. J., Mizuno, H. H., et al. (2002). Human adipose tissue is a source of multipotent stem cells. *Molecular Biology of the Cell*, *13*(12), 4279–4295.

CHAPTER THREE

Isolation and Study of Adipocyte Precursors

Christopher D. Church[*,‡], Ryan Berry[*,†,‡], Matthew S. Rodeheffer[*,†,‡,1]

[*]Section of Comparative Medicine, Yale University School of Medicine, New Haven, Connecticut, USA
[†]Department of Molecular, Cell, and Developmental Biology, Yale University School of Medicine, New Haven, Connecticut, USA
[‡]Yale Stem Cell Center, Yale University School of Medicine, New Haven, Connecticut, USA
[1]Correspondence author: e-mail address: matthew.rodeheffer@yale.edu

Contents

1. Introduction	32
2. Adipose Tissue Depots and Cell Populations	33
3. Digestion of Whole WAT for Isolation of SVF	33
4. Flow Cytometry and FACS	36
4.1 Antibody staining	36
4.2 FACS and flow cytometry	37
4.3 Cytometers/sorters	39
4.4 Flow cytometry data acquisition	39
4.5 Minimal markers for flow cytometry	39
4.6 Flow cytometry software analysis	40
5. Primary Adipocyte Precursor Cell Culture and Differentiation	41
5.1 Primary adipocyte precursor cell culture	41
5.2 Oil Red O lipid staining	42
6. Digestion of Whole WAT for Isolation of Mature Adipocytes	43
Acknowledgments	45
References	45

Abstract

White adipose tissue (WAT) is a heterogeneous tissue composed of lipid-filled adipocytes and several nonadipocyte cell populations, including endothelial, blood, uncharacterized stromal, and adipocyte precursor cells. Although lipid-filled adipocytes account for the majority of WAT volume and mass, nonadipocyte cell populations have critical roles in WAT maintenance, growth, and function.

As mature adipocytes are terminally differentiated postmitotic cells, differentiation of adipocyte precursors is required for hyperplastic WAT growth during development and in obesity. In this chapter, we present methods to separate adipocyte precursor cells from other nonadipocyte cell populations within WAT for analysis by flow cytometry or purification by fluorescence-activated cell sorting. Additionally, we provide methods to study the adipogenic capacity of purified adipocyte precursor cells *ex vivo*.

1. INTRODUCTION

Distinct white adipose tissue (WAT) depots are distributed throughout the body and include subcutaneous (inguinal) and visceral (epigonadal, retroperitoneal, omental, mesenteric) depots (Cinti, 2007, 2012). While the vast majority of WAT mass is comprised of lipid-filled mature adipocytes, the mature adipocytes account for less than half of the cells in WAT (Eto et al., 2009; Hirsch, 1979; Hirsch & Batchelor, 1976). Several stromal cell populations comprise the nonadipocyte cell populations, including endothelial, blood, and mesenchymal cell populations.

As mature adipocytes are terminally differentiated postmitotic cells, the differentiation of adipocyte precursors is necessary for the establishment of adipocyte number during development and expansion in obesity (Herberg, Döppen, Major, & Gries, 1974; Hirsch & Batchelor, 1976). In most WAT depots, the number of adipocytes is established during childhood and adolescence, with production of new adipocytes from adipocyte precursors in adulthood occurring to replace the dying adipocytes and maintain adipocyte number (Arner et al., 2010; Spalding et al., 2008).

It has long been known that adipocyte precursors reside within WAT depots as culturing of the heterogeneous mixture of WAT resident stromal cells, termed the stromal vascular fraction (SVF), results in the generation of lipid-filled adipocytes (Ng, Poznanski, Borowiecki, & Reimer, 1971). However, until recently, methodology did not allow for the purification of the adipogenic stromal cells from nonadipogenic cells within the SVF, and therefore the identity of adipocyte precursors was unknown.

Currently, the separation of adipocyte precursor populations from nonadipogenic stromal cells using a single marker is not possible (Berry & Rodeheffer, 2013), and thus the separation of these precursor cells requires multicolor flow cytometry (Rodeheffer, Birsoy, & Friedman, 2008) to exclude nonadipogenic endothelial and blood lineage cells, and enrich for mesenchymal cell populations. Recent studies identified early adipocyte progenitors, with a specific cell surface marker profile (Lin−:CD29+:CD34+:ScaI+:CD24+), that are capable of differentiating into a functional WAT depot *in vivo* that rescues hyperglycemia in a lipodystrophic AZIP mouse model (Rodeheffer et al., 2008). Moreover, cells with more limited adipogenic capacity, termed preadipocytes, are enriched in the Lin−:CD29+:CD34+:ScaI+:CD24− population. These cells are enriched for adipogenic expression markers, such as Pparγ2, and are capable of forming adipocytes when transplanted outside of the WAT microenvironment, suggesting further commitment to the adipocyte lineage (Berry & Rodeheffer, 2013),

In this chapter, we provide an overview of methods utilized to identify, analyze, and isolate these adipocyte precursor cell populations from murine WAT. We also detail protocols for isolation by fluorescent-activated cell sorting (FACS) or analysis by flow cytometry. Finally, we provide a description of the methods and conditions routinely used for the culture and adipogenic differentiation of these cell populations.

2. ADIPOSE TISSUE DEPOTS AND CELL POPULATIONS

WAT depots are composed of a heterogeneous mixture of cell populations including terminally differentiated lipid-filled adipocytes and an SVF that contains blood lineage cells, endothelial cells, immune cells, other uncharacterized stromal cells, and adipocyte precursor cells. Specific cell populations can be identified and purified based on the expression of cell surface markers (Table 3.1) and subsequently analyzed by flow cytometry or purified by FACS. Flow cytometry allows for the single cell analysis of extracellular and intracellular protein expression in complex samples in any tissue, for any markers/species for which antibodies are available. FACS allows the isolation of live cells from specific populations for *ex vivo* analysis. Adipocyte precursors can be identified by the lack of expression of lineage cell surface markers such as CD45 (blood) and CD31 (endothelial) and enriched for expression of mesenchymal stem cell markers including CD29, CD34, PdgfRα, Sca-1, and CD24 (to distinguish progenitors; Table 3.1). Additionally, antibodies recognizing cell surface markers of specific hematopoietic cell populations, such as macrophages, B cells, and T cells, can be added to assess changes in multiple SVF populations simultaneously.

3. DIGESTION OF WHOLE WAT FOR ISOLATION OF SVF

1. Individual mouse adipose depots are carefully excised and thoroughly minced with scissors (1–2 mm pieces), using sterile techniques.
2. The minced adipose tissue is digested in 0.8 mg/ml collagenase type 2 (Worthington Biochemical Corporation, NJ, USA; LS004174) in sterile Hank's Balanced Salt Solution (HBSS; diluted from 10× stock no calcium, no magnesium, no Phenol Red; available from Life Technologies, CA, USA; product number 14185-052) containing 3% bovine serum albumin (BSA), 1.2 mM calcium chloride, 1.0 mM magnesium chloride,

Table 3.1 Commonly used antibody clones and dilutions for the identification and isolation of adipose tissue cell populations

Antibody	Gene name	Cell population	Supplier	Clone	Flurochrome	Dilution[a]	Excitation (nm)	Bandpass filter	Longpass filter	Detected wavelengths
CD45	Leukocyte common antigen	Blood	eBioscience	30-F11	APC–eFluor e780	1:5000	640 nm (Red)	780/60	755 LP	755–810 nm
CD31	Pecam1; platelet/ endothelial cell adhesion molecule 1	Endothelial	eBioscience	390	PE-Cy7	1:1200	532 nm (Green)	780/60	735	750–810 nm
CD34	Cluster differentiation hematopoietic progenitor cell antigen	Mesenchymal stem cells	Biolegend	MEC14.7	Alexa Fluor 647	1:200	640 nm (Red)	660/20	N/A	650–670 nm
CD29	Integrin β1		Biolegend	HMBeta1-1	Alexa Fluor 700	1:400	640 nm (Red)	710/50	690 LP	690–735 nm

Sca1	Stem cell antigen-1 or Ly6A/E	BD Horizon	D7	V450	1:1000	405 nm (Violet)	450/50 N/A 425–475 nm
CD24	Heat-stable antigen, HSA or HsAg	BD Bioscience	M1/69	PE	1:100	532 nm (Green)	575/26 N/A 562–588 nm
				PerCP-Cy5.5	1:100	532 nm (Green)	710/50 685 LP 685–735 nm
PDGFRa	Platelet-derived growth factor receptor alpha; CD140a	Biolegend	APA5	PE	1:200	532 nm (Green)	575/26 N/A 562–588 nm

[a] Dilutions shown should be used as a guideline for initial control experiments. The correct dilution needs to be determined for each antibody, fluorochrome, and the specific cytometer/sorter used for analysis, as laser power and the filters can influence the signal versus noise.

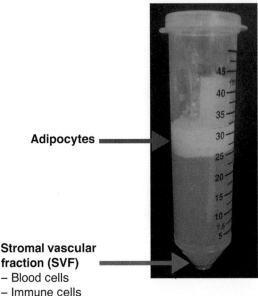

Figure 3.1 Separation of SVF from adipocytes. (See the color plate.)

and 0.8 mM zinc chloride for 75 min in a shaking water bath (120–140 rpm) including vigorous shaking by hand (for 10–20 s) after 60 min of incubation. *Approximately 5 ml of digestion buffer is used to digest one WAT depot from one mouse.*

3. Floating adipocytes are separated from the SVF by centrifugation at 300 ×g for 3 min (Fig. 3.1). *Note: This digestion protocol results in a low percentage of intact mature adipocytes. Intact mature adipocytes can be isolated through a modified digestion protocol* (Section 6).

4. The floating adipocyte fraction and supernatant is removed and the SVF pellet is resuspended in HBSS 3% BSA wash buffer and sequentially filtered through sterile 70 μm (BD Biosciences, CA, USA; product number 352350) and 40 μm (BD Biosciences, CA, USA; product number 352340) nylon mesh filters before antibody staining.

4. FLOW CYTOMETRY AND FACS

4.1. Antibody staining

1. Antibodies (Table 3.1; user-determined antibody concentrations) are diluted in HBSS with 3% BSA and the SVF is resuspended in antibody

staining solution and placed on ice in the dark for 20 min. The quantity of antibody, volume of staining solution, and incubation period should be optimized for each antibody and sample amount (e.g., approximately 500,000 cells in 100 μl of antibody staining solution).

Note: Depending on the source of tubes used for staining, it may be necessary to coat all tubes in HBSS with 3% BSA overnight at 4 °C to maintain cell viability throughout the procedure.

For maximal recovery of SVF cells through all steps, wing-bucket style centrifuges should be used to pellet the cells at the bottom of the tube.

2. An excess of HBSS 3% BSA is added to wash and then the SVF is centrifuged at $300 \times g$ for 3 min. The wash buffer is carefully removed and the SVF pellet is resuspended in HBSS with 3% BSA and subsequently filtered through a 40 μm nylon filter prior to analysis by flow cytometry or purification by FACS.

3. For FACS purification, the SVF is resuspended in FACS buffer (PBS with 0.5% BSA) with 0.5 g/ml propidium iodide (Sigma-Aldrich, MO, USA; P4170)—a fluorescent, plasma membrane impermeant molecule that intercalates into DNA—to identify and exclude dead cells. The cells are then filtered through a 40 μm filter several times until they flow easily through the filter to reduce clogging the cell sorter lines.

4. FACS purified cell populations are collected in 1.5 mL tubes that have been coated with HBSS 3% BSA (1.5 mL tubes are coated by filling the tubes with HBSS 3% BSA and incubating them at 4 °C for greater than 24 h). The buffer is removed from the tubes prior to cell collection. Live cells can be used for *in vivo* transplantation to assess lineage commitment and differentiation (Berry & Rodeheffer, 2013; Rodeheffer et al., 2008) and *in vitro* differentiation (Section 5). Additionally, cells may be sorted directly into TRIzol® LS Reagent (Life Technologies, CA, USA; product number 10296) for RNA extraction and subsequent gene expression studies, which differs from the standard TRIzol® reagent in concentration and permits processing of larger samples.

4.2. FACS and flow cytometry

The selection of multicolor fluorochrome combinations for flow cytometry can be challenging and is dependent on the specific flow cytometry system—which can have different laser and optical filter combinations to excite and properly detect a given combination of fluorochromes (Baumgarth & Roederer, 2000; Darzynkiewicz, Crissman, & Robinson, 2000; Maecker & Trotter, 2008; Ormerod, 2000; Purdue University, 2013).

The selection of fluorochrome combinations and filters can be assisted by tools such as the BD Biosciences Spectrum Guide and Fluorescence Spectrum Viewer (BD Biosciences, 2013) or Invitrogen's Flow Cytometry and data analysis tutorials (Invitrogen, Life Technologies, CA, USA, 2013). Additionally, multicolor flow cytometry requires compensation between the emission spectra of fluorochromes used in combination due to their potential overlap. Compensation is the mathematical elimination of spectral overlap (Baumgarth & Roederer, 2000; Roederer, 2001) and must be performed during multicolor flow cytometry when any two fluorochromes used have partially overlapping emission spectra. Compensation can be performed before or after data collection manually or using software-based automation. When software automation is used, it is recommended to manually check the compensation settings to ensure that calculated compensation values are correct. This becomes increasingly important as the number of fluorchromes used per sample increases as the likelihood of spectral overlap also increases, as does the risk for compensation errors (Baumgarth & Roederer, 2000). When compensation is not performed correctly, it is possible for a population that is negative for a specific antigen to appear positive for that antigen simply because the fluorescent signal from a different fluorescently conjugated antibody "bleeds" into the filter that is intended to detect the fluorescently conjugated antibody of interest.

To definitively determine if an observed fluorescent signal is derived from the fluorescently conjugated antibody of interest, a fluorescent-minus-one (FMO) control should be performed. In this control, a sample is split into two with one sample being stained with all of the antibodies in the antigen scheme. The second sample, the FMO control, is stained with all of the antibodies except for the antibody that binds to the antigen of interest. Both samples are then analyzed using the same compensation settings. If a positive signal is observed for the antigen of interest in the FMO control sample then the observed signal is the result of bleed through from the emission spectrum of a different fluorescently conjugated antibody. In this scenario, compensation settings must be properly adjusted. In a sample that it is compensated correctly, all cells in the FMO control sample will have a negative fluorescence signal for the antigen of interest.

When a sample has been stained with all antibodies and is acquired under the proper compensation settings, any cell that has fluorescent signal greater than the negative population in the FMO control can be considered positively stained for the fluorescently conjugated antibody of interest. In this manner, FMO controls can be used to ensure that any fluorescent signal above background in a multicolor flow cytometry scheme is due to staining by the antibody of interest and not from "bleed through" as a result of spectral overlap. In some cases, particularly intracellular staining, an isotype control may be a more appropriate control. In this case, samples are prepared as described above for the FMO control, but an antibody of the same isotype and conjugated fluorochrome are substituted for the antibody that recognizes the antigen of interest. Alternatively, if cell signaling events or expression repression/induction are being assayed it may be possible to use positive and negative conditions as a biological control.

4.3. Cytometers/sorters

We routinely sort our SVF samples on a BD FACSAria II and III cell sorters and analyze them on BD LSRII flow cytometers (BD Biosciences, CA, USA), each equipped with BD FACSDiva Software. Please see Table 3.1 for a recommended setup including excitation/emission optical filters based on the BD LSRII instrument.

4.4. Flow cytometry data acquisition

Specific flow cytometers and manufacturers provide software for collecting and analyzing sample data during the acquisition of experiments. We routinely start by selecting cells based on forward scatter area (FSC-A) and side scatter area (SSC-A). Live cells are subsequently gated on both SSC and FSC singlets, ensuring that the staining of individual cells is analyzed.

Note: The FSC and SSC may need to be altered to ensure that the cell population of interest is positioned on scale based on the FSC and SSC. Adipocyte precursor cells may appear larger in size than other SVF populations such as immune cells that are routinely analyzed on flow cytometers. This can be facilitated by back gating—showing the final gated population of interest within the population of its ancestors based on the hierarchy.

Data can be presented in a number of formats including dot plots and histograms that can display fluorescent intensity of one or more fluorochromes simultaneously (for examples of dot plots, see Fig. 3.2).

4.5. Minimal markers for flow cytometry

Depending on the limitation of excitation and emission detection on different flow cytometers, an alternative minimal set of cell surface markers can be used to analyze and isolate adipocyte precursors from WAT (Fig. 3.3; Table 3.2). While the use of minimal marker schemes results in reduced adipocyte precursor purity compared to the scheme described above (Table 3.2), it can still be used to enrich for adipocyte precursors.

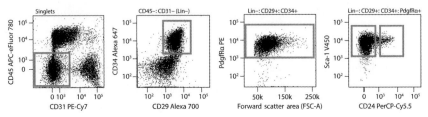

Figure 3.2 Flow cytometry gating of adipocyte precursor cells. Each plot displays only the cell population from the populations gated in the leftward plot, which is also indicated at the top of each plot.

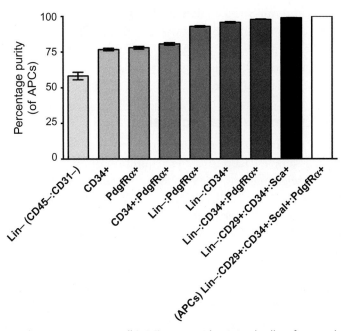

Figure 3.3 Adipocyte precursor cell (APC) purity with minimal cell surface markers. Data ($n=4$) are expressed as mean ± SEM.

4.6. Flow cytometry software analysis

Tree Star's FlowJO software package is commonly used for flow cytometry analysis and can analyze data from any flow cytometer in Windows and MacIntosh operating systems (Tree Star, 2012). FlowJo is based around "workspace" areas that include detailed graphical reports, gating, compensation, tables, and statistical analysis of parental populations. Compensation and gating templates can be applied to multiple samples within an experiment or to multiple independent experiments. Additionally,

Table 3.2 Minimal cell surface markers for the identification and isolation of adipose tissue cell populations

Antibody combination	Percentage of adipocyte precursor cells (±SEM; n = 4)
	(CD45−/CD31−/CD34+/CD29+/ScaI+: PdgfRα)
[a]Lin− (CD45−:CD31−)	58.27% ± 2.6
CD34+	76.72% ± 0.89
PdgfRα	77.97% ± 1.00
CD34+:PdgfRα	80.68% ± 0.85
[a]Lin−:PdgfRα	92.87% ± 0.47
[a]Lin−:CD34+	95.68% ± 0.49
[a]Lin−:CD34+:PdgfRα	97.69% ± 0.33
[a]Lin−:CD34+:CD29+:ScaI+	98.42% ± 0.25
[a]Lin−:CD34+:CD29+:ScaI+: PdgfRα	100%

[a]The same fluorochrome can be used for the CD45 and CD31 antibodies for exclusion of these "lineage" populations in a single channel.

high-resolution, publication quality histograms and dot plots can be exported in multiple image formats.

Another example of flow cytometry software is the BD Biosciences FACSDiva™ software package used on BD flow cytometers and sorters (BD Biosciences, NJ, USA). BD FACSDiva™ is routinely used to establish multicolor analysis templates including compensation parameters necessary for establishing multicolor templates. Additionally, BD FACSDiva™ software, like FlowJO, allows data to be expressed in detailed graphical reports, tables, and statistical population analysis. Data files from BD FACSDiva™ can be exported for further analysis and imported into FlowJo for high-quality images.

5. PRIMARY ADIPOCYTE PRECURSOR CELL CULTURE AND DIFFERENTIATION

5.1. Primary adipocyte precursor cell culture

Adipocyte precursor cells can be isolated from WAT depots of any mouse model through FACS for further study of adipogenesis *in vitro*.

1. Adipocyte precursors from subcutaneous WAT (SWAT) are isolated via FACS and plated onto carboxyl-coated 24-well plates (BD Biosciences,

CA, USA; 354775) in DMEM (ATCC, VA, USA; Cat. # 30-2002) supplemented with 10% fetal bovine serum (FBS; GIBCO, Life Technologies, CA, USA) and maintained in a 5% CO_2 atmosphere. Approximately 25,000 cells are plated per well of a 24-well plate. These cells take on a fibroblast-morphology after 3–5 days and are allowed to grow to confluence with DMEM media supplemented with 10% FBS changed every 48 h.
2. Once at confluence the cells are held for 48 h without changing the media.
3. The media is exchanged with adipocyte differentiation cocktail (MDI: 3-isobutyl-1-*m*ethylxanthine, *d*examethasone, and *i*nsulin). DMEM media supplemented with 10% FBS, 1 µg/ml insulin (Sigma-Aldrich, MO, USA; I-6634; 10,000× stock is 10 mg/ml in 0.01 M HCl), 0.25 µg/ml dexamethasone (Sigma-Aldrich, MO, USA; D4902; 10,000 × stock is 1 mg/ml in 100% ethanol), and 30 µg/ml IBMX (3-isobutyl-1-methylxanthine; Sigma-Aldrich, MO, USA; I5879; 500× stock is 15 mg/ml in 0.3 M KOH; made fresh each time).
4. Every 48 h, the media is exchanged with maintenance differentiation media (DMEM supplemented with 10% FBS and 1 µg/ml insulin). Lipid filling is observed within 48–72 h after the addition of adipocyte differentiation media in step 3. By day 7, lipid filling is complete and cells can be harvested for RNA/protein isolation or stained with Oil Red O (Section 5.2).

Note: When isolated from the inguinal SWAT, adipocyte precursor cells, positive and negative for CD24 (Rodeheffer et al., 2008) differentiate with the addition of only insulin (1 µg/ml in DMEM supplemented with 10% FBS); however, very low rates of differentiation are observed when similar cells are isolated from the visceral epigonadal depot even when induced with MDI adipocyte differentiation cocktail (Fig. 3.4B and C).

5.2. Oil Red O lipid staining

1. For lipid staining by Oil Red O, the differentiated cells are fixed with 2% formaldehyde and 0.2% glutaraldehyde in PBS for 15 min and then rinsed in PBS for 10 min, twice in water for 1 min, followed by 30 s in 60% isopropanol.
2. The cells are stained with Oil Red O (0.7% in 60% isopropanol; Electron Microscopy Sciences, PA, USA; Cat. # 26503-02) for 10 min and rinsed with 60% isopropanol for 1 min followed by water.

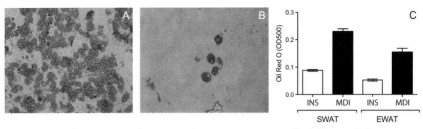

Figure 3.4 Differentiation of primary adipocyte precursor cells. (A) Day 7 Subcutaneous WAT (SWAT), (B) Day 7 Epigonadal WAT (EWAT), (C) Day 7 Oil Red O quantification. INS, insulin-only differentiation; MDI (adipogenic cocktail containing IBMX, dexamethasone, and insulin). (See the color plate.)

3. Oil Red O stained cells are directly visualized and imaged using an inverted microscope (Fig. 3.4A and B).
4. Quantification of lipid accumulation is achieved by Oil Red O extraction by lysis (100% isopropanol with 4% NP40 substitute IGEPAL CA-630, Sigma-Aldrich, MO, USA; I8896; 300 μl per well of a 24-well plate) and gentle agitation for 10 min at room temperature.
5. Following Oil Red O extraction 100 μl is transferred to a 96-well plate and absorbance measured at 490–520 nm using a plate reader or spectrophotometer (Fig. 3.4C).

6. DIGESTION OF WHOLE WAT FOR ISOLATION OF MATURE ADIPOCYTES

1. Individual adipose depots are carefully excised and minced into 2–4 mm pieces with scissors.
2. Minced WAT is digested in 1 mg/ml collagenase type 2 (Worthington Biochemical Corporation, NJ, USA; LS004174) in Krebs Ringer buffer (KRB), 1.2 mM calcium chloride, 1.0 mM magnesium chloride, 0.8 mM zinc chloride, 4 mM glucose (Sigma-Aldrich, MO, USA; G7021), and 500 nM adenosine (Sigma-Aldrich, MO, USA; A9251) for 80 min in a shaking water bath (less than 110 rpm) including gentle mixing by inversion every 20 min. Approximately 5 ml of digestion buffer is used to digest one WAT depot from one mouse.
3. The tissue digest is passed through a 200 μm nylon filter.
4. Floating adipocytes are separated from the SVF by centrifugation at $150 \times g$ for 8 min.

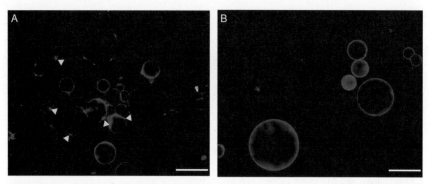

Figure 3.5 Isolation of adipocytes. (A) Intact adipocytes with single DAPI positive nuclei within a cell membrane. (B) Lipid ghosts without DAPI positive nuclei. White scale bar represents 100 μm. (See the color plate.)

5. The floating fraction is carefully transferred with a wide opening plastic transfer pipette into KRB containing 4 mM glucose and 500 nM adenosine to wash the adipocyte fraction. Excess KRB buffer is removed from below the adipocyte layer using a glass Pasteur pipette.
6. Isolation of intact adipocytes is verified by staining for plasma membrane with Cell Mask Orange (CMO; Life Technologies, CA, USA; C10045) and nuclei with DAPI (Life Technologies, CA, USA; D1306).
7. The stained adipocyte solution is transferred to a slide and immediately visualized under a fluorescent microscope for verification of purified intact adipocytes. Intact adipocytes are characterized by a CMO+ plasma membrane that contains a lipid droplet and a single DAPI+ nucleus near the CMO+ membrane (Fig. 3.5A).

The presence of excessive CMO—lipid droplets, free-floating nuclei, adipocyte ghosts (adipocyte-like structures that lack a nucleus; Fig. 3.5B) or cells with multiple nuclei suggest adipocyte lysis and/or SVF contamination. Because of the fragility of lipid-filled mature adipocytes, several isolations may be necessary to acquire a preparation of intact, purified mature adipocytes with minimal stromal cell contamination. The purified adipocyte fraction can be further studied for gene expression and metabolic studies.

8. Alternatively, the adipocytes can be lysed to yield adipocyte nuclei by resuspending the adipocyte layer in 2–4 volumes of adipocyte nuclei isolation buffer (KRB containing 0.2% NP40 substitute IGEPAL CA-630 (Sigma-Aldrich, MO, USA; I8896)), followed by vortexing for 10–20 s and placing on ice for 5 min.

9. Adipocyte nuclei are separated from cellular debris by centrifugation at $2000 \times g$ for 5 min and the careful removal of excess adipocyte nuclei isolation buffer. The adipocyte nuclei pellet is resuspended in nuclei flow cytometry buffer (KRB containing 0.02% NP40 substitute IGEPAL CA-630; Sigma-Aldrich, MO, USA) with DAPI and CMO for flow cytometry analysis. Gating on CMO−: DAPI+ nuclei allows for the analysis of purified adipocyte nuclei.

ACKNOWLEDGMENTS

We thank Elise Jeffery for protocol developments and discussion. This work was supported by American Diabetes Association Award 7-12-JF-46, DERC pilot project Grant DK045735 and NIDDK Grant DK090489 to M. S. R. and EMBO long-term postdoctoral fellowship (ALTF 132-2011 to C. C.).

REFERENCES

Arner, E., Westermark, P. O., Spalding, K. L., Britton, T., Rydén, M., Frisén, J., et al. (2010). Adipocyte turnover: Relevance to human adipose tissue morphology. *Diabetes*, *59*(1), 105–109. http://dx.doi.org/10.2337/db09-0942.

Baumgarth, N., & Roederer, M. (2000). A practical approach to multicolor flow cytometry for immunophenotyping. *Journal of Immunological Methods*, *243*(1–2), 77–97.

BD Biosciences, NJ, USA. (2013). *Multicolor flow cytometry–tools*. From http://www.bdbiosciences.com/research/multicolor/tools/index.jsp retrieved 29.03.13.

Berry, R., & Rodeheffer, M. S. (2013). Characterization of the adipocyte cellular lineage in vivo. *Nature Cell Biology*, *15*(3), 302–308. http://dx.doi.org/10.1038/ncb2696.

Cinti, S. (2007). *The adipose organ. Nutrition and health: Adipose tissue and adipokines in health and disease*. Totowa, NJ: Humana Press Inc. ISBN: 978-1-58829-721-1

Cinti, S. (2012). The adipose organ at a glance. *Disease Models & Mechanisms*, *5*(5), 588–594. http://dx.doi.org/10.1242/dmm.009662.

Darzynkiewicz, Z., Crissman, H. A., & Robinson, J. P. (2000). *Methods in cell biology Cytometry* (3rd ed.). (Vols. 63–64). San Diego: Academic press.

Eto, H., Suga, H., Matsumoto, D., Inoue, K., Aoi, N., Kato, H., et al. (2009). Characterization of structure and cellular components of aspirated and excised adipose tissue. *Plastic and Reconstructive Surgery*, *124*(4), 1087–1097. http://dx.doi.org/10.1097/PRS.0b013e3181b5a3f1.

Herberg, L., Döppen, W., Major, E., & Gries, F. A. (1974). Dietary-induced hypertrophic–hyperplastic obesity in mice. *Journal of Lipid Research*, *15*(6), 580–585.

Hirsch, J. (1979). Isotopic labeling of DNA in rat adipose tissue: Evidence for proliferating cells associated with mature adipocytes. *Journal of Lipid Research*, *20*(6), 691–704.

Hirsch, J. J., & Batchelor, B. B. (1976). Adipose tissue cellularity in human obesity. *Clinics in Endocrinology and Metabolism*, *5*(2), 299–311.

Invitrogen, Life Technologies, CA, USA (Ed.). (2013). An introduction to flow cytometry. From http://www.invitrogen.com/site/us/en/home/support/Tutorials.html retrieved 29.3.13.

Maecker, H., & Trotter, J. (2008). Selecting reagents for multicolor flow cytometry with BD™ LSR II and BD FACSCanto™ systems. Application note. *Nature Methods*, *5*, 1–8.

Ng, C. W., Poznanski, W. J., Borowiecki, M., & Reimer, G. (1971). Differences in growth in vitro of adipose cells from normal and obese patients. *Nature*, *231*(5303), 445.

Ormerod, M. G. (2000). *Flow cytometry: A practical approach*. (3rd ed.), *Issue 229 of Practical Approach Series*. Oxford University Press. ISBN-10 0-19-963824-1.
Purdue University, (Ed.). (2013). Purdue University Cytometry Laboratories. From http://www.cyto.purdue.edu/flowcyt/educate.htm retrieved 20.3.13.
Rodeheffer, M. S., Birsoy, K., & Friedman, J. M. (2008). Identification of white adipocyte progenitor cells in vivo. *Cell*, *135*(2), 240–249. http://dx.doi.org/10.1016/j.cell.2008.09.036.
Roederer, M. (2001). Spectral compensation for flow cytometry: Visualization artifacts, limitations, and caveats. *Cytometry*, *45*(3), 194–205.
Spalding, K. L., Arner, E., Westermark, P. O., Bernard, S., Buchholz, B. A., Bergmann, O., et al. (2008). Dynamics of fat cell turnover in humans. *Nature*, *453*(7196), 783–787. http://dx.doi.org/10.1038/nature06902.
Tree Star. (2012). FLOWJO Data Analysis Software for Flow Cytometry (Vol. 10, pp. 1–32). Retrieved from http://www.flowjo.com/home/manual.html retrieved 20.3.13.

CHAPTER FOUR

Imaging of Adipose Tissue

Ryan Berry[*], Christopher D. Church[†], Martin T. Gericke[‡], Elise Jeffery[§], Laura Colman[†], Matthew S. Rodeheffer[*,†,¶,1]

[*]Department of Molecular, Cell, and Developmental Biology, Yale University School of Medicine, New Haven, Connecticut, USA
[†]Section of Comparative Medicine, Yale University School of Medicine, New Haven, Connecticut, USA
[‡]Institute of Anatomy, University of Leipzig, Leipzig, Germany
[§]Department of Cell Biology, Yale University School of Medicine, New Haven, Connecticut, USA
[¶]Yale Stem Cell Center, Yale University School of Medicine, New Haven, Connecticut, USA
[1]Corresponding author: e-mail address: matthew.rodeheffer@yale.edu

Contents

1. Introduction   48
2. Imaging of Whole Mounted Adipose Tissue   48
   2.1 Preparation of slides   49
   2.2 Confocal imaging of whole mounted adipose tissue   51
3. Imaging of Sectioned Adipose Tissue   53
   3.1 Paraffin-sectioned adipose tissue   53
   3.2 Frozen sectioning   68
Acknowledgments   72
References   72

Abstract

Adipose tissue is an endocrine organ that specializes in lipid metabolism and is distributed throughout the body in distinct white adipose tissue (WAT) and brown adipose tissue (BAT) depots. These tissues have opposing roles in lipid metabolism with WAT storing excessive caloric intake in the form of lipid, and BAT burning lipid through non-shivering thermogenesis. As accumulation of lipid in mature adipocytes of WAT leads to obesity and increased risk of comorbidity (Pi-Sunyer et al., 1998), detailed understanding of the mechanisms of BAT activation and WAT accumulation could produce therapeutic strategies for combatting metabolic pathologies. As morphological changes accompany alterations in adipose function, imaging of adipose tissue is one of the most important tools for understanding how adipose tissue mass fluctuates in response to various physiological contexts. Therefore, this chapter details several methods of processing and imaging adipose tissue, including bright-field colorimetric imaging of paraffin-sectioned adipose tissue with a detailed protocol for automated adipocyte size analysis; fluorescent imaging of paraffin and frozen-sectioned adipose tissue; and confocal fluorescent microscopy of whole mounted adipose tissue. We have also provided many example images showing results produced using each protocol, as well as commentary on the strengths and limitations of each approach.

1. INTRODUCTION

Adipose tissue is distributed throughout the body in distinct "white" and "brown" adipose tissue depots. White adipose tissue (WAT) is largely composed of unilocular lipid-filled adipocytes that specialize in lipid storage, whereas brown adipose tissue (BAT) is largely composed of multilocular adipocytes that specialize in lipid burning. Although adipocytes compose the majority of WAT and BAT volume, both tissue types contain a large number of stromal cells including blood, endothelial, fibroblastic, and adipocyte precursor cells which are essential for adipose tissue function. Changes in adipose tissue morphology accompany adipose tissue development (Birsoy et al., 2011), the onset of obesity (Sun, Kusminski, & Scherer, 2011) and response to cold challenge (Seale et al., 2011), making imaging of adipose tissue a powerful tool for understanding the basic biology of adipose tissue development, maintenance, growth, and remodeling. Furthermore, imaging of adipose tissue from genetic mouse models allows for study of adipocyte precursor localization (Berry & Rodeheffer, 2013; Gupta et al., 2012; Lee, Petkova, Mottillo, & Granneman, 2012; Tang et al., 2008) and adipocyte lineage derivation (Berry & Rodeheffer, 2013; Tang et al., 2008; Tran et al., 2012), providing insight into how tissue organization allows WAT to participate in and respond to systemic metabolism. Understanding the mechanisms of WAT accumulation and BAT activation may be therapeutically beneficial as the accumulation of WAT leads to obesity and increased risk of co-morbidities (Pi-Sunyer et al., 1998). In this chapter, we will provide detailed protocols for preparing and imaging whole mount, paraffin-sectioned and frozen-sectioned adipose tissue. We will also provide discussion on the benefits and limitations of each approach to guide the application of these imaging approaches to future studies of adipose tissue biology.

2. IMAGING OF WHOLE MOUNTED ADIPOSE TISSUE

Adipose tissue that has been stained with fluorescent antibodies/dyes or isolated from fluorescent reporter mice can easily be visualized in whole mount through confocal microscopy. The advantage of imaging adipose tissue in whole mount is that it does not require fixation, processing, embedding, or sectioning. As these steps can decrease antigen recognition, deplete fluorescent signal, and lead to increased auto-fluorescence, imaging of adipose tissue in whole mount generally provides a high signal/noise ratio and allows for clear distinction of fluorescently labeled cells. This approach has recently been used by our group to perform lineage tracing of WAT by clearly differentiating mature adipocytes from stromal cells *in situ* (Berry & Rodeheffer, 2013). The disadvantage of this technique is that antigen labeling with fluorescently conjugated antibodies can be less robust than what is observed in tissue prepared for immunohistochemistry (IHC) as the antibody must permeate the tissue, but this is antibody and antigen-dependent.

2.1. Preparation of slides

Materials needed
1. Microscope slides (Thermo Scientific, MA, USA, 4951F-001)
2. Coverslips (Fischer Scientific, MA, USA, 12-545-F)
3. 10 mL syringe (Sigma-Aldrich, MO, USA, Z248029)
4. 16-gauge needle (BD Biosciences, CA, USA, 305198)
5. Fluoromount-G (Southern Biotech, AL, USA, 0100-01)
6. Rapid dry nail polish
7. Sterile PBS (Life Technologies, NY, USA, 14190-144)
8. Vasoline

Prior to starting
1. Fill 10 mL syringe with vaseline.
2. Attach 16 gauge needle to filled syringe.

Protocol
- A diagram of a completed slide prepared for imaging of whole mounted adipose tissue is shown in Fig. 4.1.
    1. Dissect adipose tissue from mouse.
    2. Cut samples into pieces that are approximately 4 mm × 4 mm × 2 mm.
    3. Stain samples with application-specific fluorescent antibodies or dyes.
- A list of commonly used stains, antibodies, and fluorescent reporter proteins along with *recommended* concentrations and staining times is provided in Table 4.1.
- We have found that fixation and permeabilization is not required for labeling of antigens with the antibodies listed in Table 4.1. However, labeling of some antigens may require fixation and/or permeabilization (Whole mount staining protocols, 2013). In our hands, incubation of small samples in mild fixative (1% paraformaldehyde) or mild detergent (0.5% Tween 20) for 30 min prior to labeling does not disrupt antigen labeling or lead to high levels of auto-fluorescence.

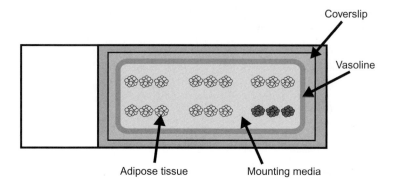
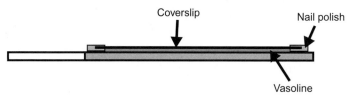

Figure 4.1 A depiction of a slide prepared for imaging of adipose tissue in whole mount. (See the color plate.)

4. Wash samples in sterile PBS.
5. Place samples onto the center of a slide.
6. Make a boundary of vaseline around the samples that is approximately 2 mm high using the vaseline filled syringe.
- The vaseline boundary should be about the same thickness (height off of the slide) as the samples.
7. Fill the inside of the vaseline boundary with Fluoromount-G.
8. Gently place a coverslip onto the slide such that it is resting on top of the vaseline boundary.
9. Lightly push down on the coverslip until the top of each sample is pressed against the underside of the coverslip.
- For best results, ensure that the coverslip is pressed down evenly on all sides, resulting in a uniform 1 mm gap between the coverslip and the slide.
10. Remove excess Fluoromount-G that has seeped out from inside the vaseline boundary after pressing down the coverslip.
11. Apply nail polish to all edges of the coverslip to adhere the coverslip to the slide.
12. Allow nail polish to dry.

Table 4.1 Commonly used fluorescent stains, antibodies, and reporter proteins for whole mount confocal imaging of adipose tissue

Antibody/ stain	Fluorochrome/ reporter protein	Company, catalog #	Excitation laser (nm)	Emission filter (nm)
N/A	dTomato	N/A	543	590–650
N/A	eGFP	N/A	488	505–540
HCS LipidTox[a]	"Green"	Invitrogen, H34475	488	505–540
HCS LipidTox[a]	"Deep Red"	Invitrogen, H34477	633	645–700
Isolectin GS-IB4[a]	Alexa Fluor 488	Invitrogen, I21411	488	505–540
Isolectin GS-IB4[a]	Alexa Fluor 647	Invitrogen, I32450	633	645–700
Cell Mask[b]	"Orange"	Invitrogen, C10045	543	565–585
CD45[c]	Alexa Fluor 647	Biolegend, 103123	633	645–700
F4/80[c]	Alexa Fluor 647	Biolegend, 123121	633	645–700
CD11b[c]	Brilliant Violet 421	Biolegend, 101235	405	420–470
CD24[c]	Brilliant Violet 421	Biolegend, 101825	405	420–470
DAPI[d]	"Blue"	Invitrogen, D1306	405	420–470

[a]Stain at 1:100 in PBS for 1 h.
[b]Stain at 1:2000 in PBS for 1 h.
[c]Stain at 1:100 for 24 h.
[d]Stain at 1:10,000 in PBS for 1 h.

2.2. Confocal imaging of whole mounted adipose tissue

- For brevity, all fluorescent molecules including fluorochromes conjugated to antibodies, fluorescent stains, and fluorescent reporter proteins are referred to as "fluorochromes" in this section.

Equipment needed
1. Leica TCS SP5 Spectral Confocal Microscope or equivalent

- For excitation, this microscope is equipped with a violet laser (405 nm), an argon laser (split into four excitation wavelengths: 458, 476, 488, and 514 nm) and a He/Ne laser (provides two excitation wavelengths: 543 and 633 nm).
- For emission detection, this microscope uses a prism spectrophotometer detection system that allows the user to set the desired emission wavelengths to be detected for each excitation laser–fluorochrome combination.

Software needed
1. Leica AF6000 digital imaging software or equivalent

Materials needed
1. Slide prepared as described in Section 2.1 containing the following samples:
    a. Sample containing all fluorochromes of interest.
    b. Sample fluorescent-minus-one (FMO) controls for each fluorochrome.
        - A FMO control is a sample that contains all fluorochromes of interest except for one of them. Example: To image adipose tissue that contains eGFP, dTomato, and Alexa Fluor 647, three FMO controls are needed.
            i. FMO control for Alexa Fluor 647 that contains only eGFP and dTomato.
            ii. FMO control for dTomato that contains only eGFP and Alexa Fluor 647.
            iii. FMO control for eGFP that contains only dTomato and Alexa Fluor 647.

Protocol
1. In the software, set excitation laser and emission filter parameters as listed in Table 4.1 for each fluorochrome of interest.
    - For best results, images are acquired through "sequential scanning" in which each fluorochrome is excited and detected in a separate scan of the tissue. Therefore, a new scan is made for each fluorochrome such that only the laser and filter needed to excite and detect that fluorochrome is active in that scan. When an image is acquired, each scan is performed sequentially and the detected signals from each scan can be overlayed to create a composite image.

2. Apply a small drop of immersion oil onto the coverslip of the slide and mount the slide onto the confocal microscope.
3. Set laser voltage and gain settings for each fluorochrome as follows:
   a. Focus on the first FMO sample using a scan that is designated to excite/detect a fluorochrome present in this FMO sample.
   b. Switch to the scan that is designated to excite/detect the fluorochrome that is *not* present in this FMO sample.
   c. Adjust laser voltage and gain settings for this scan such that the background fluorescence is barely visible.
   d. Save the laser voltage and gain settings for this scan.
   e. Repeat steps a–d to set the laser voltage and gain settings for each fluorochrome using the appropriate FMO controls.
4. Focus on a sample that contains all fluorochromes of interest.
5. Acquire the image at the desired resolution using the laser voltage and gain settings determined in step 4 for each scan. Example images of WAT acquired with the fluorochromes, excitation laser, and emission filter settings listed in Table 4.1 are shown in Fig. 4.2.

3. IMAGING OF SECTIONED ADIPOSE TISSUE

3.1. Paraffin-sectioned adipose tissue

As staining of whole mount tissue requires the antibody/stain to permeate the tissue, more robust staining can be accomplished by sectioning the tissue prior to staining. The most common method of sectioning involves tissue fixation and embedding in paraffin. Although these methods tend to mask antigens, they maintain excellent tissue morphology and therefore allow for tissue analysis following cellular staining. Additionally, antigens can be "unmasked" through an antigen retrieval step to facilitate antibody labeling. The following protocols for paraffin sectioning, hematoxylin and eosin (H&E) staining and immunohistochemical antigen labeling have been adapted from previously published protocols (Cinti, Zingaretti, Cancello, Ceresi, & Ferrara, 2001), but contain several specific modifications.

3.1.1 Preparation of slides
Materials needed
1. Microscope slides (Thermo Scientific, MA, USA, 4951F-001)
2. 5× Zinc-formalin concentrate (Anatech Ltd, MI, USA, 141)
3. Sterile PBS (Life Technologies, NY, USA, 14190-144)

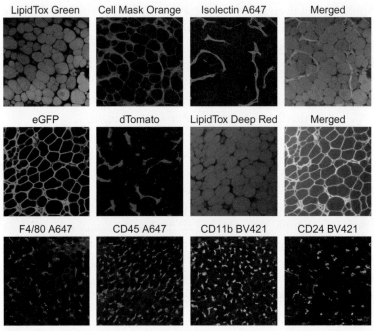

Figure 4.2 Example images of whole mounted fluorochrome-containing WAT acquired through confocal microscopy. WAT was isolated from fluorescent reporter mice (eGFP and dTomato) and/or incubated in fluorescent stains/antibodies to label lipids (LipidTox), plasma membranes (Cell Mask Orange), endothelial cells (Isolectin GS-IB$_4$), macrophages (F4/80 and CD11b), nonspecific blood lineage cells (CD45), B-cells, and adipocyte progenitor cells (CD24) as described in Table 4.1. Images were acquired with the appropriate laser/filter settings listed in Table 4.1 and laser voltage/gain settings determined by FMO controls. (See the color plate.)

4. Sodium citrate dihydrate (Avantor Performance Materials, PA, USA, 3646-01)
   - Only needed if performing antigen retrieval for IHC
5. Acetone (Avantor Performance Materials, PA, USA, 9006-01)
6. CitriSolv (Fischer Scientific, MA, USA, 22-143-975), or xylene
7. 100% Ethanol (Decon Labs, PA, USA, 2701)
8. 95% Ethanol (Decon Labs, PA, USA, 2801)
9. Biobond (Electron Microscopy Services, PA, USA, 71304)
10. Paraffin (Leica Microsystems, Germany, EM-400)
11. Embedding cassettes (Sakura Finetek USA, CA, USA, 4135)

Equipment needed
1. Embedding station (containing heating and cooling plates)

- Individual heating and cooling plates can be used in place of an embedding station.
2. Metal embedding molds (Sakura Finetek USA, CA, USA, 4124)
3. Microtome
4. Heating water bath
5. 2100 Retriever (Electron Microscopy Services, PA, USA, 62706)
   - Only needed if performing antigen retrieval for IHC.
   - There are several alternative protocols for antigen retrieval, including some that do not require specialized equipment (D. Antigen retrieval (IHC-P guide), 2013).

Prior to starting
1. Prepare 75% and 70% ethanol:
   a. Mix together 750 mL 100% EtOH and 250 mL dH$_2$O (75% EtOH).
   b. Mix together 500 mL 75% EtOH and 35 mL dH$_2$O (70% EtOH).
2. Prepare *10 mM citrate buffer* for antigen retrieval:
   a. To make 1 L, add 2.94 g sodium citrate dihydrate to 1 L dH$_2$O.
   b. pH to 6.0 using hydrochloric acid.
      - Tris/EDTA pH 9.0 buffer can be used in place of 10 mM citrate buffer and may be better suited for retrieval of some antigens (D. Antigen retrieval (IHC-P guide), 2013).
3. Prepare 1 × zinc-formalin by adding 10 mL 5 × zinc-formalin to 40 mL dH$_2$O.

Protocol
1. Dissect adipose tissue from mouse.
2. Incubate samples in 1 × zinc-formalin overnight at 4 °C.
   - For downstream labeling of phosphorylated antigens, animals must be perfused with fixative to preserve phosphorylation events.
3. Wash samples three times in sterile PBS.
4. Place samples into labeled embedding cassette.
5. Incubate cassette in 70% ethanol until tissue processing is performed.
   - Samples can be stored in 70% ethanol at 4 °C until tissue processing.

Tissue processing
- Tissue processing can be performed manually as described below or by using a histology tissue processing machine.
6. Incubate cassette in 75% ethanol for 30 min.
7. Incubate cassette in 95% ethanol for 75 min. Repeat this step a second time with fresh 95% ethanol.

8. Incubate cassette in 100% ethanol for 60 min. Repeat this step a second and third time with fresh 100% ethanol.
9. Incubate cassette in CitriSolv for 60 min. Repeat this step a second time with fresh CitriSolv.
10. Incubate cassette in melted paraffin for 60 min at 60 °C.
11. Transfer cassette to fresh melted paraffin and incubate overnight at 60 °C.
12. Transfer cassette to fresh melted paraffin and incubate for 60 min at 60 °C.

Paraffin embedding

13. Remove cassette from paraffin and place on a heating plate set to 60 °C.
14. Place metal embedding mold on a heating plate set to 60 °C.
15. Fill metal embedding mold 75% full with melted paraffin.
16. Remove samples from cassette and place in melted paraffin within the metal embedding mold in the desired orientation.
17. Place metal embedding mold on a cold plate set to 0 °C for a few seconds to harden the paraffin just enough to maintain the sample orientation.
18. Place embedding mold back on the heating plate set to 60 °C and place the embedding cassette on top of embedding mold.
19. Quickly pour melted paraffin into the embedding cassette to fill the remainder of the embedding mold and the embedding cassette.
    - The melted paraffin will seep through the holes in the bottom of the cassette to fill the mold before filling the cassette.
20. Immediately transfer the mold with cassette to the cold plate and leave the mold/cassette on the cold plate until the paraffin has fully hardened.
    - The paraffin will harden through the holes in the cassette to attach the cassette to the molded paraffin block.
21. Carefully remove the paraffin block from the mold by pulling the attached cassette away from the mold.
    - If the paraffin block is not completely hardened, the molded block may detach from the cassette, necessitating re-embedding of the samples. Therefore, if the paraffin block does not easily detach from the mold, incubate the mold at −20 °C for 30 min before attempting to detach the block (Fig. 4.3).

Sectioning

22. Prior to sectioning, coat slides with Biobond as follows:
    a. Dilute Biobond 1:50 in acetone.
    b. Incubate slides in diluted Biobond for 5 min.

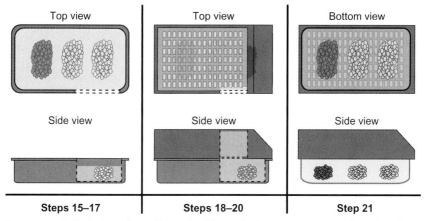

Figure 4.3 A depiction of paraffin embedding of adipose tissue. (See the color plate.)

  c. Incubate slides in dH$_2$O for 5 min.
  d. Allow slides to dry before sectioning.
23. Cut 5 μm thick paraffin sections using microtome.
24. Carefully transfer paraffin section(s) to a water-bath set to 37 °C.
   • Paraffin section(s) should float.
25. Transfer each paraffin section to a slide by submerging the slide into the water bath, placing the slide under the floating section, and raising the slide out of the water such that the section attaches to the slide as it is removed from the water.
26. Incubate the slides with sectioned adipose tissue at 40 °C overnight to dry and adhere the tissue sections to the slides.
   • For H&E staining, proceed to Section 3.1.2.
   • For IHC applications, proceed to step 27.

Deparaffinization and rehydration

27. Incubate the dried slides in CitriSolv for 5 min to dissolve the paraffin. Repeat this step a second time with fresh CitriSolv.
28. Rehydrate the sectioned tissue through sequential incubations in ethanol as described below:
   a. Incubate slides in 100% ethanol for 10 min. Repeat this step a second time with fresh 100% ethanol.
   b. Incubate slides in 95% ethanol for 10 min. Repeat this step a second time with fresh 95% ethanol.
   c. Incubate slides in 70% ethanol for 5 min. Repeat this step a second time with fresh 70% ethanol.

29. Incubate slides in PBS for 5 min. Repeat this step a second and third time with fresh PBS.
    - If you do not want to perform antigen retrieval, skip to step 33.
30. Perform antigen retrieval by running slides submerged in 10 mM citrate buffer through the 2100 Retriever.
    - This process takes 35–40 min.
    - Several methods exist for antigen retrieval (D. Antigen retrieval (IHC-P guide), 2013).
31. Remove slides from 2100 Retriever and allow slides to cool at room temperature (RT) for 15 min in 10 mM citrate buffer.
32. Incubate slides in PBS for 5 min. Repeat this step a second and third time with fresh PBS.
33. Proceed to Section 3.1.4.

3.1.2 H&E staining

H&E staining is a useful tool for the study of adipose tissue as it can display changes in tissue composition and morphology such as the formation of crown-like structures, "browning" of WAT, "whitening" of BAT, or changes in adipocyte cell size as described in Section 3.1.3.

Materials needed

1. Slides containing paraffin-sectioned adipose tissue dried overnight at 40 °C.
2. Eosin Y Alcoholic (Leica Microsystems, Germany, 3801600)
3. Shandon Gill 3 Hematoxylin (Thermo Scientific, MA, USA, 6765010)
4. Hydrochloric acid (Avantor Performance Materials, PA, USA, 9535-00)
5. Lithium carbonate (Sigma-Aldrich, MO, USA, L4283)
6. Xylene (Avantor Performance Materials, PA, USA, 8668-16)
7. 100% ethanol (Decon Labs, PA, USA, 2701)
8. 95% ethanol (Decon Labs, PA, USA, 2801)
9. Mounting media (Thermo Scientific, MA, USA, 8310-16)
10. Coverslips (Fischer Scientific, MA, USA, 12-545-F)

Prior to starting

1. Prepare a *1% lithium carbonate (saturated) solution* by adding 5 g lithium carbonate to 500 mL dH$_2$O and mixing well.
2. Prepare 70% ethanol by mixing 70 mL 100% ethanol and 30 mL dH$_2$O.
3. Prepare a *HCl–ethanol solution (0.125% HCl in 96.25% ethanol)* by adding 625 µL hydrochloric acid to 62.5 mL 70% ethanol in a 500 mL flask. Add 437 mL 100% ethanol and gently mix the solution.

Protocol
1. Incubate slides in xylene for 20 s. Repeat this step three more times using fresh xylene each time.
2. Incubate slides in 100% ethanol for 20 s. Repeat this step a second time with fresh 100% ethanol.
3. Incubate slides in 95% ethanol for 15 s.
4. Incubate slides in dH$_2$O for 15 s.
5. Incubate slides in Shandon Gill 3 Hematoxylin for 15 s. Repeat this steps four more times using fresh hematoxylin each time.
6. Incubate slides in dH$_2$O for 45 s.
7. Incubate slides in HCl–ethanol solution for 15 s.
8. Incubate slides in dH$_2$O for 15 s.
9. Incubate slides in 1% lithium carbonate solution for 15 s.
10. Incubate slides in dH$_2$O for 15 s.
11. Incubate slides in Eosin Y Alcoholic for 15 s.
12. Incubate slides in 95% ethanol for 15 s. Repeat this step a second and third time using fresh 95% ethanol.
13. Incubate slides in 100% ethanol for 15 s.
14. Incubate slides in fresh 100% ethanol for 45 s.
15. Incubate slides in xylene for 45 s.
16. Allow slides to air dry and coverslip the slides using mounting media.
17. H&E-stained sections can be imaged using bright-field microscopy to produce images as shown in Fig. 4.4A. For automated adipocyte size analysis using H&E-stained sections in Cell Profiler, proceed to Section 3.1.3.

3.1.3 Automated adipocyte size analysis using cell profiler

Adipocyte size has been shown to correlate with metabolic phenotypes such as insulin resistance (de Souza et al., 2001; Lundgren et al., 2007). Additionally, adipocyte hypertrophy is an important component of WAT accumulation in development (Birsoy et al., 2011) and obesity (Hirsch & Batchelor, 1976; Johnson, Zucker, Cruce, & Hirsch, 1971). As the vast majority of WAT volume is accounted for by lipid within mature adipocytes, H&E staining of adipocyte plasma membranes allows for adipocyte size measurements in paraffin-sectioned WAT to determine the contribution of adipocyte hypertrophy to WAT mass accumulation. The following protocol details automated adipocyte size measurements using H&E-stained WAT sections and Cell Profiler software (Carpenter et al., 2006; CellProfiler cell image analysis software, 2013).

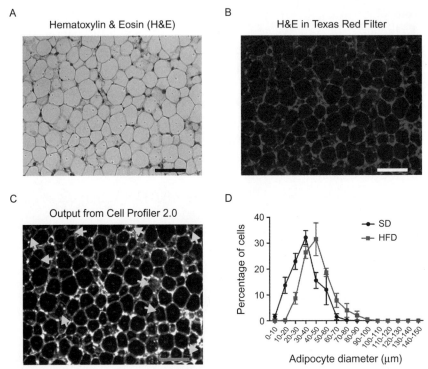

Figure 4.4 Example images and graph of Cell Profiler Adipocyte size analysis. (A) Bright-field image of H&E-stained paraffin-sectioned WAT. (B) Fluorescent image of H&E-stained paraffin-sectioned WAT using the Texas Red filter cube. (C) Image from (B) outputted from Cell Profiler 2.0 following adipocyte pixel area analysis. Yellow arrows indicate improperly gated cells that must be manually excluded from the dataset. (D) Adipocyte cell size distribution in WAT depots isolated from standard- (SD) and high-fat-diet (HFD) fed mice following Cell Profiler 2.0 analysis and conversion of measured adipocyte pixel area to adipocyte diameter as described in Section 3.1.3. (See the color plate.)

Materials needed
1. Slide containing paraffin-sectioned and H&E-stained adipose tissue.

Equipment needed
1. Zeiss Axioplan 2 fluorescent microscope with a Texas Red filter cube (560/55 excitation filter (533–588 nm); 645/75 emission filter (608–683 nm)).
    - Any fluorescent microscope with a green laser (514–568 nm) and a detection filter for red wavelengths (575–700 nm) can be used.

Software needed
1. Zeiss Axiovision 4.2 or equivalent.
2. Cell Profiler 2.0 (Carpenter et al., 2006)
3. Adipocyte membrane-specific pipeline for Cell Profiler 2.0
    - The pipeline that our lab uses to automate adipocyte size measurement in Cell Profiler 2.0 can be downloaded from the Cell Profiler Forum (CellProfiler cell image analysis software, 2013) (http://cellprofiler.org/forum/) under CellProfiler 2.0 Help → Adipocyte H&E Cell Profiler Pipeline. The pipeline file is named "Adipocyte Pipeline numbered Rodeheffer Lab.cp" and can be directly downloaded from the following page: http://cellprofiler.org/forum/viewtopic.php?f=14&t=1687&hilit=adipocyte&start=15. This pipeline is a series of commands that allows the Cell Profiler 2.0 software to recognize the plasma membrane of adipocytes in images of H&E-stained sections that are acquired as described below; number each recognized adipocyte; measure the pixel area of each adipocyte; export an image that shows the software determined adipocyte boundaries for each adipocyte; and export a .CSV spreadsheet containing the pixel area of each adipocyte. Cell Profiler 2.0 software and the listed pipeline are designed to sequentially analyze large numbers of images that are uploaded at one time, allowing for measurement of hundreds to thousands of adipocytes for comparison of cell size distribution in different adipose depots or the same depot under varying experimental conditions.

Protocol
1. Acquire low magnification ($\sim 20\times$), high-resolution ($\sim 1300 \times 1030$ pixels) RGB images of H&E-stained adipose tissue sections using the Texas Red filter cube.
    - As the tissue sections are not labeled with fluorescent molecules, this imaging approach utilizes the auto-fluorescence of the H&E-stained plasma membranes in the Texas Red emission filter upon exposure to filtered light of 533–588 nm wavelengths to acquire an image in which the adipocyte membranes are easily distinguishable from the cell interior. Example images acquired under bright-field or as-described fluorescent conditions are shown in Fig. 4.4A and B, respectively.
2. Export images acquired under fluorescent settings as .TIFF files.

3. Place all .TIFF images from the same depot or experimental condition in an "input" folder on your computer hard-drive.
4. Make an "output" folder for each depot or experimental condition.
5. Load "Adipocyte Pipeline numbered Rodeheffer Lab.cp" in Cell Profiler.
6. Select experimental "input" folder as the default input folder at the bottom of the Cell Profiler 2.0 window.
   - Image files from the input folder are shown in a window in the bottom left of the software window.
7. Select a previously made output folder as the default output folder at the bottom of the Cell Profiler 2.0 window.
8. Name the output file (.CSV file containing acquired data) at the bottom of the Cell Profiler 2.0 window.
9. Click on the "Analyze Images" button.
   - The final analysis step for each image produces a .TIFF file as shown in Fig. 4.4C in which the software designated adipocyte number and software determined cell boundaries are shown.
   - After Cell Profiler 2.0 has analyzed all of the images in the input folder, a spreadsheet in .CSV format is saved in the output folder that contains the pixel area of every adipocyte from every image in the input folder.
10. Identify structures that have been improperly gated during the analysis by manually checking each .TIFF image showing the software determined adipocyte boundaries and adipocyte number and deleting the software calculated pixel area from the .CSV file for improperly gated structures.
    - This pipeline does not allow for correction of improperly gated structures, so manual exclusion of these structures from the dataset is necessary for obtaining an accurate cell size distribution.
11. In the spreadsheet, convert pixel area to pixel diameter for all of the properly gated adipocytes using the formula: $\text{Diameter} = \sqrt{\frac{4(\text{Area})}{\pi}}$
12. Determine the pixel to micrometer scaling factor for the imaging settings used to acquire the initial images in step 1. This is usually accomplished through the imaging software or by using a stage micrometer.
13. In the spreadsheet, convert the diameter in pixels to diameter in micrometer by multiplying the diameter in pixels by the determined scaling factor.
14. For each dataset, group the cell diameters into user-determined "bins." For adipose tissue, we recommend 10 μm bins such that each

dataset can be displayed in graphical format as the percentage of adipocytes with cell diameters that fit in each bin (i.e., 0–10 μm, 10–20 μm, etc.). This allows for representation of adipocyte size data as a distribution of sizes that can be compared between depots or experimental conditions. A graph comparing the "binned" adipocyte size distribution in a WAT depot dissected from mice that have been fed either a standard diet (SD) or a high-fat diet (HFD) is shown in Fig. 4.4D.

3.1.4 IHC using horseradish peroxidase substrates

Detection of a specific antigen in paraffin-sectioned tissue is routinely performed through IHC, in which an antigen-specific antibody is directly or indirectly coupled to a reporter molecule, allowing microscopic visualization of antibody labeled antigens. The simplest IHC protocol involves antigen labeling with a fluorochrome-conjugated monoclonal antibody. This technique often produces poor results as fixation and processing of tissue can modify antigen structure and reduce antibody recognition. Additionally, this technique relies on a 1/1 ratio of antigen/reporter molecule, limiting detection of sparsely expressed antigens. When tissue preparation only requires light fixation, such as in the frozen sectioning protocol provided in Section 3.2, these limitations can be overcome by amplifying the fluorescent reporter through the "sandwich method," in which the antigen is labeled with a nonfluorescent primary antibody and the primary antibody is labeled with a fluorochrome-conjugated secondary antibody. This method amplifies the reporter signal as multiple secondary antibodies can bind one primary antibody.

Because antigen labeling is generally less robust in paraffin-sectioned tissue, protocols for paraffin IHC rely on the sandwich method and replacement of the conjugated fluorochrome with an enzyme capable of repeatedly converting a substrate molecule into a fluorescent or colored reporter to result in high reporter signal amplification. The most common antibody-conjugated enzymes are alkaline phosphatase and horseradish peroxidase (HRP), both of which can be used in combination with several reporter substrates. Below we provide protocols for antigen labeling with colored (3,3′-diaminobenzidine (DAB)) and fluorescent (tyramide signal amplification (TSA)) HRP substrates, as these substrates produce high signal amplification and low background staining in paraffin-sectioned WAT.

3.1.4.1 3,3′-Diaminobenzidine

When fluorescent antigen labeling is not feasible or desired, labeling of antigens with DAB provides a good alternative. DAB is a soluble organic compound that in the presence of hydrogen peroxide is oxidized by HRP to yield an insoluble compound that is brown in color. Therefore, DAB-labeled sections can be visualized though bright-field microscopy.

Materials needed

1. Slides containing paraffin-sectioned adipose tissue dried overnight at 40 °C
2. 30% hydrogen peroxide (Avantor Performance Materials, PA, USA, 2186-01)
3. Sterile PBS (Life Technologies, NY, USA, 14190-144)
4. Unconjugated IHC grade primary antibody
5. VECTASTAIN Peroxidase ABC kit (Vector Labs, CA, USA, PK-4000–PK-4007)
    - ABC kits are species-specific. Choose the kit that is designed for the species from which the primary antibody is derived (i.e., PK-4001 is an ABC Peroxidase Rabbit IgG kit and should be used when the primary antibody is derived from rabbit).
6. Avidin/Biotin Blocking kit (Vector Labs, CA, USA, SP-2001)
7. DAB Peroxidase Substrate kit (Vector Labs, CA, USA, SK-4100)
8. Pap Pen (Life Technologies, NY, USA, 00-8899)
9. Mounting media (Thermo Scientific, MA, USA, 8310-16)
10. Coverslips (Fischer Scientific, MA, USA, 12-545-F)

Prior to starting

1. Prepare *0.3% H_2O_2 in PBS* by mixing 1 mL 30% H_2O_2 in 100 mL PBS.
2. Prepare *Normal Blocking Serum* by adding three drops of the supplied normal serum concentrate from the VECTASTAIN Peroxidase ABC kit (yellow bottle) in 10 mL sterile PBS.
    - The blocking serum used is from the species from which the secondary antibody is derived. For VECTASTAIN Peroxidase ABC kits, the secondary antibody is derived from horse, and therefore the blocking serum is normal horse serum.

Protocol

1. Make a hydrophobic barrier around the tissue sections with the Pap Pen.
2. Incubate sections in 0.3% H_2O_2 in PBS for 15 min at RT to block endogenous peroxidase activity.
3. Incubate slides in PBS for 5 min. Repeat this step a second and third time with fresh PBS.

4. Incubate sections in normal blocking serum for 30 min at RT to prevent nonspecific antibody binding.
5. Blot away normal blocking serum and wash sections in sterile PBS.
6. Incubate sections in avidin serum from the Avidin/Biotin blocking kit for 15 min at RT to prevent nonspecific binding of the ABC complex.
7. Blot away avidin serum and wash sections with sterile PBS.
8. Incubate sections in biotin serum from the Avidin/Biotin blocking kit for 15 min at RT to prevent nonspecific binding of the ABC complex.
9. Blot away biotin serum and wash sections with sterile PBS.
10. Incubate sections in primary antibody diluted in normal blocking serum overnight at 4 °C.
    - Antibody concentration is user-determined.
11. Incubate slides in PBS for 5 min. Repeat this step a second and third time with fresh PBS.
12. Prepare biotinylated secondary antibody by adding one drop of biotinylated secondary antibody stock concentrate from the VECTASTAIN Peroxidase ABC kit (blue bottle) to 10 mL sterile PBS.
13. Incubate sections in biotinylated secondary antibody for 30 min at RT.
    - Immediately after incubation begins, prepare ABC reagent by adding two drops of Reagent A from the VECTASTAIN ABC kit to 10 mL sterile PBS. Mix solution and add two drops of Reagent B. Mix solution again and allow ABC reagent to stand at RT until step 14.
14. Incubate slides in PBS for 5 min. Repeat this step a second and third time with fresh PBS.
15. Incubate sections in ABC reagent for 30 min at RT.
16. Incubate slides in PBS for 5 min. Repeat this step a second and third time with fresh PBS.
17. Prepare DAB reagent from the DAB Peroxidase Substrate kit as follows:
    - Add two drops of Buffer Stock solution to 5 mL dH$_2$O and mix.
    - Add four drops of DAB Stock solution and mix.
    - Add two drops of Hydrogen Peroxide solution and mix.
    - Add two drops of Nickel solution and mix.
18. Incubate section in DAB reagent for 15 min at RT.
    - Incubation time needed for effective antigen staining with low background staining is antigen-dependent and user-determined, but usually varies from 1 to 30 min.
19. Incubate slides in sterile PBS for 45 min at RT.
    - Several colored HRP substrates are commercially available. Therefore, this protocol can be repeated for labeling of a second or third antigen in the same tissue section.

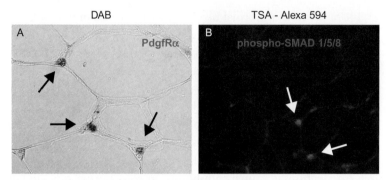

Figure 4.5 Immunohistochemistry on paraffin-sectioned WAT. (A) HRP-DAB-mediated immunolabeling of PdgfRα in paraffin-sectioned WAT. Black arrows indicate PdgfRα+ cells marked by deposition of oxidized DAB, which is brown in color. (B) HRP-TSA-mediated immunolabeling of phospho-SMADs 1,5,8 in paraffin-sectioned WAT. White arrows indicate phospho-SMAD+ cells marked by fluorescence of oxidized TSA-A594. (See the color plate.)

20. Coverslip slides using mounting media and visualize staining through bright-field microscopy.
    - An example image of DAB-labeled PdgfRα in WAT is provided in Fig. 4.5A.

3.1.4.2 Tyramide signal amplification

When fluorescent antigen labeling is desired, TSA provides high sensitivity to overcome the normal limitations of fluorescent labeling of paraffin-sectioned tissue. In TSA, inactive fluorochrome-conjugated tyramide derivatives are oxidized by HRP in the presence of hydrogen peroxide to yield highly fluorescent reporter molecules that can be visualized through fluorescent microscopy. There are a large number of fluorochrome-labeled tyramide derivatives available, and TSA-labeled sections can be counterstained with other fluorescent markers such as Isolectin GS-IB$_4$, Cell Mask, and DAPI (Table 4.1) for detailed fluorescent analysis of adipose tissue.
Materials needed
1. Slides containing paraffin-sectioned adipose tissue dried overnight at 40 °C
2. 30% hydrogen peroxide (Avantor Performance Materials, PA, USA, 2186-01)
3. Sterile PBS (Life Technologies, NY, USA, 14190-144)
4. TSA kit (Life Technologies, NY, USA, T-20911–T-30955)

- TSA kits are species-specific. Choose the kit that is designed for the species from which the primary antibody is derived (i.e., T-20925 is a TSA kit with HRP-Goat Anti-Rabbit IgG and Alexa Fluor 594 and can be used when the primary antibody is derived from rabbit). Invitrogen makes HRP-goat anti-rabbit IgG kits, HRP-goat anti-mouse IgG kits, and HRP-streptavidin kits. If the primary antibody is derived from non-rabbit or mouse species, the HRP-streptavidin kit can be used in combination with a biotin-conjugated primary antibody.
5. Unconjugated or biotinylated IHC grade primary antibody
6. Streptavidin/Biotin Blocking kit (Vector Labs, CA, USA, SP-2002)
   - Only needed if using a TSA HRP-streptavidin kit.
7. Pap Pen (Life Technologies, NY, USA, 00-8899)
8. Mounting media (Thermo Scientific, MA, USA, 8310-16)
9. Coverslips (Fischer Scientific, MA, USA, 12-545-F)

Prior to starting
1. Prepare *0.3% H_2O_2 in PBS* by mixing 1 mL 30% H_2O_2 in 100 mL PBS.
2. Prepare *tyramide stock solution*, *1% blocking reagent*, *HRP-conjugate stock solution* and *amplification buffer* per TSA kit protocols.

Protocol
1. Make a hydrophobic barrier around the tissue sections with the Pap Pen.
2. Incubate sections in 0.3% H_2O_2 in PBS for 15 min at RT to block endogenous peroxidase activity.
3. Incubate slides in PBS for 5 min. Repeat this step a second and third time with fresh PBS.
4. Incubate the sections in prepared 1% blocking reagent for 60 min at RT.
5. Blot away 1% blocking reagent and wash sections with sterile PBS.
   - If using HRP-streptavidin kit, proceed to step 6. If not, proceed to step 10.
6. Incubate sections in streptavidin serum from the streptavidin/biotin blocking kit for 15 min at RT.
7. Blot away streptavidin serum and wash sections with sterile PBS.
8. Incubate sections in biotin serum from the Streptavidin/Biotin blocking kit for 15 min at RT.
9. Blot away biotin serum and wash sections with sterile PBS.
10. Incubate sections in primary antibody diluted in 1% blocking reagent overnight at 4 °C.
    - Antibody concentration is user-determined.

11. Incubate slides in PBS for 5 min. Repeat this step a second and third time with fresh PBS.
12. Prepare a HRP-conjugate working solution by diluting the HRP-conjugate stock solution 1:100 in 1% blocking reagent.
13. Incubate sections in 100 µL of HRP-conjugate working solution for 30–60 min at RT.
14. Incubate slides in PBS for 5 min. Repeat this step a second and third time with fresh PBS.
15. Prepare a tyramide working solution by diluting the tyramide stock solution 1:100 in amplification buffer.
16. Incubate the sections in 100 µL tyramide working solution for 5–10 min at RT.
17. Incubate slides in PBS for 5 min. Repeat this step a second and third time with fresh PBS.
    - Tissue sections can now be counterstained with additional fluorescent markers such as those listed in Table 4.1.
18. Coverslip slides using mounting media and visualize staining through fluorescent microscopy using appropriate lasers and filters for each fluorescent stain.

3.2. Frozen sectioning

To circumvent complex labeling strategies needed to amplify reporter signal in paraffin-sectioned adipose tissue, less harsh processing and sectioning protocols have been developed in which adipose tissue is lightly fixed and embedded in gelatin before hardening through freezing. Frozen sectioning maintains antigen integrity and allows for simple immunofluorescent labeling of a single, or multiple, antigen(s). The frozen sectioning and immunofluorescent labeling protocols below have been adapted from previously published protocols (Crisan et al., 2008), with several specific modifications.

3.2.1 Preparation of slides
Materials needed
1. Microscope slides (Thermo Scientific, MA, USA, 4951F-001)
2. 5× Zinc-formalin concentrate (Anatech Ltd, MI, USA, 141).
3. Sterile PBS (Life Technologies, NY, USA, 14190-144)
4. Sucrose (Avantor Performance Materials, PA, USA, 4072-01)
5. Gelatin (MP Biomedicals, 960317)
6. 12-well polystyrene plate (BD Biosciences, CA, USA, 351143)

7. Scalpel
8. Small spatula

Equipment needed
1. Cryostat

Prior to starting
1. Prepare *PBS–sucrose buffer*.
    a. To make 1 L, add 150 g sucrose to 1 L PBS.
    b. Store at 4 °C.
2. Prepare *sucrose–gelatin solution*.
    a. To make 100 mL, add 15 g sucrose, and 7.5 g gelatin to 100 mL PBS.
    b. Incubate at 60 °C until the gelatin melts.
3. Prepare 1× zinc-formalin by adding 10 mL 5× zinc-formalin to 40 mL dH_2O.

Protocol
1. Dissect adipose depot(s) from mouse.
2. Incubate samples in 1× zinc–formalin at 4 °C.
   - Incubation times needed for proper fixation varies based on sample size. We recommend:
     - For small samples (~10 mg), incubate for 1–2 h.
     - For medium-sized samples (<1 g), incubate overnight.
     - For large samples (>1 g), incubate for 36–48 h.
   - For downstream labeling of phosphorylated antigens, animals must be perfused with fixative to preserve phosphorylation events.
3. Wash samples three times in PBS.
4. Incubate samples in fresh PBS overnight at 4 °C.
5. Incubate samples in PBS–sucrose buffer overnight at 4 °C.
   - Samples can be stored in PBS–sucrose buffer at 4 °C for up to 2 weeks.
6. Incubate samples in sucrose–gelatin solution for 1–2 h at 37 °C.
7. Cut samples to 1 cm × 1 cm using a scalpel.
8. Place each sample in the center of a well in a 12-well plate.
9. Fill the well half full with warm sucrose–gelatin solution from step 6.
   - Adipose tissue sample will float to the top.
10. Incubate the plate for 10–20 min at 4 °C to solidify the sucrose–gelatin solution.
11. Fill the well with sucrose–gelatin solution ensuring that the sample remains submerged.
12. Incubate the plate for 10–20 min at 4 °C to solidify the sucrose–gelatin solution.

13. Carefully detach the sucrose–gelatin block from the well using a small spatula ensuring that the sample remains entirely within the block.
14. Trim the block with a scalpel to ~2 cm × 2 cm × 2 cm ensuring that the sample remains entirely within the block.
15. Place the trimmed block back into the well and incubate the plate containing the block at −80 °C for at least 24 h.
16. Section the frozen sucrose-gelatin block at −45 °C using a cryostat.
    - We recommend a section thickness of 12–20 μm.
    - The frozen sucrose–gelatin block must be entirely frozen when sectioning. Therefore, it is essential to adhere to the following sectioning protocol:
        i. Quickly transfer the sucrose–gelatin block from −80 °C to the cryostat pre-set to −45 °C.
        ii. Allow the block to refreeze at −45 °C prior to sectioning.
        iii. Trim the frozen block through sequential 100 μm sections until the tissue is exposed.
        iv. Allow the block to refreeze at −45 °C for 15–30 min.
        v. Section the block at a thickness of 12–20 μm.
17. Gently transfer each section to a slide by slowing approaching the section with the slide until the section adheres to the slide.
    - Care must be taken to prevent mechanical alteration of the section during this transfer process.
18. Allow the section to air dry for 5 min at RT.
19. Proceed to Section 3.2.2.

3.2.2 IHC using fluorochrome-conjugated antibodies

As mentioned, frozen sectioning results in less antigen masking and allows for effective antigen labeling with fluorochrome-conjugated antibodies. In our hands, HRP-based signal amplification produces significantly higher background staining in frozen sections as compared to paraffin sections. Because of this, immunofluorescent labeling is our preferred method of staining frozen sections. Additionally, although the provided frozen sectioning protocol produces sections with excellent tissue morphology, paraffin-embedded tissue can be sectioned thinner, producing sections more suitable for H&E staining. The following protocol details immunofluorescent labeling of frozen-sectioned adipose tissue and provides example images of immunofluorescent-labeled WAT in Fig. 4.6.

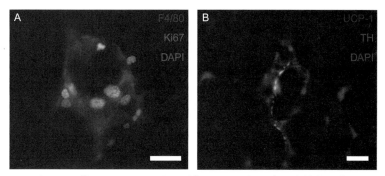

Figure 4.6 Immunofluorescent staining of frozen-sectioned WAT. (A) A classic crown-like structure stained with antibodies for F4/80 and Ki67 and counterstained with DAPI. (B) Sympathetic innervation of a beige adipocyte as shown through staining with antibodies for UCP-1 and the sympathetic nerve fiber marker tyrosine hydroxylase (TH) and counterstained with DAPI. Scale bar is 20 μm. (See the color plate.)

Materials needed
1. Slide containing frozen-sectioned adipose tissue
2. Unconjugated IHC grade primary antibody
3. Fluorochrome-conjugated secondary antibody
4. Bovine Serum Albumin, BSA (Sigma-Aldrich, MO, USA, A9647)
5. Sterile PBS (Life Technologies, NY, USA, 14190-144)
6. Triton X-100 (Sigma-Aldrich, MO, USA, X100)
7. Acetone (Avantor Performance Materials, PA, USA, 9006-01)
8. Methanol (Avantor Performance Materials, PA, USA, 9070-01)
9. Pap Pen (Life Technologies, NY, USA, 00-8899)
10. Slide staining Jar (Electron Microscopy Services, PA, USA, 71405-01)
11. Mounting media (Thermo Scientific, MA, USA, 8310-16)
12. Coverslips (Fischer Scientific, MA, USA, 12-545-F)

Prior to starting
1. Prepare *0.3% PBS-T* by dissolving 600 μL Triton X-100 in 200 mL PBS.
2. Prepare *1% BSA in PBS-T* by dissolving 1 g BSA in 100 mL 0.3% PBS-T.
3. Prepare *1:1 acetone:methanol* by adding 50 mL acetone to 50 mL methanol. Incubate at −20 °C until ice-cold.

Protocol
1. Incubate slides containing frozen-sectioned adipose tissue in ice-cold 1:1 acetone:methanol for 10 min in a slide staining jar.
2. Make a hydrophobic barrier around the tissue sections with the Pap Pen.
3. Incubate sections in 0.3% PBS-T for 5 min. Repeat this step a second and third time with fresh PBS-T.

4. Incubate sections in 1% BSA in PBS-T for 1 h at RT to block non-specific antibody binding.
5. Incubate sections in primary antibody diluted in 1% BSA in PBS-T overnight at 4 °C.
   * Antibody concentration is user-determined.
6. Incubate sections in 0.3% PBS-T for 5 min. Repeat this step a second and third time with fresh PBS-T.
7. Incubate sections in species-specific fluorochrome-conjugated secondary antibody diluted in 1% BSA in PBS-T for 60 min at RT.
   * Antibody concentration is user-determined. A 1:200 dilution is generally a good secondary antibody concentration to start with.
8. Incubate slides in 0.3% PBS-T for 5 min. Repeat this step a second and third time with fresh PBS-T.
9. Tissue can be counterstained with DAPI or other fluorescent stains. Additionally, steps 4–8 can be repeated to label other antigens, using primary antibodies derived from different species for each antigen.
10. Incubate slides in 0.3% PBS-T for 5 min. Repeat this step a second and third time with fresh PBS-T.
11. Coverslip slides using mounting media and visualize staining through fluorescent microscopy using appropriate lasers and filters for each fluorescent stain.

ACKNOWLEDGMENTS

We thank Michael Schadt, the Section of Comparative Medicine Histology Manager at Yale University for aiding in the development of adipose tissue sectioning and H&E staining protocols. We also thank Dr. Mark Bray from the Carpenter lab at the Broad Institute of MIT and Harvard for assistance in developing and refining a Cell Profiler pipeline suitable for adipocyte cell size analysis. This work was supported by American Diabetes Association Award 7-12-JF-46, DERC pilot project Grant DK045735, and NIDDK Grant DK090489 to M. S. R.; EMBO long-term postdoctoral fellowship ALTF 132-2011 to C. C.; Deutsche Forschungsgemeinschaft DFG-SFB 1052/1: "Obesity mechanisms" (project A04), Helmholtz alliance "Imaging and Curing Environmental Metabolic Disease," and Research grant from the Medical Faculty, Leipzig University to M. T. G.

REFERENCES

Berry, R., & Rodeheffer, M. S. (2013). Characterization of the adipocyte cellular lineage in vivo. *Nature Cell Biology*, *15*, 302–308.

Birsoy, K., Berry, R., Wang, T., Ceyhan, O., Tavazoie, S., Friedman, J. M., et al. (2011). Analysis of gene networks in white adipose tissue development reveals a role for ETS2 in adipogenesis. *Development*, *138*(21), 4709–4719.

Carpenter, A. E., Jones, T. R., Lamprecht, M. R., Clarke, C., Kang, I. H., Friman, O., et al. (2006). Cell profiler: Image analysis software for identifying and quantifying cell. *Genome Biology, 7*(10), R100.

CellProfiler cell image analysis software. (2013). Retrieved April 1, 2013 from http://www.cellprofiler.org/.

Cinti, S., Zingaretti, M. C., Cancello, R., Ceresi, E., & Ferrara, P. (2001). Morphologic techniques for the study of brown adipose tissue and white adipose tissue. *Methods in Molecular Biology, 155*, 21–51.

Crisan, M., Yap, S., Casteilla, L., Chen, C. W., Corselli, M., Park, T. S., et al. (2008). A perivascular origin for mesenchymal stem cells in multiple human organs. *Cell Stem Cell, 3*(3), 301–313.

D. Antigen retrieval (IHC-P guide) | Abcam. (2013). Retrieved April 1, 2013 from http://www.abcam.com/?pageconfig=resource&rid=11488.

de Souza, C. J., Eckhardt, M., Gagen, K., Dong, M., Chen, W., Laurent, D., et al. (2001). Effects of pioglitazone on adipose tissue remodeling within the setting of obesity and insulin resistance. *Diabetes, 50*(8), 1863–1871.

Gupta, R. K., Mepani, R. J., Kleiner, S., Lo, J. C., Khandekar, M. J., Cohen, P., et al. (2012). Zfp423 expression identifies committed preadipocytes and localizes to adipose endothelial and perivascular cells. *Cell Metabolism, 15*, 230–239.

Hirsch, J., & Batchelor, B. (1976). Adipose tissue cellularity in human obesity. *Clinics in Endocrinology and Metabolism, 5*, 299–311.

Johnson, P. R., Zucker, L. M., Cruce, J. A., & Hirsch, J. (1971). Cellularity of adipose depots in the genetically obese Zucker rat. *Journal of Lipid Research, 12*, 706–714.

Lee, Y. H., Petkova, A. P., Mottillo, E. P., & Granneman, J. G. (2012). In vivo identification of bipotential adipocyte progenitors recruited by β3-adrenoceptor activation and high-fat feeding. *Cell Metabolism, 15*, 480–491.

Lundgren, M., Svensson, M., Lindmark, S., Renstrom, F., Ruge, T., & Eriksson, J. W. (2007). Fat cell enlargement is an independent marker of insulin resistance and 'hyperleptinaemia'. *Diabetologia, 50*(3), 625–633.

Pi-Sunyer, F. X., Becker, D. M., Bouchard, C., Carleton, R. A., Colditz, G. A., Dietz, W. H., et al. (1998). Clinical guidelines on the identification, evaluation, and treatment of overweight and obesity in adults. NHLBI Obesity Education Initiative Expert Panel on the Identification, Evaluation, and Treatment of Obesity in Adults (US). Bethesda (MD): National Heart, Lung, and Blood Institute.

Seale, P., Conroe, H. M., Estall, J., Kajimura, S., Frontini, A., Ishibashi, J., et al. (2011). Prdm16 determines the thermogenic program of subcutaneous white adipose tissue in mice. *The Journal of Clinical Investigation, 121*(1), 96–105.

Sun, K., Kusminski, C. M., & Scherer, P. E. (2011). Adipose tissue remodeling and obesity. *The Journal of Clinical Investigation, 121*(6), 2094–2101.

Tang, W., Zeve, D., Suh, J. M., Bosnakovski, D., Kyba, M., Hammer, R. E., et al. (2008). White fat progenitor cells reside in the adipose vasculature. *Science, 322*, 583–586.

Tran, K. V., Gealekman, O., Frontini, A., Zingaretti, M. C., Morroni, M., Giordano, A., et al. (2012). The vascular endothelium of the adipose tissue gives rise to both white and brown fat cells. *Cell Metabolism, 15*(2), 222–229.

Whole mount staining protocols | Abcam. (2013). From http://www.abcam.com/index.html?pageconfig=resource&rid=11330. Retrieve date 05.10.13 (May 5th 2013).

CHAPTER FIVE

Adipose Tissue Angiogenesis Assay

Raziel Rojas-Rodriguez, Olga Gealekman, Maxwell E. Kruse, Brittany Rosenthal, Kishore Rao, SoYun Min, Karl D. Bellve, Lawrence M. Lifshitz, Silvia Corvera[1]

Program in Molecular Medicine, University of Massachusetts Medical School, Worcester, Massachusetts, USA
[1]Corresponding author: e-mail address: silvia.corvera@umassmed.edu

Contents

1. Introduction — 76
2. Materials — 77
   2.1 Medium, instruments, and culture dishes — 77
   2.2 Adipose tissue samples — 78
3. Methods — 79
   3.1 Sample preparation — 79
   3.2 Embedding procedure — 80
   3.3 Quantification of angiogenic potential — 80
4. Method Limitations — 90
Acknowledgment — 90
References — 91

Abstract

Changes in adipose tissue mass must be accompanied by parallel changes in microcirculation. Investigating the mechanisms that regulate adipose tissue angiogenesis could lead to better understanding of adipose tissue function and reveal new potential therapeutic strategies. Angiogenesis is defined as the formation of new capillaries from existing microvessels. This process can be recapitulated *in vitro*, by incubation of tissue in extracellular matrix components in the presence of pro-angiogenic factors. Here, we describe a method to study angiogenesis from adipose tissue fragments obtained from mouse and human tissue. This assay can be used to define effects of diverse factors added *in vitro*, as well as the role of endogenously produced factors on angiogenesis. We also describe approaches to quantify angiogenic potential for the purpose of enabling comparisons between subjects, thus providing information on the role of physiological conditions of the donor on adipose tissue angiogenic potential.

1. INTRODUCTION

The adipose tissue of mammals has evolved as a preferred storage site for excess calories in the form of triacylglycerol, and also as a critically important endocrine organ, which controls whole body metabolic homeostasis. One of the unique features of adipose tissue is its ability to massively expand in response to chronic positive energy balance, and to decrease in size under conditions in which stored calories are mobilized for use in other organs. Adipose tissue expansion results from an increase in adipocyte size, as well as from the differentiation of precursors into new adipocytes. As is the case for any tissue, changes in adipose tissue mass must be accompanied by parallel changes in microcirculation (Cho et al., 2007; Christiaens & Lijnen, 2010; Crandall, Hausman, & Kral, 1997). Adequate vascularization is required to deliver nutrients and oxygen to all cells in the tissue, to remove waste products, and to allow the tissue to functionally interact with the rest of the organism through the sensing and production of hormones and growth factors. In the case of adipose tissue, inadequate microcirculation could impair appropriate triglyceride storage by preventing access to circulating lipoproteins under fed conditions; it could also prevent adequate fuel delivery to other organs during fasting. Recent data suggest that adipose tissue microcirculatory alterations occur in type 2 diabetes, raising the possibility that impaired vascular development may be pathogenic in this disease (Hodson, Humphreys, Karpe, & Frayn, 2012; Hosogai et al., 2007; Michailidou et al., 2012; Pasarica et al., 2009; Rausch, Weisberg, Vardhana, & Tortoriello, 2008). In support of an important role for adipose tissue angiogenesis, increased production of pro-angiogenic factors, in particular VEGF, by adipose tissue improves whole body metabolism in high-fat diet-fed mice (Michailidou et al., 2012; Sung et al., 2013; Wree et al., 2012). Thus, investigating the mechanisms that regulate adipose tissue angiogenesis could lead to better understanding of adipose tissue function and potentially reveal new therapeutic strategies.

Research on complex processes, such as tissue vascularization, can benefit from *in vitro* models that mimic important features of the process. In the area of angiogenesis, significant information has been derived from the aorta ring assay, in which aortic rings dissected from rat or mouse thoracic aortas generate outgrowths of branching microvessels (Aplin, Fogel, Zorzi, & Nicosia, 2008; Baker et al., 2012). The formation of sprouts can be visualized by light microscopy of live cultures, and branching microvessels are composed of the same cell types that operate *in vivo*. The formation of aortic ring sprouts is regulated by endogenous and exogenously added pro- and anti-angiogenic factors. Thus, this assay has been extremely valuable in dissecting the basic mechanisms of angiogenesis.

Here we present an adaptation of the aorta ring assay, which we have used to assess the angiogenic capacity of adipose tissue from mice and humans (Gealekman et al., 2008, 2012, 2011). We have found that angiogenic capacity of adipose tissue *in vitro* reflects the physiological conditions of the donor. For example, adipose tissue angiogenesis, but not that of aorta, is enhanced in ob/ob hyperphagic mice, and in response to stimuli that increase adipose tissue growth, such as thiazolidinediones (Gealekman et al., 2008). Angiogenic capacity also differs among human adipose tissue depots and correlates with insulin sensitivity (Gealekman et al., 2011). Thus this assay can be used to dissect endogenous factors that regulate adipose tissue angiogenesis under different physiological conditions.

2. MATERIALS
2.1. Medium, instruments, and culture dishes

- Matrigel™ Basement Membrane Matrix, Growth Factor Reduced, Phenol Red-free, *LDEV-Free (BD Biosciences, San Jose, CA, catalog # 356231). Matrigel comes in bottles of 10 ml. To avoid freeze–thaw cycles and contamination, it is dispensed into 1 ml aliquots and stored at $-20\,°C$
- EBM-2 medium supplemented with EGM-2 MV (Lonza, Basel, Switzerland, BulletKit, CC-3202)
- Dulbecco's phosphate-buffered saline (DPBS) (Life Technologies Corp-Gibco®, Carlsbad, CA, catalog # 14190)
- Dispase (BD Biosciences, San Jose, CA, catalog # 354235) enzyme is used for proteolysis of Matrigel. Aliquot dispase in 1.5 ml tubes and store at $-20\,°C$. Try to avoid multiple freeze–thaw cycle
- 50 mM EDTA
- Trypsin-Versene (Lonza, Basel, Switzerland, catalog # 17-161E)
- Formaldehyde (Ted Pella, Inc., Redding, CA, catalog #18505)
- Hoechst 33258, pentahydrate (Life Technologies Corp, Carlsbad, CA, catalog # 947743)
- Triton X-100 (Sigma-Aldrich, St. Louis, MO, catalog # T-9284)
- BSA (Sigma-Aldrich, St. Louis, MO, catalog # A3059)

- 96-well flat bottom multiwell plates (BD Falcon™, Franklin Lakes, NJ, catalog # 353072)

35 mm glass-bottom culture dishes (MatTek Corporation, Ashland, MA, catalog # P35G-1.5-14-C)
- Very fine point forceps (VWR International, Radnor, PA, catalog # 25607-856)
- Micro Surgery Scissors (Integra™-Miltex®, York, PA, catalog # 17-2150)
- Sterile pipettes, tips, and conical tubes
- Three 100 mm × 20 mm petri dishes
- Round bladed disposable scalpels (FEATHER® Safety Razor Co. Ltd, Osaka, Japan, catalog # 2975#10)

2.2. Adipose tissue samples

Sample collection of adipose tissue from mice or human subjects requires IACUC and IRB approval from the institution at which the procedure will be performed.

In preliminary studies, only the epididymal adipose depot of mice was competent to form angiogenic sprouts *in vitro*, for reasons that are currently under investigation. Thus, all of our studies have been carried out with male mice on a C56BL/6 background, or on genetic variants such as the *ob/ob* mouse. For this assay, epididymal adipose tissue is dissected and processed for embedding as described in detail below.

This assay has also been used to measure angiogenesis from human adipose tissue. Explants from both subcutaneous and visceral adipose tissue depots develop capillary sprouts (Gealekman et al., 2011). However, differences in the extent of angiogenic growth have been seen between visceral and subcutaneous adipose tissue. Thus, the exact source of each adipose tissue sample (including morphometric and metabolic characteristics of human subjects and the exact location from which the adipose tissue sample is extracted) should be carefully tracked and kept as consistent as possible between subjects within a study. An additional factor that may affect the outcome of this assay is the time between the excision of adipose tissue from the patient and the embedding of the tissue pieces into Matrigel. Both the conditions in which adipose tissue samples are stored and the time between excision and embedding should be optimized and, at very least, kept consistent between the subjects being compared. For optimal results we recommend collecting adipose tissue pieces that do not exceed 1 g, and storing them at room temperature in EGM-2 MV-supplemented EBM-2 medium. The time between excision of the adipose tissue from human subject and its embedding into Matrigel should be minimized, and in our laboratory is routinely under 3 h.

3. METHODS
3.1. Sample preparation

Before starting
- Protocol should be performed in a Class II Biocabinet.
- Wear personal protective garment including the use of lab coat, double gloves, facial barrier protection, hair net, etc. Follow Human Tissue Handling Guidelines from your institution.
- Instruments for handling tissue should be previously sterilized.
- Remove aliquots of Matrigel from $-20\,°C$ and place on ice to bring it to $4\,°C$. Use 2.5 ml of Matrigel for 60 wells of a 96-well-multiwell plate.

3.1.1 Collection of mouse adipose tissue
- Male C57Bl/6 mice from 9 to 23 weeks of age, fed either normal chow or high-fat diet.
- Mice are sacrificed according to institutional protocols.
- Epididymal fat pads are harvested by dissection using iris scissors according to the proximity to the epididymis and vesicular gland, taking care not to include the internal spermatic artery/vein and caput epididymis.
- After harvesting, adipose tissue is placed into 50 ml conical tubes containing 25 ml of EGM-2 MV-supplemented EBM-2 medium in which it is stored until embedding.

3.1.2 Collection of human adipose tissue
- Tissue is obtained following IRB protocols. In our studies, human adipose tissue has been obtained from needle biopsies, bariatric surgery, or panniculectomy procedures.
- Tissue is cut into ~1 g fragments, from which large vessels and obvious connective tissue are removed using iris scissors.
- After harvesting, adipose tissue is placed into 50 ml conical tubes containing 25 ml of EGM-2 MV-supplemented EBM-2 medium in which it is stored until embedding.

3.2. Embedding procedure

1. Label 3 100 cm petri dishes as #1, 2, and 3.
2. Pipette 25 ml of EGM-2MV-supplemented EBM-2 medium in plate #1 and #3. Put 15 ml of medium in plate #2.
3. Using forceps, transfer tissue from 50 ml conical tube into plate #1 and wash it by gently moving it around the plate.
4. Using forceps and scalpel, cut the tissue into strips (Fig. 5.1). Use the forceps to hold the piece and the scalpel to make the cuts, always moving in one direction (no upward or downward movement). Rinse the strip with a gentle mix in the medium on the plate. Move the tissue strips to plate #2.
5. In plate #2, cut the strips into small slices. Maintain the size of pieces constant. As a guide for size, you can place millimeter paper underneath plate #2. Each piece should not be greater than 1 mm^3. Once cut, move the piece to plate #3. Cut approximately 75–80 slices per tissue sample.
6. Obtain a small tray and fill with ice. Put 96-well plate over ice. Maintain the plate in ice during the following steps.
7. Dispense 40 μl of Matrigel into wells to be used. Do not use the wells around the perimeter of the plate. Keep Matrigel on ice at all times.
8. Using the forceps, place one piece of adipose tissue per well.
9. After embedding, take plate out of ice. Make sure each explant is in the middle of the well. Use forceps if accommodation is needed.
10. Incubate at 37 °C in 5% CO_2 for 30 min.
11. Add 200 μl of EGM-2 MV-supplemented EBM-2 medium per well. Make sure to fill all well in the plate, including the well around the perimeter of the plate, which do not contain adipose tissue explants. Wells at the perimeter of the plate tend to evaporate faster and placing explants only in the wells located in the middle of the plate allows for maintenance of a constant level of medium in all wells containing the explants.
12. Incubate at 37 °C in 5% CO_2. 100 μl of the medium should be replaced every other day.

3.3. Quantification of angiogenic potential

The mount of capillary growth can be evaluated using several approaches. Here we describe three independent strategies to quantify angiogenic potential, which we have successfully applied to both mouse and human samples.

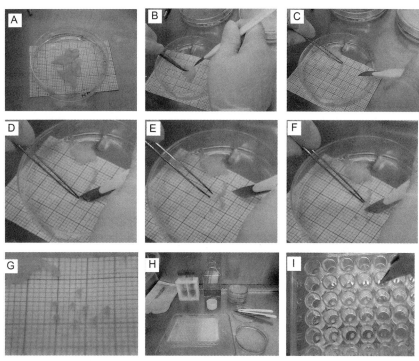

Figure 5.1 Embedding procedure. (A) Adipose tissue samples placed in a 100 cm petri dish containing 25 ml of EGM-2 MV medium. The millimeter (mm) paper placed under the petri dish is used as a size reference. (B) Sample of adipose tissue in plate #2 containing 15 ml of EGM-2 MV medium. The scalpel and forceps are used to hold the fat and cut it into strips. (C) Piece of fat strip cut from the adipose tissue sample. Using the millimeter paper reference, the fat strip is aligned in order to cut the appropriate size of each slice (explant). (D) For the first cut, it is easier to start at one of the ends of the adipose tissue strip. The forceps are used to hold the fat while the scalpel is used to cut the slice. (E) The explant is aligned with one of the quadrants in the millimeter paper to verify adequate size. (F) The rest of the strip is cut into slices. The adipose tissue is held by forceps and the cut is done by the scalpel. While handling the forceps, avoid pulling or stretching the fat, since it may damage the tissue. (G) Individual slices cut to appropriate size and verified with the millimeter paper. (H). Display of workstation in the biocabinet before starting the embedding procedure. Explants were transferred to plate #3, containing 25 ml of EGM-2 MV medium. 96-multiwell plate is kept in a tray filled with ice for the embedding steps. (I) Embedding step. After the Matrigel is dispensed, forceps are used to place the explants, one per well. The explant is positioned at the center of the well. (See the color plate.)

We recommend using at least two independent methods so correlation analysis can be performed to validate reproducibility of the data obtained.

3.3.1 Counting of capillary sprouts

A capillary sprout is defined as a branching structure of at least three cells connected to each other in a linear manner (Fig. 5.2A and B). This structure

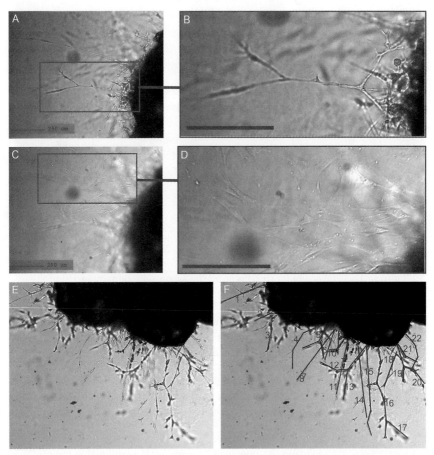

Figure 5.2 Cells emerging from mouse adipose tissue explant. (A, B) Capillary sprout emerging from embedded mouse explant, displaying characteristic linear branching structure. (C, D) Focus set to the surface of the well, where fibroblastic adherent cells can be seen emerging from the explant, observed at a different optical plane of the image. (E) Phase contrast image of the explant and the capillary sprouts 14 days post-embedding. (F) Structures shown in red highlight formations that can be considered to be sprouts. (See the color plate.)

is qualitatively different from other cell types that can also emerge from the explant, but which grow in a disorganized manner and typically adhere to the surface of the plate (Fig. 5.2C and D). To detect capillary sprouts, imaging under sufficient magnification under phase contrast is required. In our laboratory we use a Zeiss Axiovert equipped with a $10\times$ objective. Phase contrast images of embedded explants and emerging capillary sprouts from mouse adipose tissue at day 14 are shown in Fig. 5.2E, and in Fig. 5.2F, the structures that would be considered to be sprouts are delineated in red. As new sprouts emerge they become interconnected and grow into the three-dimensional volume of the Matrigel; thus, counting should be performed early during the growth period. Sprouts begin to emerge after 4–5 days postembedding of mouse epididymal fat explants, and we quantify the number of sprouts per explant at day 7 and day 14 postembedding. Because the definition of a sprout is somewhat subjective, it is necessary to have more than one individual perform the counting, and that the individuals performing the counting be blinded as to the experimental condition of the sample.

Because of the variability in the number of sprouts emerging from each explant, at least 25 explants must be analyzed for each experimental condition. The results are expressed as the average number of sprouts per explant. When studying mouse epididymal fat, there is a linear correlation between the number of explants that develop sprouts and the average number of sprouts per explant (Fig. 5.3). Thus, the percent of explants that display sprouting is an alternative way to quantify angiogenic potential.

Counting can also be performed on human explants, which form well-defined capillaries with tight junctions defining primitive lumens (Fig. 5.4). Human explants develop many more sprouts compared to mouse (Fig. 5.5). After 11 days, the sprouts become highly branched and expand into the three-dimensional volume of the Matrigel, making counting more complex. Thus, counting of human sprouts is best done at days 5–7 postembedding.

3.3.2 Digital analysis of growth area

The capillary growth can be further analyzed by measuring the area of sprouting in each well. In our laboratory, we have used a Zeiss Axio Observer Z1 microscope equipped with an automated stage. Brightfield images of each well of a 96-well plate are acquired using a $2.5\times$ objective and captured using an Andor Clara E interline CCD camera. The stage position, illumination and acquisition setting are controlled by Micro-manager open source microscopy software (Edelstein, Amodaj, Hoover, Vale, & Stuurman, 2010), using a multiwell plate plugin developed by Karl Bellve

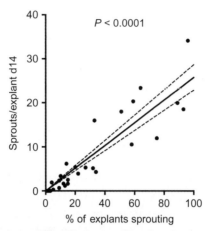

Figure 5.3 Linear correlation between the number of explants sprouting and the quantity of sprouts per explant from mouse adipose tissue. Capillary sprouting was quantified in a study of 36 mice fed normal or high-fat diet for 3–30 weeks. 25–30 explants from each mouse were embedded. The percent of the embedded explants for each mouse displaying sprouts after 14 days of culture is plotted on the x-axis. The mean number of sprouts per explant (i.e., sum of sprouts in all explants/total number of explants embedded) is plotted on the ml axis. Linear regression was calculated using PRISM software.

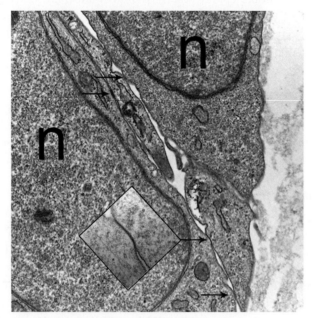

Figure 5.4 Electron micrograph of capillary sprouts from a human adipose tissue explant. Arrows identify tight junctions. n, nucleus.

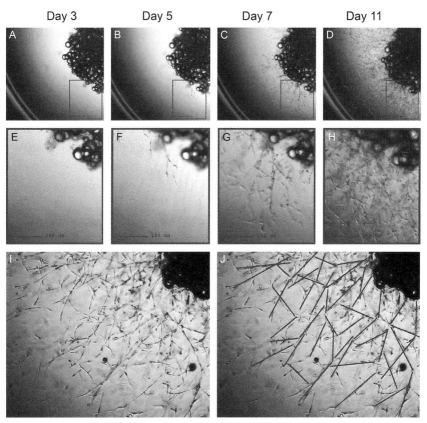

Figure 5.5 Capillary sprouting from human adipose tissue. A human explant from subcutaneous adipose tissue at days 3 (A, B), 5 (C, D), 7 (E, F), and 11 (G, H) post-embedding. Capillary sprouting begins to be observed at day 5. After day 11, the growth is highly increased (I, J), making difficult to identify all sprout formation. (See the color plate.)

and Ben Czech (http://valelab.ucsf.edu/~MM/MMwiki/index.php/Well_Plate_Plugin). The plugin is designed to handle standard SBS well plates from 24, 96, 384, and 1536 wells. The plugin will calculate the correct X, Y, and Z coordinate for each well position, and for any number of positions within each well. Imaging can then iterate through each well, starting at the first well. The plugin will move the stage down or up each column, until the last row, before moving to the next column and reversing direction. The zigzag path chosen for moving the stage is the fastest path on a Zeiss AxioObserver Z1. The starting well, or the ending well, can be selected to only work on a subregion. For our experiments, each well of a 96-well plate containing an explant is divided into four quadrants, with a 50 CCD pixel overlap. Five optical sections spaced at 150 μm are collected for each quadrant, and a montage of the quadrants is then generated for further analysis.

For further analysis, the composite image is imported into Fiji, an image processing package, which is a distribution of ImageJ, Java, Java 3D, and several plugins organized into a coherent menu structure (Schindelin et al., 2012). For calculation of the growth area, areas of explant and capillary sprout growth are delineated. Various parameters can then be extracted, including the area occupied by explant and sprouts (Fig. 5.6) as follows:

To measure areas in the image files:
1. Open Fiji
2. Open image to be analyzed
3. Visually examine in which plane the explant is more in focus. Use it for selecting the area of the explant
4. In the tools menu, select "Freehand selections"
5. Select the area of the explant:
   1. Highlight the borders of the explant with the cursor, keeping pressed the left button on the mouse. If the button is released, the area selected will be automatically closed
6. Edit > Selection > Add to Manager. Automatically, the selected area is added to ROI Manager. Keep the ROI Manager window open. ROI will allow you to select the area of growth in the same image and measure a series of parameters, including the area measurements
7. Go back to the image and select the area of growth. Make sure to examine every plane. You can move around each plane while making the selection, so all the growth is included
8. After selecting the area of growth, press "Add (t)" in the ROI Manager window
9. In the ROI Manager window, select both areas
   1. "Shift" + "↓" if needed
10. Press "Measure" in the ROI Manager window. A "Results" window will appear. It contains different kinds of measurements, including area of the explant and growth
    1. Item #1 is the area of the explant
    2. Item #2 is the area of the growth
    3. It is very important to maintain the order of measurements. First the explant, then the area of growth.
11. In the Results window, go to File > Save as

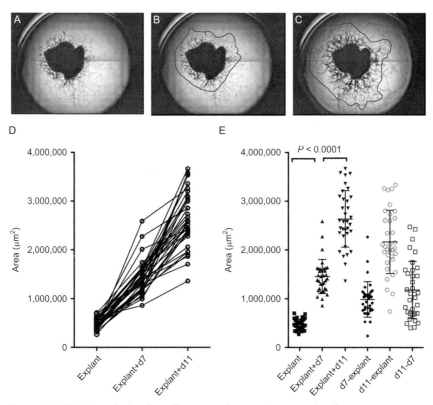

Figure 5.6 Digital analysis of capillary growth area. An example of montages generated from bright field images of quadrants of a single well from a 96-well-multiwell plate containing an explant from human omental adipose tissue. The region of the explant (A), and of capillary growth at day 7 (B), and day 11 (C) postembedding is delineated. The areas are calculated for the selected regions highlighted in red. (D) Calculated areas of 34 explants from the same tissue sample growing in the same 96-well-multiwell plotted in a before–after format, revealing linear growth in all embedded explants over the culture period. (E) Scatter plot displaying the means and standard deviation of the values obtained for each explant at each time point, and values obtained after subtracting the area of the initial explant. Paired Student's t-test between time points reveals highly significant differences, which can be used to compare angiogenic potential among different donors. (See the color plate.)

1. Suggestion: Save results file with the well name (A2, B2, etc.) and a short identification of the sample. Keep consistency in naming, in case a program is developed to help in analysis. (Examples: A2_Control_Day7.txt, B7_Sample1_Day7.txt)
12. In the ROI Manager window, select "More>>," Save

1. Save ROI file as a .zip. This will allow you to save both explant and growth area as .roi files. Suggestion: Name the .zip file with the well name (A2, B2, etc.) and a short identification of the sample. Keep consistency in naming, in case a program is developed to help in analysis. (Examples: A2_Control_Day7.zip, B7_Sample1_Day7.zip)
13. To open image files that were already measured
    1. Open Fiji
    2. Open desired image
    3. File > Open > Select file > Open ".zip" file
    4. In the ROI Manager window, select both areas
        a. "Shift" + "↓" if needed
    5. Mark "Show All"
    6. Both selected areas should appear in the active image

3.3.3 Immunofluorescence analysis of explants

Adipose tissue explants can also be analyzed by immunofluorescence, with the exception that embedding must be done on coverslips or glass-bottom culture dishes. In our laboratory we have used 35 mm glass-bottom culture dishes (MatTek Corporation). The use of glass-bottom culture dishes allows for examination of stained samples at higher magnifications. The protocol used for visualizing human Von Willebrand factor is as follows:

1. Dispense 100 μl of Matrigel to cover the bottom of each dish, and embed explants and maintain in culture as described above.
2. On the desired day, wash samples three times in DPBS by gently aspirating medium from the side of the dish and using 3 ml of DPBS for each wash.
3. Prepare fresh 4% formaldehyde solution by diluting the supplied 16% solution in DPBS at 1:3 ratio. Make sure to use chemical hood with proper ventilation at all times while handling formaldehyde.
4. Inside of chemical hood, gently aspirate DPBS from the dishes with samples and add 1 ml of freshly prepared 4% formaldehyde solution into each dish. Incubate for 15 min at room temperature with very gentle shaking.
5. Gently remove 4% formaldehyde from the dish and wash again three times in DPBS, gently shaking for 5 min in each wash step.
6. Prepare permeabilizing solution by supplementing DPBS with 0.5% Triton X-100 and 1% BSA.

7. Permeabilize and block nonspecific antibody attachment by incubating in 0.5% Triton X-100 1% BSA in DPBS for 60 min with gentle shaking. Use 2 ml of the solution per each well.
8. Remove solution and replace with primary antibody, in this case anti-human polyclonal rabbit Von Willebrand factor (Dako, catalog # A0082), diluted 1:100 in permeabilizing solution. Use at least 1 ml of antibody solution for each dish. Incubate overnight at 5 °C with gentle shaking.
9. Remove primary antibody, and wash three times with DPBS, gently shaking for 5 min between each wash step.
10. Incubate for 1 h with secondary antibodies diluted in permeabilizing solution. The secondary antibody we have used is Alexa Fluor 488 (Invitrogen; Molecular probes, catalog # 11008, dilution 1:500).
11. Wash three times with DPBS, gently shaking for 5 min between each wash step.
12. Perform nuclear counter staining by incubating for 5 min with Hoechst 33258, pentahydrate diluted 1:5000 in DPBS.
13. Wash in DPBS three times.

3.3.4 Isolated cell analysis

After completing manual counting and digital assessments of capillary growth, cells comprising capillary sprouts can be harvested from the Matrigel for further analysis by western blotting, FACS, etc. To quantitatively recover the cells from each well, the following procedure is used:
Before starting
- Dispase is used for proteolysis of Matrigel. Dispase should be aliquoted in 1.5 ml volume vials and stored at -20 °C. Avoid multiple freeze–thaw cycles. Before use, dispase is thawed at 37 °C.
- Most of the capillary cells grow in Matrigel, but some cells (e.g., fibroblasts) attach to the bottom of the plate. Trypsin-Versene is used for detaching cells that remain adherent to the well after the dispase treatment.
- Materials and tools used, including jeweler's forceps and scissors, should be sterile.
1. Carefully, remove the medium by aspirating it from the wells. Be careful not to aspirate Matrigel or explants.
2. Rinse twice with 200 µl of sterile DPBS. Discard DPBS.
    Caution: In case of human tissue, discard medium and DPBS according to your institution's safety regulations.

3. Add 50 μl of dispase to each well. Incubate at 37 °C for 1.5–2 h. *Make sure that all cells are detached and floating in the digested Matrigel by looking under the microscope after incubation.*
4. Prepare a 1:1 mixture of Trypsin-Versene and 50 mM EDTA.
5. Add 50 μl of the Trypsin-Versene/EDTA mixture to each well to stop dispase digestion and dislodge adherent cells. Gently mix by pipetting up and down. Incubate at 37 °C for 10 min.
6. Use forceps to remove remaining explant from wells. Discard the explants in biomedical waste.
7. Add 50 μl of DPBS to each well. Mix by pipetting up and down.
8. Transfer the cell suspension from all wells to a 15 ml conical tube. Fill with EGM-2 MV-supplemented EBM-2 medium to achieve a final volume of 10 ml.
9. Centrifuge the tube containing the cells at $200 \times g$ for 10 min at room temperature.
10. Make sure the cells are pelleted down. Aspirate the supernatant and resuspend the pellet in 1 ml of fresh EGM-2 MV-supplemented EBM-2 medium.
11. Obtain cell concentration and viability. For this purpose, the Cellometer Auto T4 Cell Counter and Software was used (Nexcelom Bioscience).
12. Resuspend cells for further use.

4. METHOD LIMITATIONS

1. There is significant variation between each individual explant, so quantitative analysis requires at least 25 explants per condition.
2. The capillary sprouts are defined by morphological criteria, so changes in the cellular composition, cell shape or cell motility could affect the result.
3. The Matrigel is relatively impermeable and thus penetration by large molecules such as nucleic acids, or viruses that could be used to interrogate the properties of the growing cells is difficult.
4. Immunostaining is complicated by nonspecific attachment of antibodies to Matrigel, and therefore successful results depend on the availability of high affinity antibodies and molecules expressed at relatively high levels.

ACKNOWLEDGMENT

This work was funded by the National Institutes of Health Grant DK089101 to S. C.

REFERENCES

Aplin, A. C., Fogel, E., Zorzi, P., & Nicosia, R. F. (2008). The aortic ring model of angiogenesis. *Methods in Enzymology*, *443*, 119–136.

Baker, M., Robinson, S. D., Lechertier, T., Barber, P. R., Tavora, B., D'Amico, G., et al. (2012). Use of the mouse aortic ring assay to study angiogenesis. *Nature Protocols*, *7*, 89–104.

Cho, C. H., Koh, Y. J., Han, J., Sung, H. K., Jong Lee, H., Morisada, T., et al. (2007). Angiogenic role of LYVE-1-positive macrophages in adipose tissue. *Circulation Research*, *100*, e47–e57.

Christiaens, V., & Lijnen, H. R. (2010). Angiogenesis and development of adipose tissue. *Molecular and Cellular Endocrinology*, *318*, 2–9.

Crandall, D. L., Hausman, G. J., & Kral, J. G. (1997). A review of the microcirculation of adipose tissue: Anatomic, metabolic, and angiogenic perspectives. *Microcirculation*, *4*, 211–232.

Edelstein, A., Amodaj, N., Hoover, K., Vale, R., & Stuurman, N. (2010). Computer control of microscopes using microManager. *Current Protocols in Molecular Biology* Chapter 14: Unit14 20.

Gealekman, O., Burkart, A., Chouinard, M., Nicoloro, S. M., Straubhaar, J., & Corvera, S. (2008). Enhanced angiogenesis in obesity and in response to PPARgamma activators through adipocyte VEGF and ANGPTL4 production. *American Journal of Physiology Endocrinology and Metabolism*, *295*, E1056–E1064.

Gealekman, O., Guseva, N., Gurav, K., Gusev, A., Hartigan, C., Thompson, M., et al. (2012). Effect of rosiglitazone on capillary density and angiogenesis in adipose tissue of normoglycaemic humans in a randomised controlled trial. *Diabetologia*, *55*, 2794–2799.

Gealekman, O., Guseva, N., Hartigan, C., Apotheker, S., Gorgoglione, M., Gurav, K., et al. (2011). Depot-specific differences and insufficient subcutaneous adipose tissue angiogenesis in human obesity. *Circulation*, *123*, 186–194.

Hodson, L., Humphreys, S. M., Karpe, F., & Frayn, K. N. (2012). Metabolic signatures of human adipose tissue hypoxia in obesity. *Diabetes*, *62*, 1417–1425.

Hosogai, N., Fukuhara, A., Oshima, K., Miyata, Y., Tanaka, S., Segawa, K., et al. (2007). Adipose tissue hypoxia in obesity and its impact on adipocytokine dysregulation. *Diabetes*, *56*, 901–911.

Michailidou, Z., Turban, S., Miller, E., Zou, X., Schrader, J., Ratcliffe, P. J., et al. (2012). Increased angiogenesis protects against adipose hypoxia and fibrosis in metabolic disease-resistant 11beta-hydroxysteroid dehydrogenase type 1 (HSD1)-deficient mice. *The Journal of Biological Chemistry*, *287*, 4188–4197.

Pasarica, M., Sereda, O. R., Redman, L. M., Albarado, D. C., Hymel, D. T., Roan, L. E., et al. (2009). Reduced adipose tissue oxygenation in human obesity: Evidence for rarefaction, macrophage chemotaxis, and inflammation without an angiogenic response. *Diabetes*, *58*, 718–725.

Rausch, M. E., Weisberg, S., Vardhana, P., & Tortoriello, D. V. (2008). Obesity in C57BL/6J mice is characterized by adipose tissue hypoxia and cytotoxic T-cell infiltration. *International Journal of Obesity*, *32*, 451–463.

Schindelin, J., Arganda-Carreras, I., Frise, E., Kaynig, V., Longair, M., Pietzsch, T., et al. (2012). Fiji: An open-source platform for biological-image analysis. *Nature Methods*, *9*, 676–682.

Sung, H. K., Doh, K. O., Son, J. E., Park, J. G., Bae, Y., Choi, S., et al. (2013). Adipose vascular endothelial growth factor regulates metabolic homeostasis through angiogenesis. *Cell Metabolism*, *17*(1), 61–72.

Wree, A., Mayer, A., Westphal, S., Beilfuss, A., Canbay, A., Schick, R. R., et al. (2012). Adipokine expression in brown and white adipocytes in response to hypoxia. *Journal of Endocrinological Investigation*, *35*, 522–527.

CHAPTER SIX

Quantifying Size and Number of Adipocytes in Adipose Tissue

Sebastian D. Parlee[*], Stephen I. Lentz[†], Hiroyuki Mori[*], Ormond A. MacDougald[*,†,1]

[*]Department of Molecular and Integrative Physiology, School of Medicine, University of Michigan, Ann Arbor, Michigan, USA
[†]Division of Metabolism, Endocrinology and Diabetes, Department of Internal Medicine, School of Medicine, University of Michigan, Ann Arbor, Michigan, USA
[1]Corresponding author: e-mail address: macdouga@umich.edu

Contents

1. Introduction 94
2. Technical Aspects 96
   2.1 Tissue preparation 96
   2.2 Image collection and analysis 100
   2.3 Data analysis 111
   2.4 Supporting data, results, and experimental guidance 113
3. Future Challenges and Conclusions 114
Acknowledgments 119
References 119

Abstract

White adipose tissue (WAT) is a dynamic and modifiable tissue that develops late during gestation in humans and through early postnatal development in rodents. WAT is unique in that it can account for as little as 3% of total body weight in elite athletes or as much as 70% in the morbidly obese. With the development of obesity, WAT undergoes a process of tissue remodeling in which adipocytes increase in both number (hyperplasia) and size (hypertrophy). Metabolic derangements associated with obesity, including type 2 diabetes, occur when WAT growth through hyperplasia and hypertrophy cannot keep pace with the energy storage needs associated with chronic energy excess. Accordingly, hypertrophic adipocytes become overburdened with lipids, resulting in changes in the secreted hormonal milieu. Lipids that cannot be stored in the engorged adipocytes become ectopically deposited in organs such as the liver, muscle, and pancreas. WAT remodeling therefore coincides with obesity and secondary metabolic diseases. Obesity, however, is not unique in causing WAT remodeling: changes in adiposity also occur with aging, calorie restriction, cancers, and diseases such as HIV infection. In this chapter, we describe a semiautomated method of quantitatively analyzing the histomorphometry of WAT using common laboratory equipment. With this technique, the frequency distribution of adipocyte sizes across the tissue depot and the number of total adipocytes per depot can be estimated by counting as few as 100 adipocytes per animal. In doing so, the method described herein is a useful tool for accurately quantifying WAT development, growth, and remodeling.

1. INTRODUCTION

White adipose tissue (WAT) is a dynamic and modifiable component of overall body mass in adulthood, comprising between ~3% and ~70% of total body weight (Hausman, DiGirolamo, Bartness, Hausman, & Martin, 2001). WAT develops in late (14–24 weeks) gestation in humans (Ailhaud, Grimaldi, & Negrel, 1992; Poissonnet, Burdi, & Garn, 1984) and postnatally in mice and rats (Han et al., 2011; Pouteau et al., 2008) and is unique in its potential for continuous and seemingly limitless growth, as observed in humans and animals under states of persistent energy surplus.

With obesity, the morphology and function of both individual adipocytes and whole WAT depots become altered. This process is a form of WAT remodeling. In periods of chronic positive energy imbalance (Poretsky & Ebooks Corporation, 2010), adipocytes store surplus energy as triacylglycerols, expanding in size (hypertrophy) and in number (hyperplasia) as a consequence. Adipocyte hypertrophy results from a relative increase in lipid deposition versus lipolysis (Kaartinen, LaNoue, Martin, Vikman, & Ohisalo, 1995; Reynisdottir, Ellerfeldt, Wahrenberg, Lithell, & Arner, 1994). When demand for lipid storage exceeds the capacity of existing adipocytes, the pools of adipocyte precursors (preadipocytes) compensate by dividing and differentiating into adipocytes; this process, called adipogenesis, results in adipocyte hyperplasia (Cawthorn, Scheller, & MacDougald, 2012a, 2012b; de Ferranti & Mozaffarian, 2008; Faust, Johnson, Stern, & Hirsch, 1978). Impaired adipogenesis is believed to contribute to the development of metabolic comorbidities including type 2 diabetes, because engorgement of adipocytes with excess lipids triggers pathological changes to the adipose tissue. These changes include altered adipokine (e.g., chemerin) secretion, and increased adipose tissue inflammation due to macrophage infiltration and activation. Additionally, lipids that cannot be stored in adipocytes become elevated in the circulation and deposited ectopically in the liver, muscle, and pancreas (Goralski & Sinal, 2007; Le Lay et al., 2001; Ozcan et al., 2004; Roman, Parlee, & Sinal, 2012; Suganami & Ogawa, 2010). Together these metabolic abnormalities trigger systemic insulin resistance and abnormal insulin production, the basic pathology of type 2 diabetes. Obesity, however, is not alone in causing changes to WAT. Numerous diseases, developmental stages, and genetic animal models coincide with or result in WAT remodeling. For instance hypogonadism and Cushing's syndrome result in elevated and modified adipose deposition, whereas HIV infection, cachexia, and certain parasitic infections are marked by adipose tissue wasting (Desruisseaux, Nagajyothi, Trujillo, Tanowitz, & Scherer, 2007; Lee, Pramyothin, Karastergiou, & Fried, 2013; Santosa & Jensen, 2012; Tchkonia et al., 2010; Tisdale, 1997). In addition, genetic modifications in transgenic animals including FABP4-Wnt10b, $LXR\beta^{-/-}$, $SFRP5^{-/-}$, $Timp^{-/-}$, and many others all cause marked changes in adipocyte size or number compared to controls (Gerin et al., 2005, 2009; Longo et al., 2004; Mori et al., 2012). Accordingly, quantifying the number and size of adipocytes in the development, deposition, or remodeling of WAT is essential in characterizing the phenotype of a given adipose tissue depot.

Several methods have been described for quantitative histomorphometry of WAT (Bjornheden et al., 2004; Bradshaw, Graves, Motamed, & Sage, 2003; Chen & Farese, 2002; Hirsch & Gallian, 1968; Lee, Chen, Wiesner, & Huang, 2004; Maroni, Haesemeyer, Wilson, & DiGirolamo, 1990; Okamoto et al., 2007). The basis for a number of these techniques is to disrupt the tissue with collagenase allowing for isolation of the individual adipocytes, which are subsequently stained and/or analyzed by hemocytometer or coulter counter (Bradshaw et al., 2003; Hirsch & Gallian, 1968; Maroni et al., 1990). A limitation with these methods is that collagenase digestion, mesh separation, and/or centrifugation of adipocytes (Bradshaw et al., 2003; Hirsch & Gallian, 1968; Maroni et al., 1990) may damage or exclude certain cells. In contrast, flow cytometry lacks the drawbacks of mesh separation and centrifugation, instead separating adipocytes based on adipocyte size as a reflection of internal lipid accumulation (Lee et al., 2004). Flow cytometry, however, requires specialty training and highly expensive equipment. Yet another reported approach takes advantage of lipophilic fluorophores to mark cellular lipids after which adipocytes size and number can be quantified using confocal fluorescent microscopy. In addition to the expensive microscope, a drawback with this method is the need to count 1000 adipocytes for accuracy, which is a relatively slow and tedious process (Nishimura et al., 2007). Osmium tetroxide is another powerful tool to mark cellular lipid content and has been used in conjunction with microcomputed tomography to visualize marrow adipocytes (see Chapter 7); however, this chemical is extremely toxic and can cause adipocyte swelling (Hirsch & Gallian, 1968). Finally, other investigators have described automated methods for quantifying hematoxylin and eosin (H&E) stained adipocytes. Each method, however, has a distinctive limitation minimizing their utility. For the method of Chen and Farese (2002), the inability to control for adipocytes touching the edge of the image results in an underestimation of adipocyte sizes. While the method of Osman et al. (2013) requires analysis of 16 representative fields of view per whole-slide image, which is laborious. In contrast to these previous approaches, we describe herein a quick and easy semiautomated procedure for histomorphometric analysis of paraffin-embedded adipose tissue that uses common and relatively safe laboratory reagents, basic laboratory skills, and standard laboratory equipment.

2. TECHNICAL ASPECTS

2.1. Tissue preparation

2.1.1 Obtaining and preserving tissues

Both human and rodent WAT are similar in terms of gross anatomical regions (subcutaneous or visceral depots) and morphology (Bjorndal, Burri, Staalesen, Skorve, & Berge, 2011). In humans, subcutaneous adipose depots accumulate beneath the dermis of the skin in multiple locations including the thigh, abdominal wall, and buttocks. In contrast, visceral adipose depots are located within the thorax (mediastinic) and abdomen (omental, mesenteric, perirenal, retroperitoneal, and parametrial) (Pond, Mattacks, Calder, & Evans, 1993). Rodents likewise have specific accumulation of subcutaneous and visceral WAT including epididymal, perirenal, mesenteric, pericardial, and inguinal depositions (for anatomy of mouse WAT depots see Cinti, 2005).

It is well established that increased visceral adipose tissue is highly associated with obesity-related cardiovascular and metabolic abnormalities when compared to subcutaneous depots (Brochu et al., 2001; Fujimoto et al., 1999; Kloting et al., 2010; Lau et al., 2011; Lefebvre et al., 1998; Nicklas et al., 2004). This difference coincides with a divergence in the expression of genes associated with glucose intolerance (increased *Glut4*, decreased *insulin receptor* (Lefebvre et al., 1998)) and atherosclerosis (increased *CCL4*, *CCL5*, and *IL-1β*) (Lau et al., 2011). Unexpectedly, we have also observed differences in gene expression across a WAT depot. For example, with increasing distance from the testis, expression of *sFRP1* mRNA decreases, whereas expression of adiponectin mRNA increases in epididymal WAT (Fig. 6.1). Given this heterogeneity in function and/or expression between and within adipose tissues, it is important prior to necropsy to

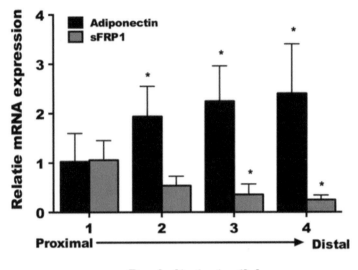

Figure 6.1 Expression of *Adiponectin* increases and *secreted frizzled-related protein 1* (*sFRP1*) decreases in epididymal adipose tissue sampled progressively from the testicle. Twelve-week-old C57BL6/J ($N=12$) mice were sacrificed and epididymal adipose tissue harvested. The depot was divided into quarters and orientation to the testicle noted. Total RNA was harvested and analyzed by RT-qPCR. Results indicate an inverse relationship with *adiponectin* expression increasing and *sFRP1* decreasing in samples more distal to the testicle. Two-way ANOVA with Bonferroni post-hoc analysis, $^*p<0.05$ when compared to the proximal quarter (area 1).

carefully consider which adipose depots, and the locations therein, will be of interest. For help in selecting the appropriate adipose depots, consult the method of Casteilla, Penicaud, Cousin, and Calise (2008).

During dissections, a standard necropsy procedure should be maintained between animals to minimize any variance in tissue collection. A wet-tissue weight is taken immediately after removal using an analytical balance. Special care is taken to remove any nonadipose-associated tissues from the depot including glands and lymph nodes. Once weighed, 100–500 mg of the adipose tissue of interest is placed in a sealable tube and covered by a ratio of >10 mL of 10% formalin per gram of WAT (Sigma-Aldrich, Cat# HT501320). Covered tissue is placed at 4 °C for 72 h, then switched to 70% ethanol (equal volume as was used for formalin) for 48 h. While extended formalin storage can result in difficulty with certain antigen detection due to "overfixation" of (Webster, Miller, Dusold, & Ramos-Vara, 2009), this has little impact on the current method as buffered formalin will not cause variable tissue hardening (Fox, Johnson, Whiting, & Roller, 1985). Of greater concern is the degradation of the tissue as a result of under-fixation and fixative procedures that lead to variable tissue "shrinkage." A standard procedure should therefore be strictly followed for WAT fixing and dehydration.

1. Identify adipose tissue depot of interest
2. Carefully dissect out adipose depot removing nonadipose-associated tissue (e.g., glands, lymph nodes)
3. Weigh depot
4. Cover 100–500 mg of tissue with >10-fold volume (mL) of 10% formalin to tissue (g) and let sit at 4 °C for a minimum of 72 h
5. Remove tissue and place in 70% ethanol for 48 h

2.1.2 Tissue embedding

Preserved tissues are then placed in labeled histology cassettes (e.g., Cat# CA95029-822, CA18000-174; VWR International, Radnor, PA, USA) and paraffin (e.g., Paraplast Tissue Embedding Medium, Cat# 1006, Sigma-Aldrich, St. Louis, MO, USA) processed (e.g., Leica ASP 300 Paraffin Tissue Processor, Leica Microbiosystem, Wetzlar, Germany). Below are the steps used for paraffin processing adipose tissue.

1. 1 h in 70% ethanol
2. 1 h in 80% ethanol
3. 1 h in 95% ethanol (×2)
4. 1 h in 100% ethanol (×2)
5. 1 h in xylene (×2)
6. 1 h in 60 °C paraffin (×3)

Once the sample is processed, the tissue is embedded into a paraffin block (e.g., Tissue-Tek Paraffin Tissue Embedder, Sakura Finetek, Torrance, CA, USA) and stored at 4 °C. Time and care should be taken in the embedding process to ensure the tissue is placed into the center of the block. This will help standardize each sample for sectioning.

Note: Allowing the tissues to remain in the 60 °C paraffin for long periods of time after processing should also be avoided to limit tissue damage.

2.1.3 Sectioning, deparaffinizing, and staining adipose tissue

Embedded tissue samples are faced off using a low profile microtome blade (e.g., Accu-Edge, Sakura Finetek, Cat# 4689, Sakura Finetek, Torrance, CA, USA) to the apex of the tissue using a paraffin microtome (e.g., Leica 2155 rotary paraffin microtome, Leica microbiosystems, Wetzlar, Germany).

Note: Samples should be kept on ice between sectioning. Colder blocks will provide cleaner sections. Similarly, a 40–42 °C water bath should be used to collect sections. When the dehydrated tissue comes into contact with water the tissue will begin to expand. Embedded adipocytes are fragile. Therefore, if the water is too warm the tissue will expand quickly tearing the delicate adipocyte membrane and leading to difficulties with downstream analysis.

Starting at the tissue apex 5×5 μm sections are made at a minimum of 100 μm intervals across the sample tissue. Using a minimum interval distance of 100 μm will ensure that each section will contain a unique sampling of adipocytes. Serial sections are recommended as this will provide matching tissue samples for additional independent analyses (e.g., immunohistochemistry). Sections are placed onto labeled electrostatically charged precleaned microscope slides (e.g., Cat# 12-550-15, Thermo Fisher Scientific, Waltham, MA, USA) and are left to dry for a minimum of 24 h.

1. Embedded tissues are faced off to tissue apex
2. 5×5 μm sections are taken at 100 μm intervals and placed on slides
3. Slides are allowed to dry for 24 h

Note: Cutting tissue seems like an inconsequential process, however paraffin-embedded adipose tissue is very delicate and membranes are prone to tearing. Extreme caution should be taken with sectioning. It is wise to perfect sectioning on nonessential adipose tissue samples before moving onto important samples.

Once dry, slides are placed into a 60 °C dry oven for 1.5 h to begin removal of paraffin, and any residual water. Tissues are then stained with H&E using the following method.

1. Slides are baked at 60 °C for 1.5 h
2. 5 min in xylene (×2)
3. 2 min in 100% ethanol (×2)
4. 2 min in 95% ethanol (×2)
5. 2 min in 70% ethanol
6. 2 min in deionized H_2O
7. Stain for 2 min in Harris' hematoxylin (Cat# HHS16, Sigma-Aldrich, St. Louis, MO, USA)
8. Rinse under running H_2O until clear
9. Three dips in a bluing solution
10. Rinse for 2 min under gently running H_2O
11. 2 min in 80% ethanol
12. 2 min in eosin (Cat# 318906, Sigma-Aldrich, St. Louis, MO, USA)
13. Four dips in 80% ethanol to rinse out excess eosin

14. 2 min in 95% ethanol
15. 2 min in 95% ethanol
16. 2 min in 100% ethanol (×2)
17. 3 min in xylene (×3)

Slides are then treated with a xylene-based permount and coverslips mounted (e.g., Cat# 2955-245, Corning Incorporated, Corning, NY, USA). Slides should be left horizontal to set for at least 24 h before putting them into a storage box.

2.2. Image collection and analysis

2.2.1 Using MetaMorph microscopy automation and image analysis software

Using a Zeiss inverted fluorescent microscope (or similar) five representative photos of each section are taken with a 40× objective using a monochrome AxioCAM MrC (or similar) camera. A semiautomated custom image analysis protocol was developed using MetaMorph image analysis software to quantify the area of the individual adipocytes. A brief description of the MetaMorph program is as follows. Based on the magnification of the microscope used, each image is first calibrated to a known distance. For a Zeiss microscope at 40× objective this equates to 0.6812 μm per pixel. Adipocyte cell membranes are enhanced by 3 pixels through erosion filtration to improve the cell boundaries. Histological images are segmented by an inclusive threshold filtration (200 low and 255 high) creating a binary mask and converting the image to a 1-bit configuration. Any adipocytes with visible lacerations to the membranes are "closed" digitally prior to continuing with the automated area quantification. To quantify the adipocyte areas, images are analyzed and adipocytes highlighted if they meet the following four criteria: (1) the adipocyte contains an area between 500 and 15,000 μm^2; (2) the adipocyte has a shape factor of 0.35–1 (a shape factor of 0 indicating a straight line and 1 a perfect circle); (3) the adipocyte has an equivalent sphere surface area between 5,000 and 1,000,000 μm^2; and (4) the adipocyte does not border the image frame. In general, 70% of total tissue adipocytes meet these criteria and are picked up automatically. An additional 10% of adipocytes meet the aforementioned criteria, but are not picked up by the automated program. Thus for our studies, size of these adipocytes was included after manually. To visualize the adipocytes more clearly and provide traceability for their respective quantified areas, each adipocyte is subsequently labeled with a unique number. The unique adipocyte number and respective area is then exported for statistical analysis.

Quantitative Histomorphometry of Adipose Tissue

The following is a detailed description of setting up and running the MetaMorph program (Molecular Devices, Sunnyvale, CA, USA) with the adipocyte analysis tool. To run this analysis both MetaMorph and Microsoft Excel (Microsoft, Albuquerque, NM, USA) software programs will be needed.

1. Open base image

    In the tool bar located at the top of the screen File → Open are selected. An image of interest is then chosen.

2. Calibrate image(s).

    The first step in analyzing the adipocytes is to calibrate the image. The process will allow the program to recognize the size of the sample based on the number of pixels and the magnification of the microscope. The following steps are applied for each image.

    Located in the tool bar at the top of the screen Measure → Calibrate Distances are chosen. This opens a "Calibration Distances" window. In the Calibration box the correct calibration name is selected. To create a new calibration, see the instructions below. "Apply to All Open Images" button is subsequently chosen.

How to add a new calibration

    An image with scale bar of a known length is opened. This image should be taken at the same magnification as the adipocyte of interest. The "+" button is selected. A new entry line item called "Calibration #x" appears in the Calibration box. The line item is selected and the "Calibrate by Region" button clicked. A new window appears called "Calibrate by Region." The steps in *BLUE* are followed. The image containing a stage micrometer or other calibration standard is selected. It should be listed in the "Image" box. If not, the "Image" box is clicked and the correct image is chosen. The "Type of Region Tool" is chosen. This will allow the calibration standard on the image to be measured. The calibration is defined by choosing "Line Rgn" (not "Rectangle Rgn"). The region position and size is adjusted to determine the calibration ratio of selected units to pixels. In the calibration window a box labeled with "Name of Calibration" (e.g., Calibration #x) will be present. A descriptive name (e.g., Zeiss 40 × AxioCam MrC, etc.) can be provided in the "Name of Calibration" box. Because each of the future images is taken at the same magnification and on the same microscope, this calibration can be used for future analyses. If any of the imaging variables change then a new calibration must be made. A value will need to be entered into the X box (Y: if rectangle) of the "Calibrated Length" window. This value is determined by the standard scale bar in the image. Typically, this will be in the micron range. The "Region Length X" (Size if rectangle, Y: if rectangle) is set by the length (in pixels) of the standard scale bar. The region is measured by clicking and holding the middle of the calibration line. The calibration line is moved so that one edge is at the start of the scale bar. The start or end of the line is measured by clicking/holding on the ends (little squares) and moving them to the desired location. The line is positioned to travel straight across the scale bar. The "Units/Pixel X" (Y: if rectangle) box will display a value for the units per pixel based on the calibrated length that was entered and the actual length of the standard scale (in pixels). For the "Units" box, the correct units are chosen by clicking on the down arrowhead. When finished, these calibrations are ready to be used by pressing "OK." These calibrations are applied to all current and future images.

3. Enhance border

To help define the membranes of the individual adipocytes each membrane was enhanced by 3 pixels. The benefit of this enhancement materializes in steps 6 and 7 when the computer algorithm identifies the individual adipocytes. This process is accomplished via the following steps.

Located within the tool bar Process → Morphological Filters → Erode → Parameters menu are selected. The following parameters must be specified.
- Filter shape: click on circle
- Diameter: 3 pixels

Once these are adjusted the APPLY button is pressed. A new Image appears named "Erode." If the "Erode" button is clicked a preferred name can be entered (Specified(Erode)) before pressing APPLY.

4. Threshold image

In this step, the images are transformed into a binary image. The purpose is to provide defined areas consisting of membrane material and empty space identified by black and white respectively. This process is accomplished via the following steps.

Located within the tool bar, Measure → Threshold Image menu is selected. The eroded image must be listed in the "Source Image" box. If it is not it can be changed by clicking on the "Source Image" and changing it to "Erode." The following parameters must also be changed. In the threshold tab the "Threshold State" should be set to inclusive. The low level should be set from 0 to high (255) so that orange color fills the white spaces. Depending on the clarity of the image these numbers may change. Keep in mind that the orange will fill the areas that are empty spaces where the lipid would be found in the adipocytes. Adjustments should be made with the low/high level to best fit the threshold to the empty space.

Next in the clip tab, the "Create Binary Mask" found in the mode section is selected. The bit depth of binary mask should then be set to 1-bit. The resulting image will be called "1-Bit Binary." If the name "1-Bit Binary" is clicked the name can be adjusted to whatever is preferred (Specified(1-Bit Binary)) before pressing APPLY. A new image appears named "1-Bit Binary" (or the selected name).

5. Separate unique adipocytes

This is a good place for the "Cut Drawing" tool to be used to separate adipocytes that have a perforated membrane and are joined to a second adipocyte. The "Cut Drawing Tool" is used (the icon looks like a pencil with red tip down) to separate two or more joined adipocytes with a black line. The "Cut Drawing Tool" is selected and then in the "1-Bit Binary" image, the left mouse button is clicked and held while a small line is used to "cut" the object into separate parts. This process is repeated as necessary.

6. Define initial areas of interest

Now that the adipocytes are defined, a computer algorithm can be used to identify the unique areas of the individual adipocytes. To accomplish this step the following steps are followed:

Located in the tool bar at the top of the screen Measure → Integrated Morphometry Analysis is opened and the Source: "1-Bit Binary" is selected.

The following characteristics should be marked off in the "Preferences" tab:

- Measure methods: Check "Measure All Regions," "Fill Holes in Objects" and "Exclude Objects Touching Edge."
- Object Drawing: Check "Draw Object Borders" and "Draw Centroid Mark."
- Miscellaneous: NONE of the options should be checked.
- Mouse click interaction: Check "Highlight Objects on Left Click" and "Add/Remove Object from Measurement Set on Double Click."
- Object standards: "Standard area: 1," "Optical Density Low Boundary: 1," "Optical Density High Boundary: 255."
- Object mask: Check "Binary-Intensity of 1 for each Object."

The following should be marked off in the "Measurement Parameters and Filters" tab.
- Area: "Display and Filter" are checked. "Comparison" is between Limit 1:500 and Limit 2:15000.
- Shape (Shape Factor): "Display and Filter" are checked. "Comparison" is between Limit 1:0.35 and Limit 2:1.000.
- Equiv (Equiv. sphere surface area): "Display and Filter" are checked. "Comparison": between Limit 1:5000, Limit 2:1,000,000.

The "Reset Current" button is subsequently pressed prior to clicking on the "Measure" button. The "1-Bit Binary" image will have an object overlay appear that colors objects green. When an object is clicked it will be highlighted in yellow. In the "Object Tab," you can see the area, shape and equivalence values for the selected object.

7. Visualize objects more clearly by creating numbered regions over the object

To provide traceability each adipocyte is labeled with a unique number. To accomplish this, the following process is followed:

The "Region Tool Properties" is selected. The icon looks like an overlapping combination of a red circle, red square, notepad and pencil located in tool bar. For the "Region color" NOTHING SHOULD BE CHECKED especially "Use Same Region for New Regions."

On the tool bar Edit → Preferences is opened.
- In the "Region Labels" tab "Draw Labels Next to Regions" should be checked.
- For the label location "Center" should be checked.
- "Draw Label Inside of Region Boundary" as well as "Draw Labels With Respect to the Image Zoom" are also checked.
- For the label color "Same as Region" is checked prior to pressing OK.

Now located in the tool bar Regions → Create Regions Around Objects is chosen.

This will put numbered regions around the objects to help identify which objects need editing.

A moment should be taken to look at the image. If it appears that there are adipocytes that are joined together that should in fact be two or more unique cells then proceed through step 8 or 9 by first removing regions (for now) by clicking Regions → Clear Regions → Yes.

If on the other hand adipocytes identified appear accurate continue to step 10.

8. Modify selected adipocytes by splitting a continuous adipocyte into smaller parts

    If an identified adipocyte is labeled as one continuous adipocyte but should in fact be 2 or more, the adipocytes can be separated using the following steps. The object that is not restricted to a defined adipocyte boundary is selected. If the yellow highlight covers two or more adipocytes then it will need to be split using the "Cut Drawing Tool" (Icon looks like a pencil with red tip down). The "Cut Drawing Tool" is chosen and then in the "1-Bit Binary" image, the left mouse button clicked and held while a small black line is drawn to "cut" the object into separate parts.

    The "Reset Current" button is pressed followed by the "Measure button." The "1-Bit Binary" image will have an object overlay appear that colors objects in green. The object can be selected to see if the object was successfully clipped in two. If not, the procedure is repeated again.

9. Modify selected objects by joining a split object into a larger object

    If a single adipocyte is being registered as two separate adipocytes they can be joined together manually. Click on an object/adipocyte that you feel is broken into smaller parts. If the yellow highlight covers a fragment of the adipocyte, then the fragments are joined by using the "Join Drawing Tool" (icon looks like a pencil with red tip up). The "Join Drawing Tool" is clicked and then in the "1-Bit Binary" image, the left mouse button is pressed and held while drawing a small red line between the fragments to "join" them into one larger object. The "Reset Current" button is then selected followed by the "Measure" button. The "1-Bit Binary" image will have an object overlay appear that colors objects in green. The object can be chosen to confirm the adipocytes have been combined successfully. If not, the procedure is repeated again.

10. Adding/removing objects that are not highlighted in green

    If the automatic computer algorithm does not register a specific adipocyte, it can be added manually by double left-clicking rapidly on the object that is not included and highlighted in green. If, on the other hand, the program identifies an object that is not an adipocyte it can be removed by similarly double left-clicking rapidly on the object.

    Note: Be patient between double clicks, it may take a moment for the new object to become highlighted/unhighlighted.

11. Logging data

    Once the image has been analyzed, the associated data need to be recorded. Recording the data is done by logging the data from objects.

    Data can be logged as objects in the Integrated Morphometry Analysis Window. The "Object Data" Tab is selected followed by the

"Configure Log" button. A new window will open called "Configure Log." The "Parameter Configurations" is opened to select the parameters that will appear in the Excel spreadsheet. Examples include image name, object number, area, shape factor, equiv. sphere surface area, etc. In the "Logging Options" tab, "Log Column Titles" is checked followed by OK. Next the "Open Log" button is chosen. A new window will open called "Open Object Log." In the "Log Measurements" section "Dynamic Data Exchange (DDE)" is pressed followed by OK. A new window will open called "Export Log Data." For the application, "Microsoft Excel" is chosen. The sheet name (typically Sheet1), starting row (typically 1) and starting column (typically 1) are selected, followed by OK. Excel will open and you will be taken to the Excel program. Click on "MetaMorph Offline Button" in the Task bar to return to MetaMorph. The Open Log button is now named "F9: Log Data." Press the "F9: Log Data Button" and the object data will be pasted into Excel. The Excel sheet is SAVED at this point to prevent data loss.

Once the data have been logged, Excel can be disconnected from MetaMorph by closing the Excel Sheet and/or Excel program. By doing so a new window appears in MetaMorph called "Link Disconnected." By pressing on "OK" Excel becomes disconnected from MetaMorph. To log any further data via this method, a connection between Excel and MetaMorph will need to be re-established.

12. Saving regions

Regions can be saved for later viewing by selecting Select Regions → Save Regions from the task bar. A new window appears to save regions to a folder. These regions can be opened at a later date by selecting Regions → Load Regions. The correct image should be opened before loading regions.

13. Adding regions manually

A region can be added manually by clicking on the "Trace Region Icon." This icon looks like an irregular shaped object in the tool bar on top panel. In the image, the first point of the adipocyte parameter is anchored by left-clicking once. The outer parameter of the adipocyte is then traced by moving along the outer edge and anchoring points by left-clicking. When the entire parameter is traced double left-clicking the mouse completes the identification of the region.

Note: The tool will stay on the "Trace Region Icon" unless an option is selected under the "Region Tool Properties." This can be removed by clicking on the Icon that looks like a combination of a "red circle, red square, notepad and pencil" in the tool bar and unchecking the box in the right-hand side that says "Revert to Locator Tool" after creating a region.

14. Logging data for one active region at a time

    Individual active regions are logged by clicking Measure → Region Measurements in the task bar. A new window called "Region Measurements" will open. The correct file should be confirmed as listed in the box on the top of the region measurement window. In the "Include" tab "Select Active Region" is chosen from the pull down menu. The parameters of interest, for example, Image name, Object Number, Area, Shape factor, Equiv. sphere surface area are chosen from the "Configure Tab." The following parameters are also selected.
    - In the "Color Channel," "Intensity" is selected.
    - In the "Display and Log" section, "Region Measurements and Summary" is checked.
    - "Log Image Calibration" and "Log Column Titles" are checked.
    - "Use Threshold for Intensity Measurements" is NOT checked.

    Once these parameters are set, the "Measurement" tab is selected followed by the "Open Log" button. A new window opens called "Open Object Log." In the "Log Measurements to Section," "DDE" is chosen followed by "OK." A new window opens called "Export Log Data." Microsoft Excel is chosen for the program. The Sheet (typically Sheet1), Starting Row (typically 1), and Starting Column (typically 1) are all named followed by selecting "OK." Excel will open. MetaMorph Offline Button is chosen in the Task bar to return to MetaMorph. The "Open Log" button is now named "F9: Log Data." The "F9: Log Data Button" is selected and the object data are subsequently pasted into Excel.

    Traced regions can be saved by selecting Select Regions → Save Regions from the tool bar. A new window appears to save regions to a folder. These regions can be opened at a later date by selecting Regions → Load Regions. The correct image must be opened prior loading the regions.

15. Removing unwanted regions.

    Unwanted regions can be removed by clicking on the regions edge. A flickering animated outline of the adipocyte will indicate it has been selected. Delete button is then pressed on the keyboard to remove the region.

16. Making a screen shot of the image with numbered regions

    Once the image is displaying all of the desired information, such as labeled regions, etc. (see steps 12–14 above to make regions or load regions into image) Edit→Duplicate→As Displayed is selected. A new window appears called "Duplicate As Displayed". Making sure the correct image is in the box under "Source Image: Destination Image," "Duplicate: Select Entire Image" is selected followed by "OK." The image is then saved by clicking on File→Save As. A new window appears "Save File Name As." A folder and name are chosen and saved as a MetaSeries Single/Multi-plane TIFF (*.tif). The Default name for all images is "Untitled." The name can be changed by clicking on the "Untitled" Button. Alternatively the image can be saved, by right-clicking in the "Duplicate as Displayed Image" and copied by selecting "Copy Image to Clipboard." This can be pasted into an Excel spreadsheet.

Steps 1–16 are repeated for all subsequent images of interest.

2.2.2 Using ImageJ image analysis software

Although using the MetaMorph semiautomated software is the most efficient way of analyzing the adipocyte histology, the price of the software is limiting for some laboratories. Accordingly, a similar manual procedure has been developed using the free open access software ImageJ (http://rsbweb.nih.gov/ij/). It should be noted that this method will be slightly slower as it requires manual identification of the adipocytes.

1. Open base image

    The image of interest is opened by clicking on File→Open in the task bar.

2. Duplicate image

    The image is duplicated by clicking the right button of the mouse on the image of interest and selecting "Duplicate Image." The name of the image is chosen.

3. Calibrate image(s)

    The first step in analyzing the adipocytes is to calibrate the image. The process allows the program to recognize the size of the sample based on the number of pixels and the magnification of the microscope. For each image the following steps are applied.

    On the tool bar Analyze→Set scale is selected. The following parameters are set and are dependent on the magnification of the image and the camera settings.
    - Distance in pixels
    - Known distance
    - Pixel aspect ratio = 1 (for most digital cameras)
    - Unit of length = µm

To determine these parameters an image with a known linear scale bar is opened. It is important that this image is taken at the same magnification as your images of interest. The "straight line" tool is selected on the tool bar. A line is drawn over the scale bar on the image, keeping the line as straight as possible. Analyze → Set Scale is selected from the task bar, opening the set scale window. The distance that was measured in pixels is displayed in the "Distance in Pixels Box." The known distance of the linear scale bar is entered into the "Known Distance Box." The pixel aspect ratio for most cameras is set to 1. The unit of length is dependent on the scale bar and is likely in microns. The "Global" box is selected to maintain the settings for all subsequent image analysis. Once the calibration settings are known they can be used for all subsequent images taken at the same magnification using the same camera.

4. Subtract background

    To help clarify the image for subsequent analysis, the backgrounds of images are removed based on a "rolling ball" algorithm (Sternberg, 1983). To process the image Process → Subtract Background is selected from the task bar. The following parameters should be set.
    - "Rolling Ball Radius" should be set to 50 pixels
    - "Light Background" should be checked
    - "Create Background (do not subtract)" should NOT be checked
    - "Sliding Paraboloid" should be checked
    - "Diable Smooth" should NOT be checked
    - "Preview" should be checked

    When all of the parameters are set "OK" is selected.

5. Remove excess noise

    Often times when photos of tissues are taken random variations in the brightness or color information occur. This "noise" is removed by selecting Process → Noise → Despeckle in the tool bar.

6. Threshold images

    The purpose of this step is to provide defined areas consisting of membrane material and empty space identified by black and white respectively. This process is accomplished via the following steps: Image → Adjust → Threshold are selected from the tool bar. The "Threshold Type" should be set to "Default" as well as "Over/Under." The box labeled "Dark Background" is selected. The first sliding bar should be set to 200 (this may vary depending on the clarity of the image) while the second sliding bar will be set to 255. Once these parameters are set "Apply" is chosen.

Note: The blue highlight should maximally cover the areas defined as the adipocyte membranes without covering the empty space that would normally contain lipid.
7. Convert image to a binary format

    Now that the areas of interest are defined, the image needs to be converted to a binary format to allow for future analysis. To convert the image Process → Binary → Make Binary are selected from the tool bar.

    Note: This process will convert the dominantly black image to a dominantly white image.
8. Enhance border

    To help define the membranes of the individual adipocytes each membrane is uniformly enhanced. The benefit of this enhancement materializes in step 8 when identifying the areas of the individual adipocytes. To enhance the border: Process → Binary → Dilate are chosen.

    Note: The processing of the images (Steps 4–8) can be automated by creating a macro. Plugins → New → Macro is selected and a new screen will appear. The following computer script is subsequently copied and pasted into this window.

    ```
    run("Subtract Background...", "rolling=50 light sliding");
    run("Despeckle");
    setAutoThreshold("Default dark");
    //run("Threshold...");
    setThreshold(200, 255);
    run("Convert to Mask");
    run("Make Binary");
    run("Dilate");
    ```

    This macro should be subsequently named and saved (File → Save). Once this macro is created, images can be easily analyzed by simply opening and calibrating the image, then by running the macro (click on Plugins → Macros → Run, opening the saved macro).
9. Measure and label adipocytes

    To measure the area of the adipocytes you will need to download the free "Measure and Label Macro" for imageJ (http://rsbweb.nih.gov/ij/plugins/measure-label.html). This macro will provide a read out of the adipocyte area and place a corresponding unique number within the center of the adipocyte. This method of measuring the area has the benefit of providing traceability for each adipocyte. Once the macro has been downloaded and installed, create a short cut key to activate the macro program. To initiate on the macro Plugin → Shortcuts → Create Shortcuts is selected in the tool bar. The "Measure and Label Macro" is chosen in the command pull down menu. A key (that is not already set, e.g., q) needs to be selected to be set as the shortcut. The "Wand" (tracing tool) located in the tool bar is subsequently chosen. Using this tool, the individual adipocytes of interest are clicked on. A yellow tracing of the inside of the adipocyte will appear. The shortcut key (e.g., q) that you assigned to the "Measure and Label Macro" is then clicked. A unique number will be placed in the center of the adipocyte, and in a pop-up results screen the corresponding area will be appear. The "Wand" tool and macro are then iteratively used until all remaining adipocytes are counted.

Note: Do not count adipocytes that are touching the border of the image.

Once all of the adipocytes have been counted, the recorded data are copied by clicking on the right button of the mouse, select all, then on copy, then pasting it into an excel spreadsheet. By clicking on File → Save as → Tiff this image can be saved for future reference.
Repeat steps 1 through 8 for all of the individual images of interest.

2.3. Data analysis

Once the areas of the adipocytes of interest are determined, a frequency distribution is calculated for a given tissue, being sure to first remove any objects that fall below an area of 350 μm^2, as these cells may be a mixture of adipocytes and stromal vascular cells. The frequency can be accomplished easily using the frequency function in Excel (=frequency(data_array, bins_array)). The size of the array bins may vary depending on the animal model; however, a distribution from 0 to 15,000 μm^2 in 500 increments is appropriate for most WAT depots. The number of total adipocytes within the distribution is subsequently calculated and used to convert the frequency to a percentage of total adipocytes counted (Fig. 6.2A). The benefit of using a frequency distribution rather than the average adipocyte area alone emerges when there is a change in adipocyte number in addition to size. In this case, while the overall average area of the adipocytes may not change, the number of adipocytes of a given size may be altered. By providing a frequency distribution these types of changes are accounted for. A comparison between two frequencies is made using a two-way ANOVA followed by a Bonferroni post hoc analysis. To ascertain whether there is a change in adipocyte number and size between animal models, a correlation between the average adipocyte volume and the weight of the adipose depot is determined. To calculate the average adipocyte volume, the radius of each individual adipocyte area is determined, assuming a circular model. This radius is used to calculate the volume (assuming a spherical model) of the given adipocyte. The average volume of a given tissue is

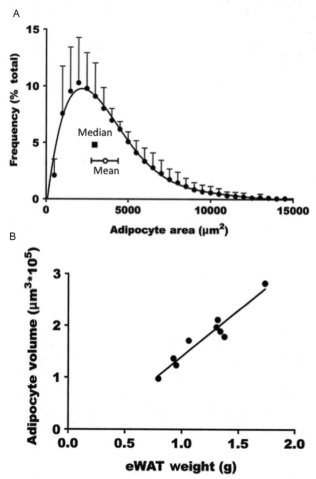

Figure 6.2 The distribution of adipocyte sizes and the linear relationship between average adipocyte volume and epididymal WAT weight. Ten-week-old C57BL6/J mice fed a 60% high-fat diet for 6 weeks were sacrificed and epididymal WAT harvested, fixed, sectioned and ~5000 adipocytes per mouse counted using the MetaMorph-based method described herein. The distribution of adipocytes within the adipose depot is calculated using the frequency function in Excel. (A) C57BL6 mice ($N=10$) have a mean adipocyte area of approximately 3600.98 ± 820 μm^2 (open circle) and a median of 2970 μm^2 (black square). (B) The relationship between average adipocyte volume weight of epididymal WAT is linear ($r^2 = 0.92$).

then calculated from these volumes. A linear correlation with a two-tailed comparison of slope and intercept is calculated and compared between different treatment groups (Fig. 6.2B). Analysis of adipocyte cellularity can then be made using either MetaMorph or ImageJ as both provide equivalent distributions (Fig. 6.3A) and average adipocyte size (Fig. 6.3B).

Quantitative Histomorphometry of Adipose Tissue

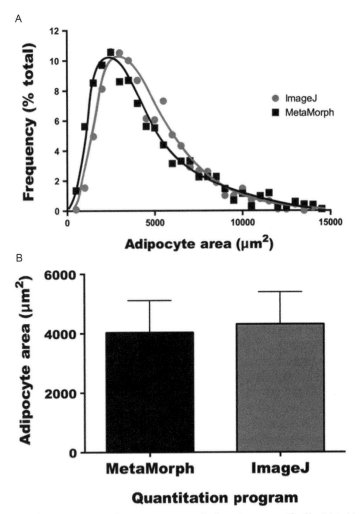

Figure 6.3 Adipocytes size and average area are similar when quantified by MetaMorph or ImageJ. Ten-week-old C57BL6/J mice ($N=5$) fed a 60% high-fat diet for 6 weeks were sacrificed and epididymal adipose tissue depots harvested, fixed, sectioned and adipocytes counted using either the MetaMorph or ImageJ method described herein. The frequency distribution of adipocyte sizes is equivalent whether analyzed by MetaMorph or ImageJ. No significant differences were found in either the frequency distribution (A, two-way ANOVA with Bonferroni post-hoc analysis) or average adipocyte area (B, Student's t-test). (See the color plate.)

2.4. Supporting data, results, and experimental guidance

A number of factors need to be taken into consideration when performing quantitative histomorphometry on adipose tissue. As was mentioned above, significant differences exist between the expression profile of adipocyte genes between tissues and within tissues (Brochu et al., 2001; Fujimoto et al., 1999; Kloting et al., 2010; Lau et al., 2011; Lefebvre et al., 1998; Nicklas et al., 2004). Analysis of adipocyte size using our described method has indicated that there are differences in the distribution of adipocyte sizes and average area between inguinal and epididymal WAT (Fig. 6.4). In contrast, with the exception of adipose tissue located directly on the testis, there were no differences in the distribution of adipocyte size or average area across the epididymal tissue depot (Fig. 6.5). As a consequence, when carrying out a comparison between treatments or genotypes in murine animal models, it is important that the subcutaneous and visceral adipose depots are considered separately and specimens taken from epididymal adipose depots should not be sampled directly from the testes.

In addition to the specific depot and location, the number of adipocytes counted and the number of animals used in the experiment are important considerations. While counting 5000 adipocytes per animal (Fig. 6.2) is possible and provides an extremely accurate distribution of the adipocyte areas across the tissue bed, it is a product of diminishing returns. By comparing the average area calculated from 3, 10, 100, 300, 500, 1000 randomly selected adipocytes repeated 10 times from an individual animal, we identified three adipocytes as the lowest evaluated number required to accurately predict average adipocyte area (Fig. 6.6A); however, counting 100 adipocytes is required to achieve a reasonably small variance around that mean (Fig. 6.6B). Counting 100 adipocytes also results in a similar distribution of adipocyte sizes compared to 300, 500, and 1000 adipocytes (Fig. 6.6C). Accordingly, our data suggest that as few as 100 adipocytes need to be counted to obtain an accurate estimation of adipocyte sizes within an adipose tissue depot. Finally, by using the frequency distribution (Fig. 6.7A) or volume of adipocytes (Fig. 6.7B) in the epididymal adipose depot, the mean adipocyte radius was calculated for mice fed a standard chow (26.7 ± 3.1 μm) or high-fat (33.8 ± 3.9 μm) diet. Based on these variances we estimated using the power equation ($1 - \beta = 0.8$, $\alpha = 0.05$) (National Research Council, 2003) that a minimum of five animals per treatment group would be required to see a ~7 μm difference from the observed mean radius (Fig. 6.7C and D), which would be comparable to the difference in average radius between the chow and high-fat diet groups.

3. FUTURE CHALLENGES AND CONCLUSIONS

While our method provides an easy and semiautomated protocol for quantitatively analyzing size and number of adipocytes within adipose tissue

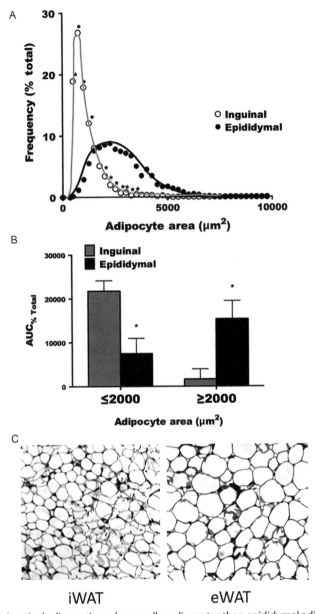

Figure 6.4 Inguinal adipose tissue has smaller adipocytes than epididymal adipose tissue. Ten-week-old C57BL6/J mice ($N=5$) fed a 60% high-fat diet for 6 weeks were sacrificed and epididymal (eWAT) and inguinal (iWAT) adipose tissue depots harvested, fixed, sectioned and adipocytes counted using the MetaMorph method described herein. The distribution of adipocyte areas indicates iWAT has greater numbers of adipocytes with areas below 2000 μm^2 and fewer adipocytes with areas greater than 2000 μm^2 when compared to eWAT (A, B). The area-under-the-curve of this distribution quantifies this elevated number of smaller (≤ 2000 μm^2) and decreased number of larger (≥ 2000 μm^2) adipocyte in iWAT. The difference in adipocyte size is evident in representative H&E stained samples of both iWAT and eWAT (C). Two-way ANOVA with Bonferroni post-hoc analysis, *$p < 0.05$ compared to respective adipocyte area increment (A and B).

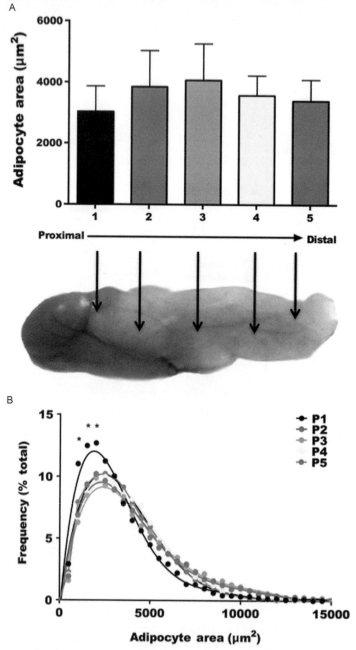

Figure 6.5 The frequency of small adipocytes but not the average adipocyte area changes in epididymal adipose tissue located on the testicle. Ten-week-old C57BL6/J mice fed a 60% high-fat diet for 6 weeks were sacrificed and epididymal WAT harvested and divided into five equal sections according to the proximity to the testicle before fixing, sectioning and counting of adipocytes (~1000 per sample) using the MetaMorph method described herein. Whereas the average size of adipocytes did not differ between samples (A), there was a higher frequency of small adipocytes (1000–2000 μm^2) directly adjacent to the testicle (labeled 1). Two-way ANOVA with Bonferroni post-hoc analysis, $^*p < 0.05$ compared to samples 2–5. (See the color plate.)

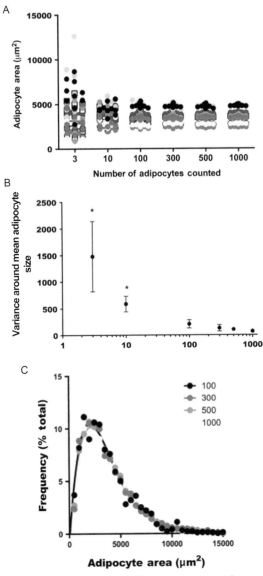

Figure 6.6 Counting the area of as few as 100 adipocytes provides an accurate distribution of adipocyte size in WAT. Ten-week-old ($N=10$) C57BL6/J mice fed a 60% high-fat diet for 6 weeks were sacrificed and epididymal WAT harvested, fixed and sectioned. For each individual animal 3, 10, 100, 300, 500 or 1000 adipocytes were randomly counted 10 times and the average adipocyte area calculated. While there was no significant difference in the mean adipocyte area (A) when counting between 3 and 1000 adipocytes, the variance around this mean is significantly greater when counting 3 or 10 adipocytes (B). The distribution of adipocyte areas, however, does not differ whether you count 100, 300, 500 or 1000 adipocytes. Accordingly counting a minimum of 100 adipocytes is sufficient to estimate mean adipocyte size, minimize variance around that mean and provide an accurate estimation of the distribution of adipocytes in WAT. One-way ANOVA with Tukey post hoc analysis, *$p<0.05$ versus 1000 adipocytes counted. (See the color plate.)

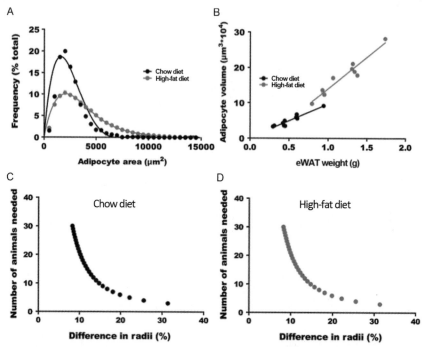

Figure 6.7 A minimum of five animals are required to detect a ~20–25% difference in mean adipocyte radius. C57BL6/J mice fed a chow diet (22-week old, $N=10$) or 60% high-fat diet (10-week old, $N=10$) for 6 weeks were sacrificed and epididymal WAT, fixed and sectioned. The distribution of adipocyte sizes was used to calculate the mean radius of adipocytes in the epididymal WAT of mice fed standard chow (26.7 ± 3.1 µm) or high-fat diet (33.8 ± 3.9 µm) (A). Average adipocyte volume was calculated and a linear relationship to weight of epididymal WAT observed (B). Based on the calculated variance in the radius of adipocytes from WAT of mice fed a standard chow (C) or high-fat (D) diet, a power equation ($1-\beta=0.8$, $\alpha=0.05$) indicates that a minimum of five mice would be required to detect a 7 µm or ~20–25% difference between treatments.

at a given time, this approach does not estimate rate of adipocyte differentiation or death within the tissue. Adipose tissue is in a constant state of flux with 1–5% of mouse adipocytes turning over daily (Rigamonti, Brennand, Lau, & Cowan, 2011) and ~8% of human adipocytes turning over every year (Spalding et al., 2008). Consequently, in addition to changes in adipocyte size and number, a future challenge will be to incorporate an easy method to evaluate rate of formation and turnover of adipocytes. However, the method described herein should have utility for both basic and clinical research, given the tremendous interest in the development, metabolism, and endocrine functions of adipose tissue development under physiological and pathophysiological conditions.

ACKNOWLEDGMENTS

This work was supported by grants to O. A. M. from the National Institutes of Health (DK095705, DK062876, and DK092759) and by a Fulbright Scholar's Award. S. D. P. was supported by a training grant from the University of Michigan Training Program in Organogenesis (T32-HD007505). Morphological analysis was supported by the Morphology and Image Analysis Cores of the Michigan Gastrointestinal Peptide Center (P30 DK034933) and the Michigan Diabetes Research Center (P30 DK020572).

Disclosure Statement: The authors have nothing to disclose.

REFERENCES

Ailhaud, G., Grimaldi, P., & Negrel, R. (1992). Cellular and molecular aspects of adipose tissue development. *Annual Review of Nutrition*, *12*, 207–233. http://dx.doi.org/10.1146/annurev.nu.12.070192.001231.

Bjorndal, B., Burri, L., Staalesen, V., Skorve, J., & Berge, R. K. (2011). Different adipose depots: Their role in the development of metabolic syndrome and mitochondrial response to hypolipidemic agents. *Journal of Obesity*, *2011*, 490650. http://dx.doi.org/10.1155/2011/490650.

Bjornheden, T., Jakubowicz, B., Levin, M., Oden, B., Eden, S., Sjostrom, L., et al. (2004). Computerized determination of adipocyte size. *Obesity Research*, *12*(1), 95–105. http://dx.doi.org/10.1038/oby.2004.13.

Bradshaw, A. D., Graves, D. C., Motamed, K., & Sage, E. H. (2003). SPARC-null mice exhibit increased adiposity without significant differences in overall body weight. *Proceedings of the National Academy of Sciences of the United States of America*, *100*(10), 6045–6050. http://dx.doi.org/10.1073/pnas.1030790100, PII: 1030790100.

Brochu, M., Tchernof, A., Dionne, I. J., Sites, C. K., Eltabbakh, G. H., Sims, E. A., et al. (2001). What are the physical characteristics associated with a normal metabolic profile despite a high level of obesity in postmenopausal women? *The Journal of Clinical Endocrinology and Metabolism*, *86*(3), 1020–1025.

Casteilla, L., Penicaud, L., Cousin, B., & Calise, D. (2008). Choosing an adipose tissue depot for sampling: Factors in selection and depot specificity. *Methods in Molecular Biology*, *456*, 23–38. http://dx.doi.org/10.1007/978-1-59745-245-8_2.

Cawthorn, W. P., Scheller, E. L., & MacDougald, O. A. (2012a). Adipose tissue stem cells meet preadipocyte commitment: Going back to the future. *Journal of Lipid Research*, *53*(2), 227–246. http://dx.doi.org/10.1194/jlr.R021089.

Cawthorn, W. P., Scheller, E. L., & MacDougald, O. A. (2012b). Adipose tissue stem cells: The great WAT hope. *Trends in Endocrinology and Metabolism*, *23*(6), 270–277. http://dx.doi.org/10.1016/j.tem.2012.01.003.

Chen, H. C., & Farese, R. V., Jr. (2002). Determination of adipocyte size by computer image analysis. *Journal of Lipid Research*, *43*(6), 986–989.

Cinti, S. (2005). The adipose organ. *Prostaglandins, Leukotrienes, and Essential Fatty Acids*, *73*(1), 9–15. http://dx.doi.org/10.1016/j.plefa.2005.04.010, PII: S0952-3278(05)00054-2.

de Ferranti, S., & Mozaffarian, D. (2008). The perfect storm: Obesity, adipocyte dysfunction, and metabolic consequences. *Clinical Chemistry*, *54*(6), 945–955. http://dx.doi.org/10.1373/clinchem.2007.100156, PII: clinchem.2007.100156.

Desruisseaux, M. S., Nagajyothi, Trujillo, M. E., Tanowitz, H. B., & Scherer, P. E. (2007). Adipocyte, adipose tissue, and infectious disease. *Infection and Immunity*, *75*(3), 1066–1078. http://dx.doi.org/10.1128/IAI.01455-06.

Faust, I. M., Johnson, P. R., Stern, J. S., & Hirsch, J. (1978). Diet-induced adipocyte number increase in adult rats: A new model of obesity. *The American Journal of Physiology*, *235*(3), E279–286.

Fox, C. H., Johnson, F. B., Whiting, J., & Roller, P. P. (1985). Formaldehyde fixation. *The Journal of Histochemistry and Cytochemistry*, *33*(8), 845–853.

Fujimoto, W. Y., Bergstrom, R. W., Boyko, E. J., Chen, K. W., Leonetti, D. L., Newell-Morris, L., et al. (1999). Visceral adiposity and incident coronary heart disease in Japanese-American men. The 10-year follow-up results of the Seattle Japanese-American Community Diabetes Study. *Diabetes Care*, *22*(11), 1808–1812.

Gerin, I., Dolinsky, V. W., Shackman, J. G., Kennedy, R. T., Chiang, S. H., Burant, C. F., et al. (2005). LXRbeta is required for adipocyte growth, glucose homeostasis, and beta cell function. *The Journal of Biological Chemistry*, *280*(24), 23024–23031. http://dx.doi.org/10.1074/jbc.M412564200.

Gerin, I., Louis, G. W., Zhang, X., Prestwich, T. C., Kumar, T. R., Myers, M. G., Jr., et al. (2009). Hyperphagia and obesity in female mice lacking tissue inhibitor of metalloproteinase-1. *Endocrinology*, *150*(4), 1697–1704. http://dx.doi.org/10.1210/en.2008-1409.

Goralski, K. B., & Sinal, C. J. (2007). Type 2 diabetes and cardiovascular disease: Getting to the fat of the matter. *Canadian Journal of Physiology and Pharmacology*, *85*(1), 113–132. http://dx.doi.org/10.1139/y06-092, PII: y06-092.

Han, J., Lee, J. E., Jin, J., Lim, J. S., Oh, N., Kim, K., et al. (2011). The spatiotemporal development of adipose tissue. *Development*, *138*(22), 5027–5037. http://dx.doi.org/10.1242/dev.067686.

Hausman, D. B., DiGirolamo, M., Bartness, T. J., Hausman, G. J., & Martin, R. J. (2001). The biology of white adipocyte proliferation. *Obesity Reviews*, *2*(4), 239–254.

Hirsch, J., & Gallian, E. (1968). Methods for the determination of adipose cell size in man and animals. *Journal of Lipid Research*, *9*(1), 110–119.

Kaartinen, J. M., LaNoue, K. F., Martin, L. F., Vikman, H. L., & Ohisalo, J. J. (1995). Beta-adrenergic responsiveness of adenylate cyclase in human adipocyte plasma membranes in obesity and after massive weight reduction. *Metabolism*, *44*(10), 1288–1292, PII: 0026-0495(95)90031-4.

Kloting, N., Fasshauer, M., Dietrich, A., Kovacs, P., Schon, M. R., Kern, M., et al. (2010). Insulin-sensitive obesity. *American Journal of Physiology Endocrinology and Metabolism*, *299*(3), E506–515. http://dx.doi.org/10.1152/ajpendo.00586.2009, PII: ajpendo.00586.2009.

Lau, F. H., Deo, R. C., Mowrer, G., Caplin, J., Ahfeldt, T., Kaplan, A., et al. (2011). Pattern specification and immune response transcriptional signatures of pericardial and subcutaneous adipose tissue. *PLoS One*, *6*(10), e26092. http://dx.doi.org/10.1371/journal.pone.0026092, PII: PONE-D-11-00779.

Lee, Y. H., Chen, S. Y., Wiesner, R. J., & Huang, Y. F. (2004). Simple flow cytometric method used to assess lipid accumulation in fat cells. *Journal of Lipid Research*, *45*(6), 1162–1167. http://dx.doi.org/10.1194/jlr.D300028-JLR200.

Lee, M. J., Pramyothin, P., Karastergiou, K., & Fried, S. K. (2013). Deconstructing the roles of glucocorticoids in adipose tissue biology and the development of central obesity. *Biochimica et Biophysica Acta*. pii: S0925–4439(13)00191–9. http://dx.doi.org/10.1016/j.bbadis.

Lefebvre, A. M., Laville, M., Vega, N., Riou, J. P., van Gaal, L., Auwerx, J., et al. (1998). Depot-specific differences in adipose tissue gene expression in lean and obese subjects. *Diabetes*, *47*(1), 98–103.

Le Lay, S., Krief, S., Farnier, C., Lefrere, I., Le Liepvre, X., Bazin, R., et al. (2001). Cholesterol, a cell size-dependent signal that regulates glucose metabolism and gene expression in adipocytes. *The Journal of Biological Chemistry*, *276*(20), 16904–16910. http://dx.doi.org/10.1074/jbc.M010955200, PII: M010955200.

Longo, K. A., Wright, W. S., Kang, S., Gerin, I., Chiang, S. H., Lucas, P. C., et al. (2004). Wnt10b inhibits development of white and brown adipose tissues. *The Journal of Biological Chemistry*, *279*(34), 35503–35509. http://dx.doi.org/10.1074/jbc.M402937200.

Maroni, B. J., Haesemeyer, R., Wilson, L. K., & DiGirolamo, M. (1990). Electronic determination of size and number in isolated unfixed adipocyte populations. *Journal of Lipid Research*, *31*(9), 1703–1709.

Mori, H., Prestwich, T. C., Reid, M. A., Longo, K. A., Gerin, I., Cawthorn, W. P., et al. (2012). Secreted frizzled-related protein 5 suppresses adipocyte mitochondrial metabolism through WNT inhibition. *The Journal of Clinical Investigation*, *122*(7), 2405–2416. http://dx.doi.org/10.1172/JCI63604.

National Research Council (2003). *Guidelines for the care and use of mammals in neuroscience and behavioral research*. Washington: The National Academies Press.

Nicklas, B. J., Penninx, B. W., Cesari, M., Kritchevsky, S. B., Newman, A. B., Kanaya, A. M., et al. (2004). Association of visceral adipose tissue with incident myocardial infarction in older men and women: The Health, Aging and Body Composition Study. *American Journal of Epidemiology*, *160*(8), 741–749. http://dx.doi.org/10.1093/aje/kwh281, PII: 160/8/741.

Nishimura, S., Manabe, I., Nagasaki, M., Hosoya, Y., Yamashita, H., Fujita, H., et al. (2007). Adipogenesis in obesity requires close interplay between differentiating adipocytes, stromal cells, and blood vessels. *Diabetes*, *56*(6), 1517–1526. http://dx.doi.org/10.2337/db06-1749.

Okamoto, Y., Higashiyama, H., Inoue, H., Kanematsu, M., Kinoshita, M., & Asano, S. (2007). Quantitative image analysis in adipose tissue using an automated image analysis system: Differential effects of peroxisome proliferator-activated receptor-alpha and -gamma agonist on white and brown adipose tissue morphology in AKR obese and db/db diabetic mice. *Pathology International*, *57*(6), 369–377. http://dx.doi.org/10.1111/j.1440-1827.2007.02109.x, PII: PIN2109.

Osman, O. S., Selway, J. L., Kepczynska, M. A., Stocker, C. J., O'Dowd, J. F., Cawthorne, M. A., et al. (2013). A novel automated image analysis method for accurate adipocyte quantification. *Adipocyte*, *2*(3), 160–164. http://dx.doi.org/10.4161/adip.24652.

Ozcan, U., Cao, Q., Yilmaz, E., Lee, A. H., Iwakoshi, N. N., Ozdelen, E., et al. (2004). Endoplasmic reticulum stress links obesity, insulin action, and type 2 diabetes. *Science*, *306*(5695), 457–461. http://dx.doi.org/10.1126/science.1103160, PII: 306/5695/457.

Poissonnet, C. M., Burdi, A. R., & Garn, S. M. (1984). The chronology of adipose tissue appearance and distribution in the human fetus. *Early Human Development*, *10*(1–2), 1–11.

Pond, C. M., Mattacks, C. A., Calder, P. C., & Evans, J. (1993). Site-specific properties of human adipose depots homologous to those of other mammals. *Comparative Biochemistry and Physiology Comparative Physiology*, *104*(4), 819–824.

Poretsky, L., & Ebooks Corporation. (2010). Principles of Diabetes Mellitus. Retrieved from http://www.msvu.ca:2048/login?url=http://www.msvu.eblib.com/EBLWeb/patron?target=patron&extendedid=P_511708_0& (pp. 1 online resource (868 pp)).

Pouteau, E., Turner, S., Aprikian, O., Hellerstein, M., Moser, M., Darimont, C., et al. (2008). Time course and dynamics of adipose tissue development in obese and lean Zucker rat pups. *International Journal of Obesity*, *32*(4), 648–657. http://dx.doi.org/10.1038/sj.ijo.0803787.

Reynisdottir, S., Ellerfeldt, K., Wahrenberg, H., Lithell, H., & Arner, P. (1994). Multiple lipolysis defects in the insulin resistance (metabolic) syndrome. *The Journal of Clinical Investigation*, *93*(6), 2590–2599. http://dx.doi.org/10.1172/JCI117271.

Rigamonti, A., Brennand, K., Lau, F., & Cowan, C. A. (2011). Rapid cellular turnover in adipose tissue. *PLoS One*, *6*(3), e17637. http://dx.doi.org/10.1371/journal.pone.0017637.

Roman, A. A., Parlee, S. D., & Sinal, C. J. (2012). Chemerin: A potential endocrine link between obesity and type 2 diabetes. *Endocrine*, *42*(2), 243–251. http://dx.doi.org/10.1007/s12020-012-9698-8.

Santosa, S., & Jensen, M. D. (2012). Effects of male hypogonadism on regional adipose tissue fatty acid storage and lipogenic proteins. *PLoS One*, *7*(2), e31473. http://dx.doi.org/10.1371/journal.pone.0031473.

Spalding, K. L., Arner, E., Westermark, P. O., Bernard, S., Buchholz, B. A., Bergmann, O., et al. (2008). Dynamics of fat cell turnover in humans. *Nature*, *453*(7196), 783–787. http://dx.doi.org/10.1038/nature06902, PII: nature06902.

Sternberg, S. R. (1983). Biomedical Image Processing. *Computer*, *16*(1), 22–34. http://dx.doi.org/10.1109/mc.1983.1654163.

Suganami, T., & Ogawa, Y. (2010). Adipose tissue macrophages: Their role in adipose tissue remodeling. *Journal of Leukocyte Biology*, *88*(1), 33–39. http://dx.doi.org/10.1189/jlb.0210072, PII: jlb.0210072.

Tchkonia, T., Morbeck, D. E., Von Zglinicki, T., Van Deursen, J., Lustgarten, J., Scrable, H., et al. (2010). Fat tissue, aging, and cellular senescence. *Aging Cell*, *9*(5), 667–684. http://dx.doi.org/10.1111/j.1474-9726.2010.00608.x.

Tisdale, M. J. (1997). Biology of cachexia. *Journal of the National Cancer Institute*, *89*(23), 1763–1773.

Webster, J. D., Miller, M. A., Dusold, D., & Ramos-Vara, J. (2009). Effects of prolonged formalin fixation on diagnostic immunohistochemistry in domestic animals. *The Journal of Histochemistry and Cytochemistry*, *57*(8), 753–761. http://dx.doi.org/10.1369/jhc.2009.953877.

CHAPTER SEVEN

Use of Osmium Tetroxide Staining with Microcomputerized Tomography to Visualize and Quantify Bone Marrow Adipose Tissue *In Vivo*

Erica L. Scheller[*,1], Nancy Troiano[†,1], Joshua N. VanHoutan[‡,1], Mary A. Bouxsein[§], Jackie A. Fretz[†], Yougen Xi[†], Tracy Nelson[†], Griffin Katz[†], Ryan Berry[¶], Christopher D. Church[¶], Casey R. Doucette[‖], Matthew S. Rodeheffer[¶], Ormond A. MacDougald[*], Clifford J. Rosen[‖], Mark C. Horowitz[†,2]

[*]Department of Molecular and Integrative Physiology, University of Michigan, Ann Arbor, Michigan, USA
[†]Department of Orthopaedics and Rehabilitation, Yale University School of Medicine, New Haven, Connecticut, USA
[‡]Department of Internal Medicine, Endocrinology, Yale University School of Medicine, New Haven, Connecticut, USA
[§]Center for Advanced Orthopedic Studies, Beth Israel Deaconess Medical Center and Harvard Medical School, Boston, Massachusetts, USA
[¶]Section of Comparative Medicine, Yale University School of Medicine, New Haven, Connecticut, USA
[‖]Maine Medical Center Research Institute, Scarborough, Maine, USA
[1]These authors contributed equally to this work.
[2]Corresponding author: e-mail address: mark.horowitz@yale.edu

Contents

1. Introduction	124
1.1 Accumulation and distribution of MAT	126
1.2 Metabolic function of MAT	127
1.3 Origin of the marrow adipocyte	128
1.4 Summary	129
2. Materials	129
2.1 Mice	129
2.2 Dissecting tools	130
2.3 Glassware and plastic ware	130
2.4 Preparation of decalcification solution	130
2.5 Preparation of osmium tetroxide staining solution	130
3. Methods	131
3.1 Isolation of mouse long bones	131
3.2 Fixation of the bone	132
3.3 Decalcification of the bones	132

3.4 Osmium staining of the decalcified bones   133
3.5 Microcomputed tomography   134
4. Summary   137
Acknowledgments   137
References   137

Abstract

Adipocytes reside in discrete, well-defined depots throughout the body. In addition to mature adipocytes, white adipose tissue depots are composed of many cell types, including macrophages, endothelial cells, fibroblasts, and stromal cells, which together are referred to as the stromal vascular fraction (SVF). The SVF also contains adipocyte progenitors that give rise to mature adipocytes in those depots. Marrow adipose tissue (MAT) or marrow fat has long been known to be present in bone marrow (BM) but its origin, development, and function remain largely unknown. Clinically, increased MAT is associated with age, metabolic diseases, drug treatment, and marrow recovery in children receiving radiation and chemotherapy. In contrast to the other depots, MAT is unevenly distributed in the BM of long bones. Conventional quantitation relies on sectioning of the bone to overcome issues with distribution but is time-consuming, resource intensive, inconsistent between laboratories and may be unreliable as it may miss changes in MAT volume. Thus, the inability to quantitate MAT in a rapid, systematic, and reproducible manner has hampered a full understanding of its development and function. In this chapter, we describe a new technique that couples histochemical staining of lipid using osmium tetroxide with microcomputerized tomography to visualize and quantitate MAT within the medullary canal in three dimensions. Imaging of osmium staining provides a high-resolution map of existing and developing MAT in the BM. Because this method is simple, reproducible, and quantitative, we expect it will become a useful tool for the precise characterization of MAT.

1. INTRODUCTION

"If the marrow were a nicely isolated organ like the spleen its study would certainly become much less time consuming" (Oehlbeck, Robscheit-Robbins, & Whipple, 1932). This quote from 1932 emphasizes a problem that has been faced by bone biologists for decades; analysis of the bone marrow (BM) requires one to first deal with the bone. This explains why most of the work in the late 1800s and early 1900s was anatomical in nature and relied on large specimens from human cadavers. Bone of this size could be sectioned for visual comparison without major disruption of the marrow elements. Optimization of decalcification protocols for downstream histological analysis from the late 1920s to early 1940s expanded our appreciation of the cellular content and morphology of the BM, including its tendency to contain a large number of adipocytes (Kramer & Shipley, 1927; Lillie, 1944). Distribution of the marrow adipose tissue (MAT) in the skeleton is a tightly regulated process while its origin and function remain largely unknown.

Quantitation of MAT in mice has historically been accomplished by counting adipocyte "ghosts" in serial histological sections of paraffin or plastic embedded bone. This method is time-consuming, resource intensive, and subject to significant variation because of interlaboratory variation and because MAT is not evenly distributed throughout the medullary canal (Fig. 7.1). If adequate analysis is not performed, traditional sectioning

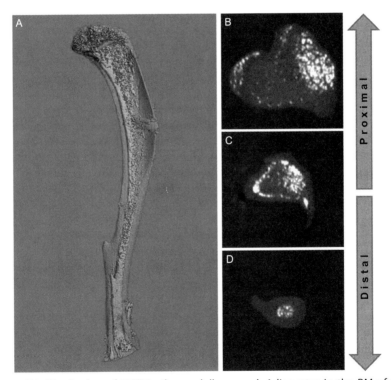

Figure 7.1 Distribution of MAT in the medullary canal. Adipocytes in the BM of the mouse are unevenly distributed throughout the medullary canal. They are most densely clustered in the epiphyses. In the metaphysis and diaphysis, adipocytes are most numerous near the central vascular canal and adjacent to the cortical bone. (A) Three-dimensional reconstruction of a 16-week-old C3H mouse osmium-stained tibia. Light blue, bone; white, MAT. (B–D) Transverse sections of the same bone from more proximal (toward the knee) to distal (toward the ankle) showing defined regional clustering of the marrow adipocytes (white). (See the color plate.)

methods can easily miss changes in MAT volume and/or distribution. In species ranging in size from rat to humans indirect imaging techniques, including computed tomography and magnetic resonance (MR) spectroscopy, have been applied with success (Bredella et al., 2009; Demontiero, Li, Thembani, & Duque, 2011; Regis-Arnaud et al., 2011). Although MR has been tried in isolated mouse femurs, quantitation of fat verses water is very imprecise (C.J. Rosen, unpublished). Thus, the inability to quantitate MAT in a rapid, systematic, and reproducible manner in a variety of mouse models has hampered a full understanding of MAT development, distribution, and function. To overcome this limitation, in this chapter we present a simple method that couples histochemical staining of lipid using osmium tetroxide with microcomputerized tomography (micro-CT) for rapid three-dimensional quantification of MAT.

1.1. Accumulation and distribution of MAT

Since 1882 it has been well documented that in early childhood the BM exists in a predominantly red or hematopoietic state (Custer, 1932). In addition, it is now known that this same BM, in addition to being hematopoietic, is also osteogenic. MAT infiltration accelerates shortly after birth in the distal portions of the appendicular skeleton before developing in more proximal areas (Emery & Follett, 1964). For example, in humans the BM of the middle phalanges of the toes is completely converted to MAT by 12 months of age (Emery & Follett, 1964). This process results in filling of peripheral bones with "yellow" fatty marrow (i.e., hands, feet, tibia) while central bones retain red hematopoietic/osteogenic marrow throughout life (i.e., pelvis, ribs, lumbar vertebrae). By age 25 human BM is approximately 70% MAT, at which point its rate of accumulation slows markedly but continues throughout life (Custer & Ahlfeldt, 1932). In human adults, males generally have more MAT than females (Shillingford, 1950). This is also true in inbred strains of mice. However, the amount of MAT particularly in the long bones is strain-dependent (M.C. Horowitz, C.J. Rosen, & C.R. Farber, unpublished). Consistent with human findings, in 1889 Ranvier recorded the presence of a MAT gradient in the vertebrae of rats ranging from predominantly cellular, red marrow in the lumbar vertebrae to fatty, yellow marrow in the tail (Huggins & Blocksom, 1936; Ranvier, 1889). This historical finding has since been reobserved in mice and used as the basis for studies of MAT regulation of bone formation and hematopoiesis (Naveiras et al., 2009; Wronski, Smith, & Jee, 1981).

MAT may exist in at least two forms. The first population, "constitutive" MAT (cMAT), corresponds with areas of early MAT formation such as the distal tibia, feet, or caudal vertebrae. The second type, "regulated" MAT (rMAT), tends to develop with aging and is located interspersed with hematopoietic cells at sites such as the distal femur and proximal tibia in mice or the proximal femur and lumbar vertebrae in humans. The underpinning of this observation was published in 1976 when it was determined histologically that populations of cMAT fail to stain with performic acid–Schiff (PFAS) while their regulated counterparts readily display the characteristic red staining pattern (Tavassoli, 1976b). This implies that not only is the lipid composition different, the two populations also have differential abilities to respond to regulatory cues linked to induction of hematopoiesis. Like osmium staining, the PFAS reaction relies on the oxidation of carbon–carbon double bonds in unsaturated lipids indicating that rMAT, at least in rabbits, contains a higher proportion of unsaturated fatty acids (FAs) than cMAT. As both populations of adipocytes contain unsaturated FAs, it is likely that longer staining times with PFAS (as with osmium) would result in staining of the majority of the marrow adipocytes. It is possible that with time rMAT can mature into cMAT. This may explain observations that MAT maturation with age is associated with a decrease in the level of unsaturated FAs and implies that it is more difficult to shed MAT in favor of hematopoiesis as we age (Yeung et al., 2005).

1.2. Metabolic function of MAT

During the early 1900s, the rabbit emerged as the main animal model for the study of MAT due to its high proportion of fatty marrow and ease of MAT access and manipulation. The existence of saturated FAs within BM was documented in the early 1930s, followed by polyunsaturated FAs in the 1940s, and a more detailed breakdown that corresponds with recent results was published in 1954 (Evans, Riemenschneider, & Herb, 1954). During the 1960s, an interest in comparing the metabolic function of the marrow adipocyte to the peripheral white adipocyte began to emerge. This is reflected by publications such as the dissertation of Shore in 1969 entitled "Bone Marrow as an Adipose Tissue" reaffirming the thought that MAT may represent a metabolically relevant depot (Shore, 1969). Subsequent work in the 1970s revealed that the marrow fat cell in rabbits is four to six times smaller in volume than that of the perinephric white adipose tissue (WAT) adipocyte (Bathija, Davis, & Trubowitz, 1979;

Trubowitz & Bathija, 1977). This is in accordance with our measurements comparing adipocyte size in MAT and gonadal WAT in mice; however, when compared to subcutaneous WAT, the MAT adipocyte was larger in size (O.A. MacDougald, unpublished).

Metabolic function was examined by measuring rates of ^{14}C-palmitate turnover in rabbit MAT and perinephric WAT (Trubowitz & Bathija, 1977). Injection of labeled palmitate, a free FA, into the circulation allows analysis of its rate of incorporation into triglyceride and is a measure of triglyceride synthesis by the adipose tissue. Incorporation of ^{14}C-palmitate was five times greater in MAT than in perinephric WAT likely reflecting an increase in the number of adipocytes per tissue volume (Trubowitz & Bathija, 1977). Strikingly, fasting of rabbits for up to 3 weeks did not diminish the FA esterification capacity of the MAT, while that of the perinephric WAT decreased by as much as 60% (Bathija et al., 1979). In his seminal work with fasted rabbits, Tavassoli found that after 10 days of food restriction electron microscopic analysis revealed marked breakdown of the central lipid vacuole in WAT adipocytes while MAT cells of the distal tibia failed to undergo fat mobilization (Tavassoli, 1974). We have observed the same phenomenon at the light microscopic level in mice fasted for 24 h when comparing MAT from caudal vertebrae to gonadal WAT and subcutaneous WAT (O.A. MacDougald, unpublished). Of note, both of these regions (distal tibia and caudal vertebrae) are cMAT depots. This observation has been supported in recent years by the finding that unlike peripheral WAT, MAT has the capacity to increase in states of calorie restriction and conditions such as anorexia nervosa (Bredella et al., 2009; Devlin et al., 2010; Fazeli et al., 2013). These findings suggest that while MAT readily esterifies and stores FA as triglyceride, similar to metabolically active WAT, it does not respond in the same way to nutritional status.

1.3. Origin of the marrow adipocyte

Elegant ultrastructural studies of developing adipocytes in BM and peripheral WAT in 1976 demonstrated that the cellular origin of BM adipose tissue differs from that of extramedullary WAT (Tavassoli, 1976a). While the WAT progenitors were characterized as "fibroblast-like" with abundant rough endoplasmic reticulum and a close association with large amounts of collagen, the MAT progenitor only occasionally contained profiles of rough ER and was not associated with collagen (Tavassoli, 1976a). Recently, WAT progenitors have been identified *in vitro* and *in vivo* as a

WAT-resident stromal cell population with the cell surface phenotype Pdgfrα$^+$ Lin$^-$ CD29$^+$ CD34$^+$ Sca1$^+$ CD24$^+$ (Berry & Rodeheffer, 2013; Rodeheffer, Birsoy, & Friedman, 2008). Consistent with the hypothesis from 1976, this WAT-specific progenitor population is not present within mouse BM (M.C. Horowitz & M.S. Rodeheffer, unpublished). This reinforces the notion that MAT arises from a unique progenitor population that has yet to be fully identified. However, using lineage tracing, the BM adipocyte progenitor also expresses Pdgfrα (M.C. Horowitz & M.S. Rodeheffer, unpublished).

1.4. Summary

Osmium tetroxide was first introduced by Palade (1952) for his classic work identifying the structure and function of cell organelles using the electron microscope. Osmium tetroxide (osmic acid) is one of the oldest fat stains and unsaturated FAs like oleic acid have been traditionally considered to be responsible for the reduction reaction. Osmium tetroxide is soluble in fats and forms a black reduction compound with them by the addition to the double carbon-to-carbon bonds. In a manner similar to Palade, we reasoned that because osmium is a heavy metal and therefore radiodense, it will appear opaque to CT and as a result can be visualized.

In recent decades, accumulation of MAT has been strongly associated with metabolic disease and development of osteoporosis (Fazeli et al., 2013). However, the time-consuming nature of MAT quantification and analysis in mice has limited our understanding of its origin and physiology. The osmium staining technique for quantification of MAT in three dimensions is an exciting new tool that will help to increase our understanding of the distribution and development of MAT and allow us to further characterize its relationship to bone formation, hematopoiesis, and systemic metabolism.

2. MATERIALS

2.1. Mice

Strains C57BL/6 (B6), C3H/HeJ (C3H) BALB/c, AZIP lipodystrophic mice (Moitra et al., 1998), FVB (background strain for AZIP), and $Ebf1^{-/-}$ (B6 × 129 background) (Fretz et al., 2010) have been analyzed. Mice from 4 to 52-weeks old have been used successfully.

Note: Although we have not tested all mouse strains, strain B6 has the lowest bone density while C3H has the highest bone density of all mouse strains. Efficient staining is dependent on penetration of the osmium into the medullary canal. The penetration is partially dependent on the amount of decalcification. Because we have successfully stained C3H, we believe any mouse strain can be assessed using this method.

Due to the poor penetration of osmium, staining of marrow adipocytes in larger bones (i.e., rat and rabbit) has several limitations. If a decalcified rat bone is stained with osmium, even for several days, only the outermost perimeter of marrow adipocytes is stained. This phenomenon is also observed when attempting to stain rabbit bones. For rats, this can be overcome by taking a 0.8–1 mm horizontal slice of bone and marrow after decalcification in your area of interest. If this slice is stained in 1% osmium tetroxide under vacuum for 24 h, the osmium stain will penetrate approximately 0.5 mm from each direction and is then suitable for CT analysis. This allows for three-dimensional marrow fat quantification in a defined region of larger bones.

2.2. Dissecting tools

Operating scissors 51/2″ straight #70-2070, Biomedical Research Instruments, Rockville, MD

Operating scissors 41/2″ straight #70-2010, Biomedical Research Instruments, Rockville, MD

2.3. Glassware and plastic ware

Shandon Cassette l Biopsy, #1000969, Thermo Fisher Scientific, Kalamazoo, MI; 20 ml disposable glass scintillation vials.

2.4. Preparation of decalcification solution

Prepare in advance

41.4 g EDTA (ethylenediamine tetraacetic acid, disodium salt dehydrate, Fisher #S311-500) + 4.4 g NaOH

QS to 1000 ml dH$_2$O (gentle heating to dissolve EDTA); adjust pH to 7.0–7.4

2.5. Preparation of osmium tetroxide staining solution

IMPORTANT osmium is very dangerous. *ALL WORK* with osmium *MUST* be done in a properly functioning fume hood. *DO NOT* get osmium on your skin—wear gloves. *VAPORS are DANGEROUS* and will damage your mucous membranes and your corneas (lachrymator).

Osmium tetroxide 2% aqueous solution from Polysciences Inc, Warrington, PA (#23311). The osmium tetroxide comes in 5 ml sealed glass vials. The vials *MUST* be opened in a fume hood.

5% potassium dichromate (5 g of potassium dichromate in 100 ml distilled water)

Prepare fresh each time you stain:

Mixed in glass scintillation vials.

1 part 5% potassium dichromate (stabilizing agent; included to prevent the reduction of the osmium).

1 part 2% osmium tetroxide

Final concentration 1%

3. METHODS
3.1. Isolation of mouse long bones
3.1.1 Mouse dissection

Male or female mice of any age can be used. We have examined mice as young as weaning age (21–28 days) to as old as 6 months. We routinely use mice 6–16 weeks old.

1. Euthanize mice by CO_2 asphyxiation and cervical dislocation; or by other methods acceptable to your animal use committee.
2. Wet the hair of the euthanized mouse with 70% ethanol to flatten the hair. Using scissors (operating scissors 5 1/2″ straight #70-2070, Biomedical Research Instruments, Rockville, MD) make a single cut through the hair and skin just behind the head, on the neck and strip-off the skin the length of the body, pulling the skin over the hind legs and feet.
3. Using scissors, and as a guide, the flexion of the leg at its attachment to the body, cut deep into the pelvic joint (femur–pelvis), keeping the femoral head intact, freeing the leg from the rest of the body. Cut off the foot just below the ankle and place in a 100 mm sterile tissue culture or Petri dish.
4. Remove as much soft tissue (tendons, ligaments, and muscle) as possible using scissors (operating scissors 4 1/2″ straight #70-2010, Biomedical Research Instruments, Rockville, MD). This is important, because the less soft tissue that remains attached to the bone, the less superfluous

tissue that will be stained and thus visualized by osmium. It is important, at this point, to keep the knee joint intact.

   Note: It has been our experience that using a Kimwipe or gauze to manipulate the soft tissue, especially the muscle, facilitates the removal of the tissue. Not all the tissue can be removed.
5. Using a scalpel with a number 15 blade, separate the tibia from the femur.

   Note: It has been our experience that inserting the blade into the knee joint and manipulating the blade side to side with light pressure ensures the joint separates without cutting off the epiphysis and damaging the growth plate.

Note: Bones are NEVER to be frozen.

3.2. Fixation of the bone

1. Place bones in tissue processing cassettes (Shandon Cassette 1 Biopsy, #1000969, Thermo Fisher Scientific, Kalamazoo, MI). Place the cassettes in a large plastic or glass beaker, and immediately immerse the bones in 10% neutral buffered formalin (Fisher #SF100-4). The volume of formalin should be sufficient to more than cover the cassettes. For small numbers of bones, a tibia and femur can be fixed in a small tube in 2.0 ml of formalin (volume of formalin 10× the bone volume (BV)). Fix the bones overnight at 4 °C with gentle agitation.

   Note: It is important to exclude organic compounds (e.g., ethanol) in the fixative due to their lipid soluble capacity.
2. The next day, pour off the formalin, rinse 1× with cool tap water, and then wash the fixed bones for 1 h in running cool tap water.

   Note: Although we have not done a side-by-side comparison, it is our experience that washing out the formalin the next day gives excellent results compared to leaving the bones in formalin for long periods of time. We recommend not leaving the bones in formalin for more than 3 days; longer periods of time result in more brittle bones.

3.3. Decalcification of the bones

1. Remove the cassettes from the wash, shake off the excess water, and immerse in the EDTA decalcification solution (decal). The volume of EDTA should be sufficient to more than cover the cassettes.

2. Decal is done at 4 °C with gentle agitation. After the first 24 h, pour off the decal and exchange with fresh decal.
3. The decal is then changed every 3–4 days until the bones are ready (~14 days) for staining.

    Note: 14 days in decal is appropriate for most mouse bones (tibia and femur). Decal is done to remove mineral to allow optimal penetration of the osmium. Because different mouse strains have different bone densities (amounts of mineral/BV), it may be necessary to adjust the time in decal. It has been our experience that 10 days is a minimum and we have not had the bones in decal past 17 days. By 17 days, the bones are flexible to the touch.
4. Pour off the decal, rinse the cassettes 1 × in cool tap water and then wash the cassettes for 1 h in cool running tap water.

3.4. Osmium staining of the decalcified bones

IMPORTANT, osmium tetroxide is very dangerous. *ALL WORK* with osmium *MUST* be done in a fume hood. *DO NOT* get osmium tetroxide on your skin—wear gloves. Wear a laboratory coat and no open toed shoes. *VAPORS are DANGEROUS* and will damage your mucous membranes and your corneas (lachrymator).

1. Remove the decalcified bones from the cassettes using forceps. For the tibia, cut off the bone just proximal to the ankle joint and place the bone in plastic screw-cap CryoTube vials (Thermo Scientific #363401, Denmark) with the cut-end up (proximal tibia facing down in the tube). Glass vials can also be used. For the femur, cut off the bone just below the femoral head and place in the vial with the cut-end up (distal femur facing down in the tube). A maximum of four bones can be placed in a single tube.

    Note: Placing the bone with the cut end facing up is done to allow maximal penetration of the osmium. If analysis of the distal tibia or proximal femur is required, the end of the bones can be left intact. However, staining time may have to be increased.
2. Incubate the bones with the osmium tetroxide for 48 h at room temperature IN THE FUME HOOD.
3. Use forceps to retrieve the bones from the staining vials and place bones in new cassettes and wash under cool running tap water for 2 h.
4. Dispose of osmium liquid waste and glass or plastics contaminated with osmium as recommended by your institution.

Note: A careful histological examination of the osmium-stained sections reveals staining only in the adipocytes. No staining could be seen in the smaller hematopoietic cells. This suggests the vast majority of the osmium staining is focused in the adipocytes.

3.5. Microcomputed tomography

Micro-CT was performed in water with energy of 55 kVp, an integration time of 500 ms, and a maximum isometric voxel size of 10 µm (the "high"-resolution setting with a 20 mm sample holder) using a micro-CT-35 (Scanco Medical, Bruttisellen, Switzerland). We increase throughput by using a 20 mm diameter sample holder and placing bones evenly in rows, separated by moist cotton, in seven standard drinking straws arrayed six around the perimeter and one in the center of the tube. Using this setup, at least 28 bones can be scanned at one time in batch mode. When creating volumes of interest (VOI), the interface between the decalcified bone and the MAT is usually apparent, facilitating placement of graphical objects. When segmentation is applied, MAT can be visualized unencumbered by the surrounding bone (Fig. 7.2A). To determine the position of the MAT within the medullary canal and to determine its change in volume, the bone can be overlaid (Fig. 7.2B). Using this method, striking differences in MAT can be seen over the length of the femur between B6 (Fig. 7.3A) and C3H (Fig. 7.3B) mice. The MAT is so prevalent in C3H that it fills the

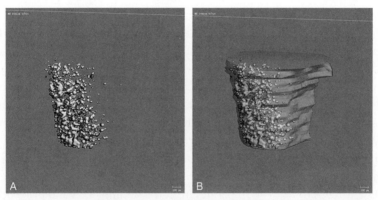

Figure 7.2 Osmium-stained and micro-CT-imaged MAT from Ebf1 null tibia. (A) Osmium-stained and micro-CT image of the proximal tibia from a 4-week-old *Ebf1*$^{-/-}$ mouse (no bone overlay). *Ebf1*$^{-/-}$ mice have very high MAT. (B) MAT from (A) with the bone overlaid. (See the color plate.)

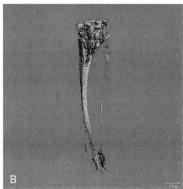

Figure 7.3 Osmium-stained and micro-CT-imaged MAT from C57BL/6 and C3H/HeJ. (A) Osmium-stained and micro-CT image of the tibia from 15-week-old B6 mice (no bone overlay). (B) Osmium-stained and micro-CT image of the tibia from 15-week-old C3H mice (no bone overlay). (See the color plate.)

fibula (thin line to the right of the tibia). In our hands, B6 mice have the lowest MAT while C3H have the highest of any of the strains we have tested.

Scanning the entire length of the bone is time consuming, although may be necessary depending on the experimental requirements. To reduce scanning time, we have selected four VOI: (1) above the growth plate in the secondary center of ossification, (2) just below the growth plate, (3) down the shaft in the metaphysis, and (4) in the diaphysis. Four VOI within the marrow space, extending 230 μm, were defined by their position along the z-axis (Fig. 7.4). The first VOI began 460 μm from the proximal end of the tibia, and each subsequent VOI began 1.2 mm distal to the previous VOI. The fractional volume of osmium-stained fat within the VOIs was determined using the Scanco 3D bone morphometry evaluation program with a threshold of 420 (1/1000 scale) and Gaussian filtering ($\sigma = 0.8$, support $= 1$). The data can be expressed as adipocyte volume/total volume (AV/TV) (Fig. 7.5). This volumetric adipocyte measurement is analogous to the volumetric bone measurement, BV/TV.

Note: These VOIs are based on the literature and our experience (Gimble, Zvonic, Floyd, Kassem, & Nuttall, 2006; Meunier, Aaron, Edouard, & Vignon, 1971; Moore & Dawson, 1990). As MAT increases in the medullary canal of the long bones it accumulates in a directed fashion. For the tibia, from the proximal end distally and for the femur, from the distal end proximal.

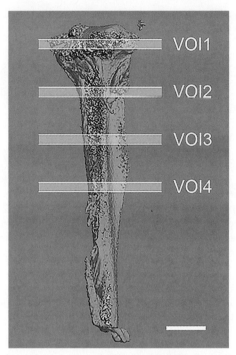

Figure 7.4 Positioning of the volumes of interest. VOIs in mouse femur. (See the color plate.)

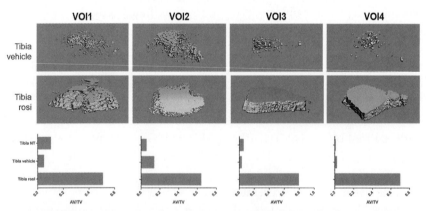

Figure 7.5 Osmium-stained and micro-CT image of MAT from the femur of C57BL/6 mice fed a diet-containing rosiglitazone or control diet. B6 mice were fed a control diet or a diet-containing rosiglitazone for 8 weeks. Rosiglitazone is a PPARγ agonist and as such a potent inducer of MAT. The tibia was collected, stained with osmium tetroxide, and imaged with micro-CT. On control diet (vehicle) MAT is highest in VOI2 and decreases as you move distally down the shaft (VOIs 3 and 4). In contrast, mice on a rosiglitazone diet have very high MAT over the length of the tibial shaft. (See the color plate.)

Because the bones are fixed in buffered formalin, they can be imaged by micro-CT to obtain cortical and trabecular bone measurements. Once accomplished, the bones can then be decalcified and osmium stained. This allows for a direct comparison, in the same bone, to measure bone and MAT parameters (i.e., medullary canal volume).

Our variability from experiment to experiment when comparing the same bones from the same strain is low. This is especially true when all of the processing (fixation, decalcification, osmium staining, and imaging) is done by the same group of investigators. We attribute the variability we do see to small changes in the processing and individual animal variability (even with the same bones from other mice of the same strain, in the same experiment). Larger variability is seen when the mice are manipulated to change MAT (e.g., high-fat diet, calorie restriction, irradiation). It is unlikely that the variability is due to the imaging component of the method. This suggests that with experience, a single group of investigators will have less variability (practice makes perfect).

4. SUMMARY

This chapter presents a new method for the visualization and quantification of MAT. This method is based on the specific staining of lipid by osmium tetroxide, which is detected and quantitated by micro-CT. This method relies on the decalcification of bone and has been successfully used on mouse long bones. The vast majority of the staining is in mature adipocytes not in other BM cells. In addition to quantitation of BM adipocytes, this method provides a 3D map of the MAT throughout the medullary canal and above the growth plate in the secondary center of ossification. This approach is rapid, reproducible, and quantitative.

ACKNOWLEDGMENTS

We thank Dr. Douglas Adams, University of Connecticut Health Science Center, Farmington, CT for his valuable input and Michelle Lynch, University of Michigan, Ann Arbor, MI for helpful discussions. The Horowitz laboratory is partially supported by NIH Grants DK092759, AR046032, and the Department of Orthopaedics and Rehabilitation, Yale University School of Medicine. The MacDougald laboratory is partially supported by NIH Grants DK092759, DK062876, and DK95705.

REFERENCES

Bathija, A., Davis, S., & Trubowitz, S. (1979). Bone marrow adipose tissue: Response to acute starvation. *American Journal of Hematology*, *6*, 191–198.

Berry, R., & Rodeheffer, M. S. (2013). Characterization of the adipocyte cellular lineage in vivo. *Nature Cell Biology*, *15*, 302–308.

Bredella, M. A., Fazeli, P. K., Miller, K. K., Misra, M., Torriani, M., Thomas, B. J., et al. (2009). Increased bone marrow fat in anorexia nervosa. *The Journal of Clinical Endocrinology and Metabolism, 94*, 2129–2136.

Custer, R. P. (1932). Studies on the structure and function of bone marrow. Part I. *The Journal of Laboratory and Clinical Medicine, 17*, 951–960.

Custer, R. P., & Ahlfeldt, F. E. (1932). Studies on the structure and function of bone marrow II. *The Journal of Laboratory and Clinical Medicine, 17*, 960–962.

Demontiero, O., Li, W., Thembani, E., & Duque, G. (2011). Validation of noninvasive quantification of bone marrow fat volume with microCT in aging rats. *Experimental Gerontology, 96*, 435–440.

Devlin, M. J., Cloutier, A. M., Thomas, N. A., Panus, D. A., Lotinun, S., Pinz, I., et al. (2010). Caloric restriction leads to high marrow adiposity and low bone mass in growing mice. *Journal of Bone and Mineral Research, 25*, 2078–2088.

Emery, J. L., & Follett, G. F. (1964). Regression of bone-marrow haemopoiesis from the terminal digits in the foetus and infant. *British Journal of Haematology, 10*, 485–489.

Evans, J. D., Riemenschneider, R. W., & Herb, S. F. (1954). Fat composition and in vitro oxygen consumption of marrow from fed and fasted rabbits. *Archives of Biochemistry and Biophysics, 53*, 157–166.

Fazeli, P. K., Horowitz, M. C., Macdougald, O. A., Scheller, E. L., Rodeheffer, M. S., Rosen, C. J., et al. (2013). Marrow fat and bone—New perspectives. *The Journal of Clinical Endocrinology and Metabolism, 98*, 935–945.

Fretz, J. A., Nelson, T., Xi, Y., Adams, D. J., Rosen, C. J., & Horowitz, M. C. (2010). Altered metabolism and lipodystrophy in the Ebf1-deficient mouse. *Endocrinology, 151*, 1611–1621.

Gimble, J. M., Zvonic, S., Floyd, Z. E., Kassem, M., & Nuttall, M. E. (2006). Playing with bone and fat. *Journal of Cellular Biochemistry, 98*, 251–266.

Huggins, C., & Blocksom, B. H., Jr. (1936). Changes in outlying bone marrow accompanying a local increase of temperature within physiological limits. *The Journal of Experimental Medicine, 64*, 253–274.

Kramer, B., & Shipley, P. G. (1927). Decalcification of bone in acid free Solutions. *Science, 66*, 484–485.

Lillie, R. D. (1944). Studies on the decalcification of bone. *The American Journal of Pathology, 20*(2), 291–296.

Meunier, P., Aaron, J., Edouard, C., & Vignon, G. (1971). Osteoporosis and the replacement of cell populations of the marrow by adipose tissue. A quantitative study of 84 iliac bone biopsies. *Clinical Orthopaedics, 80*, 147–154.

Moitra, J., Mason, M. M., Olive, M., Krylov, D., Gavrilova, O., Marcus-Samuels, B., et al. (1998). Life without white fat: A transgenic mouse. *Genes & Development, 12*, 3168–3181.

Moore, S. G., & Dawson, K. L. (1990). Red and yellow marrow in the femur: Age-related changes in appearance at MR imaging. *Radiology, 175*, 219–223.

Naveiras, O., Nardi, V., Wenzel, P. L., Hauschka, P. V., Fahey, F., & Daley, G. Q. (2009). Bone-marrow adipocytes as negative regulators of the haematopoietic microenvironment. *Nature, 460*, 259–263.

Oehlbeck, L. W. F., Robscheit-Robbins, F. S., & Whipple, G. H. (1932). Marrow hyperplasia and hemoglobin reserve in experimental anemia due to bleeding. *The Journal of Experimental Medicine, 56*, 425–428.

Palade, G. E. (1952). A study of fixation for electron microscopy. *The Journal of Experimental Medicine, 95*, 285.

Ranvier, L. A. (1889). *Traite technique d'histologie*. p. 264. Paris: F. Savy.

Regis-Arnaud, A., Guiu, B., Walker, P. M., Krause, D., Ricolfi, F., & Ben Salem, D. (2011). Bone marrow fat quantification of osteoporotic vertebral compression fractures:

Comparison of multi-voxel proton MR spectroscopy and chemical-shift gradient-echo MR imaging. *Acta Radiologica, 52,* 1032–1036.
Rodeheffer, M. S., Birsoy, K., & Friedman, J. M. (2008). Identification of white adipocyte progenitor cells in vivo. *Cell, 135,* 240–249.
Shillingford, J. P. (1950). The red bone marrow in heart failure. *Journal of Clinical Pathology, 3,* 24–39.
Shore, L. S. (1969). *Bone marrow as an adipose tissue.* Philadelphia: Hahnemann Medical College.
Tavassoli, M. (1974). Differential response of bone marrow and extramedullary adipose cells to starvation. *Experientia, 30,* 424–425.
Tavassoli, M. (1976a). Ultrastructural development of bone marrow adipose cell. *Acta Anatomica (Basel), 94,* 65–77.
Tavassoli, M. (1976b). Marrow adipose cells. Histochemical identification of labile and stable components. *Archives of Pathology & Laboratory Medicine, 100,* 16–18.
Trubowitz, S., & Bathija, A. (1977). Cell size and palmitate-1-14c turnover of rabbit marrow fat. *Blood, 49,* 599–605.
Wronski, T. J., Smith, J. M., & Jee, W. S. (1981). Variations in mineral apposition rate of trabecular bone within the beagle skeleton. *Calcified Tissue International, 33,* 583–586.
Yeung, D. K., Griffith, J. F., Antonio, G. E., Lee, F. K., Woo, J., & Leung, P. C. (2005). Osteoporosis is associated with increased marrow fat content and decreased marrow fat unsaturation: A proton MR spectroscopy study. *Journal of Magnetic Resonance Imaging, 22,* 279–285.

CHAPTER EIGHT

Brown Adipose Tissue in Humans: Detection and Functional Analysis Using PET (Positron Emission Tomography), MRI (Magnetic Resonance Imaging), and DECT (Dual Energy Computed Tomography)

Magnus Borga[*,†], Kirsi A. Virtanen[‡], Thobias Romu[*,†], Olof Dahlqvist Leinhard[*,§], Anders Persson[*,§,¶], Pirjo Nuutila[‡,∥], Sven Enerbäck[#,1]

[*]Center for Medical Image Science and Visualization, Linköping University, Linköping, Sweden
[†]Department of Biomedical Engineering, Linköping University, Linköping, Sweden
[‡]Turku PET Centre, University of Turku, Turku, Finland
[§]Department of Medical and Health Sciences, Linköping University, Linköping, Sweden
[¶]County Council of Östergötland, Linköping, Sweden
[∥]Department of Endocrinology, Turku University Hospital, Turku, Finland
[#]Department of Clinical and Medical Genetics, Institute of Biomedicine, University of Gothenburg, Gothenburg, Sweden
[1]Corresponding author: e-mail address: sven.enerback@medgen.gu.se

Contents

1. Introduction 142
2. Positron Emission Tomography 143
   2.1 Introduction 143
   2.2 Tracers for BAT metabolic imaging and principles of modeling 144
   2.3 Discovery of BAT in adult humans using PET 146
3. Magnetic Resonance Imaging 147
   3.1 Introduction 147
   3.2 Image acquisition 148
   3.3 Phase-sensitive reconstruction 149
   3.4 Intensity inhomogeneity correction and image calibration 149
   3.5 Visualization 149
4. Dual Energy Computed Tomography 150
   4.1 Introduction 150
   4.2 Projection-based two material decomposition 152
   4.3 BAT characterization 153

4.4 CT scanner protocol for BAT   154
4.5 Post processing   156
References   156

Abstract

If the beneficial effects of brown adipose tissue (BAT) on whole body metabolism, as observed in nonhuman experimental models, are to be translated to humans, tools that accurately measure how BAT influences human metabolism will be required. This chapter discusses such techniques, how they can be used, what they can measure and also some of their limitations. The focus is on detection and functional analysis of human BAT and how this can be facilitated by applying advanced imaging technology such as positron emission tomography, magnetic resonance imaging, and dual energy computed tomography.

1. INTRODUCTION

The newly established role of brown adipose tissue (BAT) in humans—as a metabolically active tissue—is based on the fact that we and other groups have identified BAT at paracervical and supraclavicular locations in healthy adults using positron emission tomography (PET) (Cypess et al., 2009; Saito et al., 2009; van Marken Lichtenbelt et al., 2009; Virtanen et al., 2009). These studies showed a severalfold increase in uptake of the glucose analog ^{18}F-fluoro-deoxy-glucose (^{18}F-FDG) in response to cold temperatures. Biopsy specimens of identified tissues were used to assess mRNA and protein levels, which together with typical morphology confirmed BAT identity and made it clear that adult humans have metabolically active BAT (Virtanen et al., 2009). Interestingly, human BAT activity was significantly lower in obese as compared to lean subjects and also significantly negatively correlated to body mass index (BMI) and percent body fat (Cypess et al., 2009; van Marken Lichtenbelt et al., 2009). Moreover, BAT was positively related to resting metabolic rate. It was also noted that women had larger BAT depots and ^{18}F-FDG uptake activity, and that BAT mass positively correlated with younger age. As an example, one of the subjects studied had an estimated supraclavicular BAT depot of approximately 60 g. It was concluded that, if fully activated, this would burn an amount of energy equivalent to approximately 4.1 kg of WAT over the course of a year (Virtanen et al., 2009). Thus, it is likely that in humans, activated BAT has the potential to contribute substantially to energy expenditure.

Although we know what ultimately causes obesity—it results when energy intake exceeds energy expenditure—today, the only proven therapy is bariatric surgery. This is a major abdominal procedure that, for reasons that remain largely unknown and which cannot be explained solely by a reduction in ventricular volume, significantly reduces energy intake. Due to cost and limited availability, however, this procedure will most likely be reserved for only a small fraction of those who stand to gain from effective antiobesity treatment. Clearly, alternative ways to treat obesity are needed. Another way to combat excessive accumulation of WAT would be to increase energy expenditure. Rodents, hibernators, and humans all have a specialized tissue—BAT—with the unique capacity to regulate energy expenditure by a process called adaptive thermogenesis. This process depends on the expression of uncoupling protein 1 (UCP1), which is a unique marker for BAT. UCP1 is an inner mitochondrial membrane protein that short-circuits the mitochondrial proton gradient, so that oxygen consumption is no longer coupled to ATP synthesis. As a consequence, heat is generated. It has been demonstrated that mice lacking Ucp1 are severely compromised in their ability to maintain normal body temperature when acutely exposed to cold and they are also prone to become obese. Moreover, it is also known that, in mice, BAT protects against diet-induced obesity, insulin resistance, and type 2 diabetes. This is, at least in part, based on prevention of excessive accumulation of triglyceride in nonadipose tissues like muscle and liver. Ectopic triglyceride storage at these locations is associated with initiation of insulin resistance and, ultimately, development of type 2 diabetes. To collect evidence of a similar role for BAT in humans we need new methods to measure both the *mass* and metabolic *activity* of this tissue enabling accurate assessments of energy turnover of this tissue both under normal and stimulated conditions. This chapter sets out to give an overview of some of the new ways in which modern imaging technology can be used for this purpose.

2. POSITRON EMISSION TOMOGRAPHY

2.1. Introduction

PET is an imaging technique that enables direct and quantitative observation of tissue radioactivity over time *in vivo*. An almost unlimited number of natural substrates (e.g., glucose and fatty acid), substrate analogs, and drugs can be labeled for use with PET. In general, PET makes it possible to isolate the organ of interest from surrounding tissues and, by mathematical modeling, the quantification of metabolic processes within the target tissue. PET is noninvasive and allows imaging of several organs during the same session in humans. PET combined with tracer kinetic models measures blood flow, membrane transport, and metabolism noninvasively and quantitatively. Thus, PET appears to provide comprehensive insights into molecular processes occurring in tissues *in vivo*.

2.2. Tracers for BAT metabolic imaging and principles of modeling

PET is based on the use of short-lived positron emitting radioisotopes (Table 8.1). Nearly, any substrate and drug can be labeled without changing their physicochemical or pharmacological properties (Phelps et al., 1979). Due to rapid decay of the positron emitters most PET tracers have to be synthesized on site. The tracer is injected intravenously to the subject or inhaled as a gas after which it distributes throughout the body via the bloodstream and enters into tissues and organs. Positrons emitted from the nucleus collide with their counterparts, electrons, and the masses of the two particles are converted into two photons emitted simultaneously in opposite directions. PET imaging is based on the detection of these paired photons in coincidence. Physiological and pharmacological phenomena and biological parameters can be estimated *in vivo* by mathematical modeling of the tissue and blood time activity curves obtained. PET is a highly sensitive and specific imaging tool for studies of BAT physiology (Table 8.1).

Table 8.1 PET tracers used in human studies of BAT with quantitative modeling

	Tracer	Half-life (min)
Perfusion (Muzik, Mangner, & Granneman, 2012; Orava et al., 2011)	$[^{15}O]H_2O$	2
Glucose uptake (Orava et al., 2011; Virtanen et al., 2009)	$[^{18}F]FDG$	109
Free fatty acid metabolism (Ouellet et al., 2012)	$[^{18}F]FTHA$	109
Oxidative metabolism (Muzik et al., 2012)	$[^{15}O]O_2$	2
(Ouellet et al., 2012)	$[^{11}C]acetate$	20

$[^{18}F]FDG$, fluorine-18-labeled 2-fluoro-2-deoxy-D-glucose; $[^{18}F]FTHA$, fluorine-18-labeled 6-thia-hepta-decanoic acid.

PET is currently the only technique that permits noninvasive quantification of regional tissue perfusion, glucose and fatty acid metabolism, protein synthesis, and oxygen consumption. To this purpose, sequential images are acquired after tracer injection to follow the distribution of the tracer in the system of interest. Regions are drawn over the organ of interest to obtain tissue-specific time activity curves. These curves, together with corresponding radioactivity levels in plasma (input to the system), are used in tracer kinetic modeling.

The glucose tracer [^{18}F]FDG is a fluorine-18-labeled glucose analog, which is transported into the cell and phosphorylated (Phelps et al., 1979; Sokoloff et al., 1977). In contrast to glucose, it cannot enter glycolysis, being trapped for the duration of most PET studies; thus, initial steps in glucose uptake and metabolism can be traced with this radiopharmaceutical. [^{18}F]FTHA is a long-chain fatty acid analog and inhibitor of fatty acid metabolism; after transport into the mitochondria it undergoes initial steps of β-oxidation and is thereafter trapped in the cell (DeGrado, Coenen, & Stöcklin, 1991; Ebert et al., 1994).

Graphical methods have the advantage of being independent of compartment number and configuration. Gjedde–Patlak analysis was first developed for the assessment of irreversible tracer transfer (Patlak & Blasberg, 1985). A graph is generated by plotting:

$$\frac{C_t(t)}{C_p(t)} \text{ versus } \frac{\int_0^t C_p(t) \mathrm{d}t}{C_p(t)}$$

where C_t is the tissue radioactivity at each sampling time point (t) and C_p is the plasma radioactivity. When irreversible influx of tracer occurs, the two variables describe a linear relationship after a few minutes of equilibration. The influx constant is then given by the slope of the linear fit of the data, excluding the first few values. If reversible uptake occurs, metabolite loss can be corrected for by using the linearization method described by Patlak and Blasberg (1985). The quantification of the fate of a natural compound, for example, glucose by use of its analog, FDG, requires the introduction of a conversion term, denominated lumped constant, representing the ratio of the metabolic rates of the tracer and the tracee (Sokoloff et al., 1977). Relative to FDG, this numerical value has been shown to approximate unity in skeletal muscle (Peltoniemi et al., 2000), myocardium (Ng, Soufer, & McNulty, 1998), and adipose tissue (Virtanen et al.,

2001), and to hold constant across metabolic disease states. Therefore, it has been assumed to be 1.0 also for human BAT (Virtanen et al., 2009).

$[^{15}O]H_2O$ is an ideal tracer for the measurement of blood flow as it is chemically inert and freely diffusible. The kinetics of the perfusion tracer $[^{15}O]H_2O$ have been validated experimentally and clinically (Iida et al., 1988; Ruotsalainen et al., 1997) and applied to the study of BAT as well (Muzik et al., 2012; Orava et al., 2011). Oxygen consumption may be determined directly in BAT using $^{15}O_2$-PET (Muzik et al., 2012). This tracer is given by inhalation and requires on-site blood radioactivity measurement or the use of image-derived input function for the modeling.

2.3. Discovery of BAT in adult humans using PET

The association between the prevalence of supraclavicular ^{18}FDG accumulation and cold outdoor temperatures was reported (Cohade, Mourtzikos, & Wahl, 2003). This tracer accumulation during diagnostic clinical PET studies was problematic for clinicians. When hybrid PET/computed tomography (CT) scanners became available, the precise localization of FDG accumulation became possible and gave further evidence for active BAT in humans (Nedergaard, Bengtsson, & Cannon, 2007). However, confirmation and direct evidence of BAT in adults took place after functional imaging was combined with biopsies taken from the active hot spots in 2009 (van Marken Lichtenbelt et al., 2009; Virtanen et al., 2009).

Since 2009, many research and clinical laboratories have actively studied BAT using ^{18}FDG-PET or PET/CT, but results have been discordant. Although the graphical method is validly used for the assessment of glucose metabolic rate, BAT activity has often been estimated using standard uptake values and static late scanning or simply using counts per time unit. The main drawbacks with these semiquantitative methods are that interindividual differences between tracer distribution volume and changes in its relative distribution and the current plasma glucose concentration are not taken into account. If an obese subject is studied, the distribution of the tracer between different tissues is not similar as compared to a lean one although the dose is adjusted with total body weight. Mathematical modeling becomes mandatory when the effects of hormones and nutrients on brown fat activity are studied. During insulin stimulation muscle glucose uptake is up to sixfold enhanced and covers 60–70% of the whole body glucose utilization (Nuutila et al., 1992). Therefore, ^{18}FDG is cleared from the plasma into muscles, its availability in plasma is decreased and "counts" are lower in BAT.

However, by modeling FDG uptake over time, we were able to demonstrate that insulin is a potent stimulator of BAT activity (Orava et al., 2011). Only dynamic data analysis with input information gives reliable values of BAT activity for comparisons between groups or subjects in interventions.

Cold activation enhances hBAT perfusion significantly when studied using ^{15}O-H$_2$O-PET (Muzik et al., 2012; Orava et al., 2011). Insulin stimulation during euglycemia increases BAT glucose uptake fivefold in line with whole body insulin sensitivity but without any increment in perfusion (Orava et al., 2011). When Muzik and coworkers used ^{15}O$_2$ to assess directly oxygen consumption during cold stimulation, MRO$_2$ values were shown to increase (Muzik et al., 2012). Based on this, the authors estimated that thermogenesis of supraclavicular BAT depots account <20 kcal/day. However, they found significant differences in total energy expenditure between BAT+ and BAT− subjects (Muzik et al., 2012). This might suggest that even smaller BAT depots need to be included in the analyses to properly estimate human BAT activity. Although PET and PET/CT or PET/MRI scanners are ideal for functional imaging, they are not updated for the optimal assessment of tiny regions of BAT, which are widely spread over the body. The resolution of PET scanners has improved but it is still between 1 and 4 mm. Also the concept of brite or beige adipose tissue means that we are analyzing data that contains a mixture of white and brown adipocytes.

3. MAGNETIC RESONANCE IMAGING

3.1. Introduction

While PET detects the metabolic activity of BAT, magnetic resonance imaging (MRI) and dual energy computed tomography (DECT) can be used for imaging the morphology of the BAT organ itself. Importantly, BAT that is not necessarily metabolically active can also be imaged. Furthermore, since MRI does not use ionizing radiation, as opposed to PET and CT, it is more suitable for large prospective studies on healthy volunteers.

A number of different contrast mechanisms have been used for detecting BAT with MRI, for example, fat–water fraction determined by chemical shift imaging (Lunati et al., 1999); fat–water fraction in combination with T1 relaxation time (Hamilton, Bydder, Smith, Nayak, & Hu, 2011); or intermolecular zero-quantum coherence (Branca & Warren, 2011). The contrast mechanism used here is similar to that used by Lunati et al. and also in the recent work by Hu, Smith, Nayak, Goran, and Nagy (2010), Hu, Tovar, Pavlova, Smith, and Gilsanz (2012), and Hu et al. (2013) and is based on the higher water content in BAT compared to white adipose tissue (WAT). The composition of WAT is typically around 90% lipid, whereas BAT has a significantly lower lipid fraction. By using multiecho chemical shift imaging (Dixon, 1984), the fat and water signals can be separated based on the difference in their resonance frequencies. Hence, the relative fat fraction can be estimated, and used for detecting BAT. The fat–water ratio itself is, however, not a reliable measure since this ratio is very noise sensitive in areas of low signal strength, for example, in air. In order to only consider anatomically relevant voxels, we use an absolute image intensity scale where WAT is used as internal reference. Using these calibrated fat images, adipose tissue with any given fat content can be detected.

Multiecho chemical shift imaging also enables detection of $T2^*$ relaxation time of the water and lipid signals (O'Regan et al., 2008; Yu et al., 2007). $T2^*$ quantification enables correction of signal loss due to $T2^*$ relaxation and also provides an additional parameter for characterization of BAT tissue (Hu et al., 2013). Khanna & Branca showed in a rat model the potential of using $T2^*$-sensitive imaging to detect BAT activity. In this study, a well-localized signal drop was observed in activated BAT caused by a strong increase in oxygen consumption and consequent increase in blood deoxyhemoglobin levels leading to a drop of the $T2^*$ relaxation time (Khanna & Branca, 2012).

Below follows a more detailed description of the different steps in the proposed imaging method.

3.2. Image acquisition

Images were acquired using an Ingenia 3.0T MRI scanner (Philips) from the upper part of thorax planned axially with the lower end of the image stack at the level of the aortic arch. Phased array coils were used for signal reception. The subjects were asked to breath shallow to minimize respiratory motion in the upper part of the thorax, thereby enabling free breathing image acquisition. This allows longer acquisition times with simultaneous coverage of supraclavicular and intrascapular BAT with high resolution and high signal-to-noise ratio. A spoiled three-dimensional gradient echo sequence was used with a 12-echo read out. The first echo was recorded at 1.15 ms and the following using 1.15 ms echo spacing. The repetition time was 26 ms and the flip angle was 10°. The obtained voxel dimension was $1.76 \times 1.76 \times 1.75$ mm^3 with a field of view (FOV) of $380 \times 302 \times 138$ mm^3. Parallel imaging acceleration, using a sensitivity-encoding (SENSE) factor of 2.5, resulted in a scan time of 1 min and 51 s.

If image acquisition with volume coverage in the lower thorax is desired, a lower number of echoes and repetition time and/or a reduced FOV coverage may be used to decrease the scan time and thereby enabling breath-hold acquisition to avoid motion artifacts.

3.3. Phase-sensitive reconstruction

Fat and water image volumes were reconstructed using the 3D inverse gradient method (Rydell et al., 2008, 2007) using the complex-valued image volumes from the first two echoes. As opposed to region-growing methods, which are essentially one-dimensional algorithms acting in two or three dimensions, the inverse gradient method is a true multidimensional method, which can benefit from the connectivity in three-dimensional data. The inverse gradient method generates a pair of phase-unwrapped real-valued in-phase (IP) and opposed-phase (OP) images. To resolve the fat–water classification problem, that is, to decide the correct sign of the OP image, the method described in Romu et al. (2011) was used.

The images were corrected for the $T2^*$ decay effect by fitting $T2^*$ decay parameters for fat and water signals respectively using a model similar to that described by O'Regan et al. (2008).

3.4. Intensity inhomogeneity correction and image calibration

In order to correct for image intensity inhomogeneities caused by various physical imperfections in the imaging system, such as variations in B0 field and coil sensitivity, WAT was used as internal reference (Leinhard, Johansson, Rydell, Borga, & Lundberg, 2008). Voxels representing pure WAT were identified using thresholding of the quotient between the fat image and the IP image at 0.9. Then a bias field was interpolated from the identified pure WAT voxels using a multiscale interpolation technique (Romu, Borga, & Leinhard, 2011). The image volumes were corrected using the inverse of this estimated bias field. The voxel values in these calibrated fat images show the *relative fat content* (RFC), which is one in WAT and zero in voxels with no adipose tissue. Note that the RFC is different from the more commonly used "fat fraction" (Reeder, Hu, & Sirlin, 2012).

3.5. Visualization

Since there is no narrow fat–water fraction interval that is generally applicable for BAT, we use a color coding for the fat fraction that is overlaid on the fat image, as shown in Fig. 8.1. Hence, the hue represents the fat

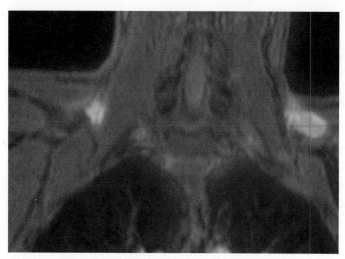

Figure 8.1 An example PET/MR image of a young male with significant activation of supraclavicular adipose tissue during cold exposure. The light color indicates high uptake of ^{18}FDG. (See the color plate.)

fraction and the brightness represents the fat signal. Finally, the water image and the colored fat image are added in order to supply more anatomical information. In the resulting image, yellow represents WAT; absence of color represents absence of fat, and a reddish color represent possible BAT. Examples of this visualization are shown in Figs. 8.2 and 8.3. In these figures, yellow indicates subcutaneous WAT, while red indicate possible BAT in the intrascapular and supraclavicular depots.

4. DUAL ENERGY COMPUTED TOMOGRAPHY

4.1. Introduction

In CT imaging, materials having different chemical compositions can be represented by the same, or very similar, Hounsfield Units (HU), making the differentiation and classification of different types of tissues extremely challenging (Macovski, Alvarez, Chan, Stonestrom, & Zatz, 1976). In addition, the accuracy with which material concentration can be measured is degraded by the presence of multiple tissue types (Alvarez & Macovski, 1976). The reason for these difficulties in differentiating and quantifying tissue types is that the measured HU of a voxel is related to the linear attenuation coefficient $\mu(E)$, which is not unique for any given material but is a

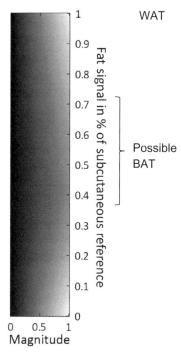

Figure 8.2 Relative fat content (RFC) color map used on the fat images. (See the color plate.)

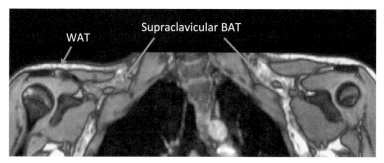

Figure 8.3 Coronal section from an adult male subject with indications of supraclavicular BAT. (See the color plate.)

function of the material composition, the photon energies interacting with the material, and the mass density of the material. In DECT, an additional attenuation measurement is obtained at a second energy, allowing the differentiation of two materials (Fig. 8.4). Although medical X-ray tubes

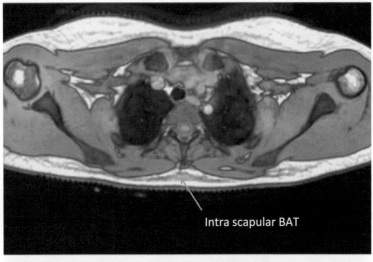

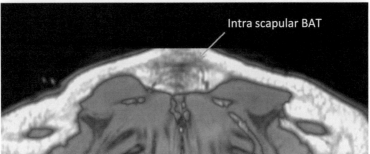

Figure 8.4 Axial and coronal sections from a male subject with indication of intrascapular BAT. (See the color plate.)

generate polychromatic spectra, the general principle remains valid. Thus, DECT can be defined as the use of attenuation values acquired with different energy spectra, and the known changes in attenuation between the two spectra, in order to differentiate and classify tissue composition.

4.2. Projection-based two material decomposition

Dual energy methods for CT were first investigated by Alvarez and Macovski in 1976 (Kalender, Klotz, & Suess, 1987; Macovski et al., 1976). They demonstrated that even with polychromatic X-ray spectra, one can still separate the measured attenuation coefficients into their contributions from the photoelectric effect and Compton scattering processes. In the 1980s, a modified commercial CT system (Siemens DR) used fast tube voltage switching to allow alternate projection measurements at the low and high tube potentials (Alvarez & Macovski, 1976). However, the tube current could not be increased quickly enough for the low tube potential setting to achieve comparable noise levels in both the low and high kilovoltage (kV) datasets. Recently, the simultaneous acquisition of dual energy data has been introduced using multi detector CT with two X-ray tubes and rapid peak kilovoltage (kVp) switching (gemstone spectral imaging) (Flohr et al., 2006; Ho et al., 2009; Johnson et al., 2007; Marin et al., 2009; Petersilka, Bruder, Krauss, Stierstorfer, & Flohr, 2008). Two major advantages of DECT are material decomposition and the elimination of misregistration artifacts. The simplest method for combining the dual energy image data for the purpose of material differentiation is to perform a linearly weighted image subtraction. The low tube potential images (typically 80 kV) are multiplied by a weighting factor and subtracted from the high voltage images (140 kV) to suppress or enhance a specific material (Kalender, Klotz, & Kostaridou, 1988). Reconstructed CT images acquired using the two different tube potentials can be processed with a three-material decomposition algorithm (Johnson et al., 2007). The principle behind this technique is illustrated in Fig. 8.5, where the typical HU of three materials of known densities are plotted on a graph where the y-axis is the HU at 140 kV and the x-axis is the HU at 80 kV. Ideally, the three materials should be sufficiently different as to create a triangle in this plot. Thereafter, corresponding HU pairs from the low- and high-energy images are mapped onto the calibration diagram. Depending on their position in the diagram, the composition of a certain material is determined. The voxels can be color-coded according to the relative composition of certain materials, or specific materials can be either suppressed or enhanced, depending on the desired clinical application.

4.3. BAT characterization

Recent studies have showed that the CT HUs of BAT are significantly greater under activated conditions than under nonactivated conditions in both patients and animal models (Baba, Jacene, Engles, Honda, & Wahl, 2010; Hu, Chung, Nayak, Jackson, & Gilsanz, 2011). Several reasons could explain the greater X-ray attenuation. BAT has unique cytological characteristics comprised of multiple lipid droplets surrounded by significant amounts of intracellular water. Brown adipocytes are also densely packed with mitochondria and are heavily vascularized and innervated. The greater

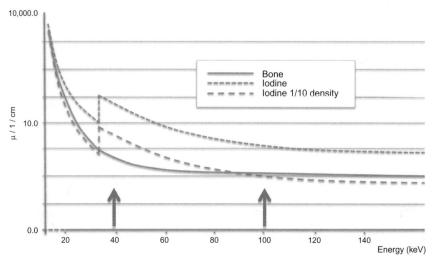

Figure 8.5 Assuming the use of mono energetic X-rays, at approximately 100 keV, the same linear attenuation coefficients are measured for bone and iodine. Data acquired at approximately 40 keV allows the differentiation of the two materials, regardless of their respective densities. (See the color plate.)

proportion of nonlipid components in metabolically active BAT versus any other inactive adipose tissues (e.g., WAT) is the cause for its greater tissue density and more positive HUs. Due to significant overlap of HU values, it currently does not seem possible to separate metabolically inactive BAT from inert WAT based on HUs alone (see example in Fig. 8.7). To overcome this problem, and to distinguish different types of fat tissues from each other, some physical property that differs between the tissues needs to be measured. Conventional attenuation measurements based on a single photon energy spectrum have limited value in this respect. When the attenuation is measured at two energies, however, their values are not exactly proportional to each other and BAT can be separated from WAT (Zachrisson et al., 2010; Fig. 8.6).

4.4. CT scanner protocol for BAT

Currently, three systems are capable of the nearly simultaneous acquisition of dual energy data during a single breath hold (Flohr et al., 2006; Marin et al., 2009; Petersilka et al., 2008): 64-slice dual source CT (Definition, Siemens Medical Systems; Erlangen, Germany), 128-slice dual-source CT

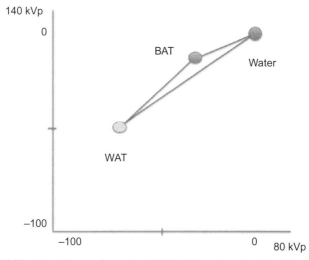

Figure 8.6 Calibration diagram based on WAT, BAT, and water. (See the color plate.)

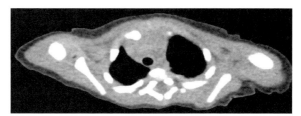

Figure 8.7 CT examination of a 1-year-old child with biopsy-verified BAT. Single energy dataset (120 kVp). BAT measures 10HUs higher that WAT indicating activated BAT (Baba et al., 2010). Using this approach, it is not possible to visualize the BAT organ.

(Definition Flash, Siemens Medical Systems), and high-definition 64-MDCT (Discovery 750 HD, GE Healthcare; Milwaukee, WI, USA). In the two systems available from Siemens, the two tubes (tubes A and B) use different kVp (80 and 140 kVp), and in high-definition 64-MDCT, the kVp switches from 80 to 140 kVp in <0.5 ms. Below follows a more detailed description of the different steps in the proposed imaging method using a 128-slice dual source CT (Definition Flash, Siemens Medical Systems). Typical Dual energy scanner settings to acquire DECT data for fat characterization are: 140 sn and 80 kVp simultaneous acquisition, ref. mAs 461 for BMI <20. 140 sn and 100 kVp simultaneous acquisition, ref. mAs 230 for BMI >20. Collimation 32×0.6 mm. Pitch 0.7, rotation

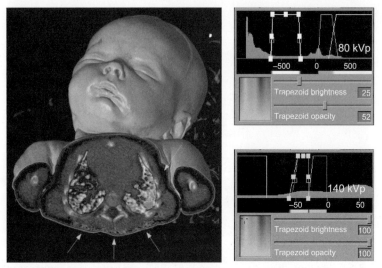

Figure 8.8 The same child as in Fig. 8.7. Datasets acquired at 80 and 140 kVp were loaded into a Siemens multimodality workstation with volume rendering 3D software (Inspace) in merged mode. For each dataset, voxels were mapped with green color for BAT and black color for WAT. Settings; WAT: 140 kVp, −50HU, 80 kVp, −75HU and for BAT: 140 kVp, −20HU, 80 kVp, −35HU. BAT can be separated from WAT. (See the color plate.)

time 0.5 s. Slice thickness 0.6 mm, reconstruction interval 0.6 mm, kernel Q30f medium soft, FOV <33.6 cm, depending on the patient size. DECT examinations should be avoided for individuals with BMI >30.

4.5. Post processing

Post processing of DECT scan data can be performed either directly in the CT scanner or in a dedicated workstation with software designed for managing raw data or DECT image data. An example of such post processing is shown in Fig. 8.8.

REFERENCES

Alvarez, R. E., & Macovski, A. (1976). Energy-selective reconstructions in X-ray computerized tomography. *Physics in Medicine and Biology, 21,* 733–744.

Baba, S., Jacene, H. A., Engles, J. M., Honda, H., & Wahl, R. L. (2010). CT Hounsfield units of brown adipose tissue increase with activation: Preclinical and clinical studies. *Journal of Nuclear Medicine, 51,* 246–250.

Branca, R. T., & Warren, W. S. (2011). In vivo brown adipose tissue detection and characterization using water-lipid intermolecular zero-quantum coherences. *Magnetic Resonance in Medicine, 65,* 313–319.

Cohade, C., Mourtzikos, K. A., & Wahl, R. L. (2003). "USA-Fat": Prevalence is related to ambient outdoor temperature-evaluation with 18F-FDG PET/CT. *Journal of Nuclear Medicine*, *44*, 1267–1270.

Cypess, A. M., Lehman, S., Williams, G., Tal, I., Rodman, D., Goldfine, A. B., et al. (2009). Identification and importance of brown adipose tissue in adult humans. *The New England Journal of Medicine*, *360*(15), 1509–1517.

DeGrado, T. R., Coenen, H. H., & Stöcklin, G. (1991). 14(R,S)-[18F]fluoro-6-thia-heptadecanoic acid (FTHA): Evaluation in mouse of a new probe of myocardial utilization of long chain fatty acids. *Journal of Nuclear Medicine*, *32*(10), 1888–1896.

Dixon, W. T. (1984). Simple proton spectroscopic imaging. *Radiology*, *153*, 189–194.

Ebert, A., Herzog, H., Stöcklin, G. L., Henrich, M. M., DeGrado, T. R., Coenen, H. H., et al. (1994). Kinetics of 14(R,S)-fluorine-18-fluoro-6-thia-heptadecanoic acid in normal human hearts at rest, during exercise and after dipyridamole injection. *Journal of Nuclear Medicine*, *35*(1), 51–56.

Flohr, T. G., McCollough, C. H., Bruder, H., Petersilka, M., Gruber, K., Süss, C., et al. (2006). First performance evaluation of a dual-source CT (DSCT) system. *European Radiology*, *16*, 256–268.

Hamilton, G., Bydder, M., Smith, D. L., Jr., Nayak, K. S., & Hu, H. H. (2011). Proper ties of brown and white adipose tissue measured by 1H MRS. *Journal of Magnetic Resonance Imaging*, *34*, 468–473.

Ho, L. M., Yoshizumi, T. T., Hurwitz, L. M., Nelson, R. C., Marin, D., Toncheva, G., et al. (2009). Dual energy versus single energy MDCT: Measurement of radiation dose using adult abdominal imaging protocols. *Academic Radiology*, *16*, 1400–1407.

Hu, H. H., Chung, S. A., Nayak, K. S., Jackson, H. A., & Gilsanz, V. (2011). Differential CT attenuation of metabolically active and inactive adipose tissues—Preliminary findings. *Journal of Computer Assisted Tomography*, *35*(1), 65–71.

Hu, H. H., Smith, D. L., Jr., Nayak, K. S., Goran, M. I., & Nagy, T. R. (2010). Identification of brown adipose tissue in mice with fat-water IDEAL-MRI. *Journal of Magnetic Resonance Imaging*, *31*, 1195–1202.

Hu, H. H., Tovar, J. P., Pavlova, Z., Smith, M. L., & Gilsanz, V. (2012). Unequivocal identification of brown adipose tissue in a human infant. *Journal of Magnetic Resonance Imaging*, *35*, 938–942.

Hu, H. H., Yin, L., Aggabao, P. C., Perkins, T. G., Chia, J. M., & Gilsanz, V. (2013). Comparison of brown and white adipose tissues in infants and children with chemical-shift-encoded water-fat MRI. *Journal of Magnetic Resonance Imaging*, *38*, 885–896. http://dx.doi.org/10.1002/jmri.24053.

Iida, H., Kanno, I., Takahashi, A., Miura, S., Murakami, M., Takahashi, K., et al. (1988). Measurement of absolute myocardial blood flow with H215O and dynamic positron-emission tomography. Strategy for quantification in relation to the partial-volume effect. *Circulation*, *78*, 104–115.

Johnson, T. R., Krauss, B., Sedlmair, M., Grasruck, M., Bruder, H., Morhard, D., et al. (2007). Material differentiation by dual energy CT: Initial experience. *European Radiology*, *17*, 1510–1517.

Kalender, W. A., Klotz, E., & Kostaridou, L. (1988). An algorithm for noise suppression in dual energy CT material density images. *IEEE Transactions on Medical Imaging*, *7*, 218–224.

Kalender, W. A., Klotz, E., & Suess, C. (1987). Vertebral bone mineral analysis: An integrated approach with CT. *Radiology*, *164*, 419–423.

Khanna, A., & Branca, R. T. (2012). Detecting brown adipose tissue activity with BOLD MRI in mice. *Magnetic Resonance in Medicine*, *68*(4), 1285–1290. http://dx.doi.org/10.1002/mrm.24118. (Epub 2012 Jan 9).

Leinhard, O. D., Johansson, A., Rydell, J., Borga, M., & Lundberg, P. (2008). Intensity inhomogeneity correction in two point dixon imaging. In *Proceedings of the international society for magnetic resonance in medicine annual meeting (ISMRM'08)*, Vol. 16, (p. 1519).

Lunati, E., Marzola, P., Nicolato, E., Fedrigo, M., Villa, M., & Sbarbati, A. (1999). In vivo quantitative lipidic map of brown adipose tissue by chemical shift imaging at 4.7 Tesla. *Journal of Lipid Research*, 40, 1395–1400.

Macovski, A., Alvarez, R. E., Chan, J. L., Stonestrom, J. P., & Zatz, L. M. (1976). Energy dependent reconstruction in X-ray computerized tomography. *Computers in Biology and Medicine*, 6, 325–336.

Marin, D., Nelson, R. C., Samei, E., Paulson, E. K., Ho, L. M., Boll, D. T., et al. (2009). Hypervascular liver tumors: Low tube voltage, high tube current multidetector CT during late hepatic arterial phase for detection—Initial clinical experience. *Radiology*, 251, 771–779.

Muzik, O., Mangner, T. J., & Granneman, J. G. (2012). Assessment of oxidative metabolism in brown fat using PET imaging. *Frontiers in Endocrinology*, 3, 1–7.

Nedergaard, J., Bengtsson, T., & Cannon, B. (2007). Unexpected evidence for active brown adipose tissue in adult humans. *American Journal of Physiology, Endocrinology and Metabolism*, 293, E444–E452.

Ng, C. K., Soufer, R., & McNulty, P. H. (1998). Effect of hyperinsulinemia on myocardial fluorine-18-FDG uptake. *Journal of Nuclear Medicine*, 39(3), 379–383.

Nuutila, P., Koivisto, V. A., Knuuti, J., Ruotsalainen, U., Teräs, M., Haaparanta, M., et al. (1992). Glucose-free fatty acid cycle operates in human heart and skeletal muscle in vivo. *The Journal of Clinical Investigation*, 89(6), 1767–1774.

Orava, J., Nuutila, P., Lidell, M. E., Oikonen, V., Noponen, T., Viljanen, T., et al. (2011). Different metabolic responses of human brown adipose tissue to activation by cold and insulin. *Cell Metabolism*, 14(2), 272–279.

O'Regan, D. P., Callaghan, M. F., Wylezinska-Arridge, M., Fitzpatrick, J., Naoumova, R. P., Hajnal, J. V., et al. (2008). Liver fat content and $T2^*$: Simultaneous measurement by using breath-hold multiecho MR imaging at 3.0 T—Feasibility. *Radiology*, 247(2), 550–557.

Ouellet, V., Labbé, S. M., Blondin, D. P., Phoenix, S., Guérin, B., Haman, F., et al. (2012). Brown adipose tissue oxidative metabolism contributes to energy expenditure during acute cold exposure in humans. *The Journal of Clinical Investigation*, 122(2), 545–552. http://dx.doi.org/10.1172/JCI60433.

Patlak, C. S., & Blasberg, R. G. (1985). Graphical evaluation of blood-to-brain transfer constants from multiple-time uptake data. Generalizations. *Journal of Cerebral Blood Flow and Metabolism*, 5, 584–590.

Peltoniemi, P., Lönnroth, P., Laine, H., Oikonen, V., Tolvanen, T., Grönroos, T., et al. (2000). Lumped constant for [^{18}F]fluorodeoxyglucose in skeletal muscles of obese and nonobese humans. *American Journal of Physiology, Endocrinology and Metabolism*, 279, E1122–E1130.

Petersilka, M., Bruder, H., Krauss, B., Stierstorfer, K., & Flohr, T. G. (2008). Technical principles of dual source CT. *European Journal of Radiology*, 68, 362–368.

Phelps, M. E., Huang, S.-C., Hoffman, E. J., Selin, C., Sokoloff, L., & Kuhl, D. E. (1979). Tomographic measurement of local cerebral glucose metabolic rate in humans with (F-18) 2-fluoro-2-deoxy-D-glucose: Validation of method. *Annals of Neurology*, 6, 371–388.

Reeder, S. B., Hu, H. H., & Sirlin, C. B. (2012). Proton density fat-fraction: A standardized MR-based biomarker of tissue fat concentration. *Journal of Magnetic Resonance Imaging*, 36, 1011–1014.

Romu, T., Borga, M., & Leinhard, O. D. (2011). MANA—Multi scale adaptive normalized averaging. In *Proceedings of the 2011 IEEE international symposium on biomedical imaging: From nano to macro* (pp. 361–364).

Romu, T., Leinhard, O. D., Forsgren, M. F., Almer, S., Dahlström, N., Kechagias, S., et al. (2011). Fat water classification of symmetrically sampled two-point dixon images using biased partial volume effects. In *Proceedings of the international society for magnetic resonance in medicine annual meeting (ISMRM'11)*.

Ruotsalainen, U., Raitakari, M., Nuutila, P., Oikonen, V., Sipilä, H., Teräs, M., et al. (1997). Quantitative blood flow measurement of skeletal muscle using oxygen-15-water and PET. *Journal of Nuclear Medicine, 38*, 314–319.

Rydell, J., Johansson, A., Leinhard, O. D., Knutsson, H., Farnebäck, G., Lundberg, P., et al. (2008). Three dimensional phase sensitive reconstruction for water/fat separation in MR imaging using inverse gradient. In *Proceedings of the international society for magnetic resonance in medicine annual meeting (ISMRM'08)* (p. 1519).

Rydell, J., Knutsson, H., Pettersson, J., Johansson, A., Farnebäck, G., Leinhard, O. D., et al. (2007). Phase sensitive reconstruction for water/fat separation in MR imaging using inverse gradient. In *Lecture notes in computer science: Vol. 4791. Medical image computing and computer-assisted intervention—MICCAI 2007* (pp. 210–218).

Saito, M., Okamatsu-Ogura, Y., Matsushita, M., Watanabe, K., Yoneshiro, T., Nio-Kobayashi, J., et al. (2009). High incidence of metabolically active brown adipose tissue in healthy adult humans: Effects of cold exposure and adiposity. *Diabetes, 58*(7), 1526–1531.

Sokoloff, L., Reivich, M., Kennedy, C., Des Rosiers, M. H., Patlak, C. S., Pettigrew, K. D., et al. (1977). The [14C]deoxyglucose method for the measurement of local cerebral glucose utilization: Theory, procedure, and normal values in the conscious and anesthetized albino rat. *Journal of Neurochemistry, 28*, 897–916.

van Marken Lichtenbelt, W. D., Vanhommerig, J. W., Smulders, N. M., Drossaerts, J. M., Kemerink, G. J., Bouvy, N. D., et al. (2009). Cold-activated brown adipose tissue in healthy men. *The New England Journal of Medicine, 360*, 1500–1508.

Virtanen, K. A., Lidell, M. E., Orava, J., Heglind, M., Westergren, M., Niemi, T., et al. (2009). Functional brown adipose tissue in healthy adults. *The New England Journal of Medicine, 360*, 1518–1525.

Virtanen, K. A., Peltoniemi, P., Marjamäki, P., Asola, M., Strindberg, L., Parkkola, R., et al. (2001). Human adipose tissue glucose uptake determined using [(18)F]-fluoro-deoxyglucose ([(18)F]FDG) and PET in combination with microdialysis. *Diabetologia, 44*(12), 2171–2179.

Yu, H., McKenzie, C. A., Shimakawa, A., Vu, A. T., Brau, A. C., Beatty, P. J., et al. (2007). Multiecho reconstruction for simultaneous water-fat decomposition and $T2^*$ estimation. *Journal of Magnetic Resonance Imaging, 26*(4), 1153–1161.

Zachrisson, H., Engström, E., Engvall, J., Wigström, L., Smedby, Ö., & Persson, A. (2010). Soft tissue discrimination ex vivo by dual energy computed tomography. *European Journal of Radiology, 75*, 124–128.

CHAPTER NINE

Analyzing the Functions and Structure of the Human Lipodystrophy Protein Seipin

M.F. Michelle Sim[*], Md Mesbah Uddin Talukder[†], Rowena J. Dennis[*], J. Michael Edwardson[†], Justin J. Rochford[‡,1]

[*]Institute of Metabolic Science, Addenbrooke's Hospital, University of Cambridge Metabolic Research Laboratories, Cambridge, United Kingdom
[†]Department of Pharmacology, University of Cambridge, Cambridge, United Kingdom
[‡]Rowett Institute of Nutrition and Health, Institute of Medical Sciences, University of Aberdeen, Aberdeen, United Kingdom
[1]Corresponding author: e-mail address: j.rochford@abdn.ac.uk

Contents

1. Introduction	162
2. Technical Aspects	163
2.1 Immunoprecipitation and Western blotting of seipin and associated proteins	163
2.2 Bimolecular fluorescence studies of seipin-interacting proteins	166
2.3 Analysis of seipin by atomic force microscopy	169
3. Discussion	173
References	174

Abstract

Disruption of the gene *BSCL2*, which encodes the protein seipin, causes severe generalized lipodystrophy in humans with a near complete absence of adipose tissue. Moreover, cell culture studies have demonstrated that seipin plays a critical cell-autonomous role in adipocyte differentiation. These observations reveal seipin as a critical regulator of human adipose tissue development; however, until recently very little has been known about the potential molecular functions of this intriguing protein. Despite significant recent interest in the function of seipin, our understanding of its molecular role(s) remains limited. The topology of seipin and lack of evidence for any enzymatic domains or activity indicate that it may act principally as a scaffold for other proteins or play a structural role in altering membrane curvature and/or budding. Work in this area has been hampered by several factors, including the lack of homology that might imply testable functions, the poor availability of antibodies to the endogenous protein and the observation that this hydrophobic ER membrane-resident protein is difficult to analyze by standard Western blotting techniques. Here we summarize some of the techniques we have applied to investigate the association of seipin with a recently identified binding partner, lipin 1. In addition, we describe the use of atomic force microscopy (AFM) to image oligomers of the seipin protein. We believe that AFM will offer a valuable tool to examine the association of candidate binding proteins with the seipin oligomer.

1. INTRODUCTION

Syndromes of congenital lipodystrophy are rare but offer valuable opportunities to gain insight into human adipose tissue development. These monogenic disorders reveal critical roles for the affected genes in adiposity, providing starting points for defining critical pathways in human fat development (Rochford, 2010). One of the most interesting in this regard is the gene *BSCL2*, which encodes the protein seipin. Disruption of *BSCL2* causes severe congenital generalized lipodystrophy (CGL) where affected individuals almost completely lack any adipose depots (Magre et al., 2001; Rochford, 2010). The molecular function of seipin remains poorly defined but it has been shown to play a critical, cell-autonomous role in adipogenesis (Chen et al., 2012, 2009; Payne et al., 2008). This implies that the study of seipin could provide novel insight regarding tissue-selective pathways regulating human adipogenesis. Given the importance of maintaining adipocyte function and turnover for human metabolic health, including in common obesity, such insights could be extremely valuable in identifying new therapeutic targets for improving prevalent metabolic disease.

Human seipin occurs as either a long or short form, the former including an additional 64 amino acid extension to the N-terminus. Although three transcripts encoding seipin were initially identified by Northern blotting, it is believed that all are capable of encoding both long and short forms of the protein (Magre et al., 2001). Seipin and its orthologs localize to the endoplasmic reticulum (ER) membrane with N- and C-termini exposed to the cytosol and two transmembrane domains so that a central loop domain is exposed to the lumen of the ER (Lundin et al., 2006).

Several studies in yeast and non-adipose cells have suggested that seipin may affect lipid droplet size, morphology and inheritance, although the molecular mechanism underlying these phenomena remains unclear (Binns, Lee, Hilton, Jiang, & Goodman, 2010; Fei, Li, et al., 2011; Fei, Shui, et al., 2011; Fei et al., 2008; Szymanski et al., 2007; Wolinski, Kolb, Hermann, Koning, & Kohlwein, 2011). Mouse embryonic fibroblasts lacking seipin have been shown to exhibit unrestrained lipolysis during adipogenesis, loss of lipid initially stored in these cells, and reduced adipogenic gene expression (Chen et al., 2012). Recently, we demonstrated that seipin can bind the phosphatidic acid (PA) phosphatase lipin 1, which itself is required for adipogenesis and adipose tissue development in mice (Reue, 2009; Sim, Dennis, et al., 2013; Sim, Talukder, et al., 2013). It is currently unclear which, if any, of these phenomena are mechanistically linked.

Technical challenges in reliably detecting seipin by Western blotting, and concerns regarding the specificity of potential interactions with candidate seipin binding proteins, are likely to have contributed to the failure to uncover more detail regarding the role of seipin in adipogenesis. Here, we describe some of the methods we have found useful in investigating the potential roles of seipin in adipocyte development.

2. TECHNICAL ASPECTS

2.1. Immunoprecipitation and Western blotting of seipin and associated proteins

Endogenous seipin has been difficult to examine as most available antibodies have proven ineffective in Western blotting although, along with others, we have detected endogenous murine seipin in immunofluorescence studies (Garfield et al., 2012; Ito, Fujisawa, Iida, & Suzuki, 2008). Indeed, we are aware of only one report, using a polyclonal antibody from Abnova (cat. no. H00026580-A02), of Western blotting endogenous seipin in samples from mouse tissues (Cui et al., 2012). For this reason, many studies have used tagged forms of seipin to aid detection. Here we use human seipin tagged at the N-terminus with a 3xFLAG tag (in the pCMV vector) which allows us to elute immunoprecipitated (IP) proteins by competition with FLAG peptide.

The protocol below describes the coimmunoprecipitation of seipin with lipin 1, from transiently transfected HEK293 cells (Sim, Dennis, et al., 2013; Sim, Talukder, et al., 2013). The ease of transfection and high expression levels makes these cells well-suited to initial examination of candidate seipin-interacting partners. In the example blot shown in Fig. 9.1A both short and long forms of seipin have been coexpressed and precipitated with either lipin 1α or lipin 1β. Singly transfected controls are included to verify that no nonspecific pull-down occurs; specifically in this case that overexpressed lipin 1 does not immunoprecipitate in the absence of seipin. We find that the ER transmembrane protein calnexin serves as a useful

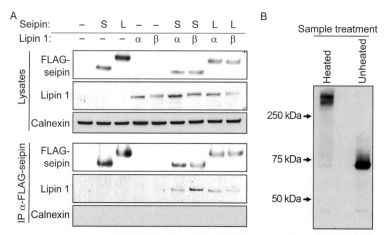

Figure 9.1 Coimmunoprecipitation of seipin and lipin 1. (A) HEK293 cells were transfected with either empty vector (−), or constructs encoding FLAG-tagged short (S) or long (L) forms of human seipin in the absence or presence of lipin 1α (α) or lipin 1β (β). Samples were extracted and FLAG-seipin immunoprecipitated (IP). IP and cell lysate samples were analyzed as described in the text. Calnexin expression is shown as a loading control for cell lysates, while its absence from the IP samples indicates that immunoprecipitated seipin does not nonspecifically coprecipitate with other ER membrane proteins. (B) A Myc-tagged long form of human seipin was expressed in HEK293 cells and identical lysate samples heated for 5 min at 95 °C or left unheated as indicated prior to separation by SDS-PAGE and analysis by Western blotting using anti-myc antibodies.

loading control for lysate samples. In addition, blotting the IP samples for calnexin allows us to verify that these samples are not contaminated with other integral ER membrane proteins.

We have also used this method to immunoprecipitate seipin with binding partners from differentiating 3T3-L1 preadipocytes stably expressing FLAG-seipin, including endogenous lipin 1 (Sim, Dennis, et al., 2013; Sim, Talukder, et al., 2013). When using 3T3-L1 cells, lysis was performed in a reduced volume of 200 μl per well of a 6-well dish to increase protein concentration. Seipin-expressing 3T3-L1 cells were generated by retroviral infection (Sim, Dennis, et al., 2013; Sim, Talukder, et al., 2013). Although not described here, anti-FLAG immunoprecipitates from FLAG-seipin-expressing 3T3-L1 adipocytes have also been subjected to proteomic analysis yielding further, subsequently validated, candidate binding partners (data not shown). Overall, our data suggest that this protocol offers a robust method for the immune-isolation of seipin and its associated proteins.

As has been reported previously, seipin forms high molecular weight complexes if heated prior to separation on SDS-PAGE (Fei, Li, et al.,

2011; Fei, Shui, et al., 2011; Sim, Dennis, et al., 2013; Sim, Talukder, et al., 2013). Hence it is critical to omit the typical boiling of either lysates or IP samples when analyzing seipin by Western blotting. An example blot obtained where an identical sample of lysate from cells transfected with a myc-tagged wild-type seipin construct has been either heated or not prior to separation on SDS-PAGE is shown in Fig. 9.1B.

2.1.1 Procedure

HEK293 cells were transfected in 6-well dishes by adding 200 μl of OptiMEM media (Life Technologies, Paisley, UK) containing 1 μg DNA and 3 μl of Fugene6 reagent (Promega, Southampton, UK) per well. Cells were lysed 48 h after transfection in 400 μl of n-octyl-β-D-glucopyranoside (ODG) lysis buffer comprising 50 mM ODG, 50 mM Tris, pH 6.8, 150 mM NaCl, 1 mM EDTA supplemented with protease inhibitors (Complete EDTA-free, Roche Applied Science, Burgess Hill, UK) and phosphatase inhibitors (PhosSTOP, Roche). All subsequent steps were performed on ice. Cells were sonicated using a Diagenode Bioruptor at medium intensity for two cycles of 30 s, incubated on ice for 20 min, centrifuged at $16,000 \times g$ for 10 min at 4 °C and supernatants retained.

Thirty microliters of dispersed anti-FLAG-agarose beads (Sigma-Aldrich, Gillingham, UK) per immunoprecipitation was aliquoted into 1.5 ml Eppendorfs and equilibrated with two washes of 500 μl of ODG lysis buffer, each followed by centrifugation at $8200 \times g$ for 30 s at 4 °C. Excess lysis buffer was then removed and lysate containing 1 mg of protein was added to the FLAG-agarose beads with ODG lysis buffer to make a final volume of 800 μl. Tubes were then gently rotated for 2 h at 4 °C. Following centrifugation ($8200 \times g$ for 30 s at 4 °C) supernatants were removed and beads washed three times with 500 μl of ODG lysis buffer. All excess lysis buffer was removed after the final wash.

FLAG-seipin was eluted from beads by addition of 100 μl of 200 ng/μl 3xFLAG peptide (Sigma-Aldrich) in TBS (50 mM Tris–HCl, pH 7.4, 150 mM NaCl) and incubated with gentle rotation for 30 min at 4 °C. Beads were then centrifuged and supernatant transferred to a new tube. Protein lysates and IP samples were stored at -80 °C if not used immediately.

An appropriate volume of lithium dodecyl sulfate (LDS) sample buffer (Invitrogen, Paisley, UK, NP-0007) was added to 20 μg of lysate and 20 μl of IP samples. As noted above, lysate and IP samples to be probed for seipin were not heated prior to loading to prevent aggregation. Samples to be Western blotted for all other proteins were boiled at 95 °C for 5 min.

Samples were separated by SDS-PAGE and transferred onto nitrocellulose membranes and Western blotted using antibodies to FLAG (Sigma-Aldrich, F1804) lipin 1 (from Dr Siniossoglou, CIMR, Cambridge, UK) or calnexin (Abcam, Cambridge, UK, ab13504) followed by horse radish peroxidase (HRP)-linked secondary antibodies. Proteins were visualized by ECL (GE Healthcare Life Sciences, Little Chalfont, UK).

2.2. Bimolecular fluorescence studies of seipin-interacting proteins

Bimolecular fluorescence complementation (BiFC) allows for visualization of protein interactions within intact cells. It avoids the nonspecific gain of binding or disruption of weak interactions that might occur during cell lysis and the mixing of different cellular compartments, as may occur when assessing interaction by coimmunoprecipitation. BiFC is based on the reconstitution of a fluorescent protein reporter, in this case yellow fluorescent protein (YFP), when two candidate interacting proteins come into contact. To achieve this, constructs encoding proteins of interest are each fused to different halves of the fluorescent protein. Individually, the fusion proteins do not fluoresce. However when the two candidate proteins interact, the YFP fusions become associated and correctly fold together to give a fluorescent signal. The technique provides a valuable means of confirming interactions identified through coimmunoprecipitation but also allows one to determine the subcellular location of the interaction in either live or fixed cells.

There are strengths and limitations to BiFC compared to other commonly used interaction-dependent fluorescent techniques, notably, fluorescence resonance energy transfer (FRET). The formation of the intact fluorophore in BiFC is irreversible. This makes it valuable for examining weak or transient interactions between proteins and increases the possibility of detecting an interaction which may occur only under certain conditions. However, it cannot be used to study the loss of an interaction in response to a given stimulus. In addition, the formation of the fluorophore is not rapid and so BiFC is not applicable for the study of acutely inducible interactions in real time. In such cases, FRET would clearly offer a more useful means of analysis.

BiFC is most frequently reported in easy-to-transfect cells, which provide high levels of expression such that results may be achieved rapidly. However, we wished to examine the interaction between seipin and lipin 1 in a more relevant model system and the protocol below describes a method optimized specifically for differentiating 3T3-L1 preadipocytes. Example images from the experiment described are shown in Fig. 9.2.

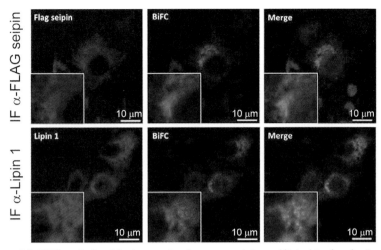

Figure 9.2 BiFC analysis of the interaction between seipin and lipin 1 in developing adipocytes. Differentiating 3T3-L1 adipocytes were cotransfected with the Seipin-C-Yn construct with a FLAG tag incorporated at the N-terminus and the Lipin-C-Yc construct. Seipin-C-Yn was detected using antibodies to the FLAG (upper panels), Lipin-C-Yc was detected using an anti-lipin 1 antibody (lower panels). Centre panels show the reconstituted YFP fluorescence, reporting the interaction of seipin and lipin in each case. (See the color plate.)

2.2.1 Procedure

Plasmid constructs were generated to express seipin fused to the N-terminus of YFP (amino acids 1–155, Yn) and lipin 1 fused to the C-terminus of YFP (amino acids 155–238, Yc). A flexible linker region was inserted between the protein of interest and the fluorophore sequence to allow for sufficient freedom of movement of the two halves of YFP and facilitate formation of the BiFC fluorophore (Kerppola, 2006). It may be necessary to generate and test all combinations of N- and C-terminal fusions of the proteins of interest to identify one that appropriately juxtaposes the half-YFP proteins. In this instance fusion of Yn to the C-terminus of seipin (Seipin-C-Yn) and Yc to the C-terminus of lipin 1 (Lipin-C-Yc) generated a YFP signal. Other fusions serve as useful controls, for example, in this case the lack of signal when a fusion of Yc to the N-terminus of lipin 1 (Lipin-N-Yc) further implies that the fluorescence observed with Sepin-C-Yn and Lipin-C-Yc reports a specific interaction.

We confirmed that the subcellular localization of the proteins is not affected by the presence of the fluorophore fusions using immunofluorescence. Seipin was detected via a FLAG tag incorporated at the N-terminus of the Seipin-C-Yn construct (Fig. 9.2 upper panels). Lipin-C-Yc was detected using an anti-lipin 1 antibody (Fig. 9.2 lower panels).

3T3-L1 preadipocytes were seeded onto glass coverslips in 12-well cell culture dishes pre-coated with 0.1% poly-L-lysine. Preadipocytes were cultured in DMEM containing 10% heat-inactivated (65 °C, 30 min) newborn calf serum supplemented with 100 units/ml Penicillin, 0.1 mg/ml streptomycin, 7 mM L-glutamine, nonessential amino acids and 0.85 mM sodium pyruvate. Two days after reaching confluence cells were induced to differentiate in DMEM containing 10% fetal bovine serum (FBS) supplemented with 100 units/ml penicillin, 0.1 mg/ml streptomycin, 7 mM L-glutamine, MEM nonessential amino acids and 0.85 mM sodium pyruvate (DMEM/FBS) supplemented with 0.5 mM 3 Isobutyl-1-methyl-xanthine, 1 μM dexamethasone and 1 μM insulin. All tissue culture reagents were from Sigma-Aldrich.

On day 3 post-induction differentiating adipocytes were fed with DMEM/FBS containing 1 μM insulin, then transiently transfected with BiFC constructs using Lipofectamine LTX (Invitrogen). For each well of a 12-well dish a total of 1 μg DNA (0.5 μg each of the Yn and Yc fusion constructs) was mixed with 200 μl OptiMEM media and 1 μl Lipofectamine LTX PLUS reagent and incubated for 5 min before addition of 2 μl of Lipofectamine LTX reagent. Reagents were incubated at room temperature for 30 min, then added to the cells. 3T3-L1 adipocytes are generally difficult to transfect and this method typically transfects around 10–15% of cells. Although low, this transfection efficiency is sufficient for fluorescence analysis of multiple individual cells in each experiment. As controls, cells were also transfected with either Seipin-C-Yn and the Yc fragment or Lipin-C-Yc and the Yn fragment to verify that these do not give any reconstituted YFP fluorescence.

For the cell type and fusion proteins used, it is necessary to optimize incubation times for expression, protein interaction and subsequent assembly of the two half-YFP polypeptides to generate a fluorescently active protein (Kerppola, 2008). To achieve optimal fluorescence, a series of temperature shifts is required to encourage fluorophore maturation. In differentiating 3T3-L1 cells following transfection, cells were incubated at 37 °C for 24 h, at 32 °C for a further 20 h then at 30 °C for 4 h. While lowering temperatures can improve fluorophore maturation, it may affect cell viability and protein interactions or result in nonspecific interactions. Hence, exposures to lower temperatures prior to observation should be limited.

Cells were rinsed once in cold phosphate buffered saline (PBS) and fixed in 4% formaldehyde for 15 min. Following three brief washes in PBS, cells may be mounted immediately or probed with antibodies. For detection with anti-FLAG, anti-lipin 1 or anti-calnexin antibodies, cells were permeabilized by incubation for 5 min in PBS containing 0.5% saponin then washed for 3 × 5 min with PBS. Cells were blocked in PBS 0.1% Tween 20 (PBST) containing 1% BSA then incubated with primary antibodies to FLAG (1:500), lipin 1 (1:500) or calnexin (1:500) diluted in blocking buffer. Cells were washed 3 × 5 min in PBST containing 0.1% BSA, then incubated for 30 min with appropriate AlexaFluor highly cross-adsorbed secondary antibodies (Molecular Probes, Paisley, UK) at 1:1000 in blocking buffer. The secondary antibodies were chosen with excitation and emission spectra distinct from those of the BiFC fluorophore.

2.3. Analysis of seipin by atomic force microscopy

Analysis of the yeast ortholog of seipin, Fld1p, using velocity sedimentation, gel filtration and electron microscopy has indicated that this protein forms a homo-oligomer of nine subunits in a toroidal arrangement (Binns et al., 2010). In the same study a mutant form of Fld1p carrying the single amino acid substitution, G228P, was analyzed. G228P in the yeast Fld1p mimics the pathogenic A212P mutation in the human protein that causes CGL. This experiment revealed that G228P Fld1p does not form the 9-subunit homo-oligomers, suggested that this mutation, and by inference A212P-seipin, may prevent the assembly of seipin into this oligomeric structure (Binns et al., 2010).

To investigate the architecture of human seipin, we have employed atomic force microscopy (AFM) (Sim, Dennis, et al., 2013; Sim, Talukder, et al., 2013). AFM is a scanning probe technique that provides nanometer-scale resolution without the requirement for extensive sample preparation. The atomic force microscope works by scanning a fine pyramidal tip back and forth over a surface in a raster pattern. The tip is situated at the end of a micro-engineered cantilever (Fig. 9.3A). As the tip is deflected by interactions with objects on the sample surface, the cantilever also deflects, and the magnitude of the deflection is registered by the change in direction of a laser beam that is reflected off the end of the cantilever and detected by an array of photomultipliers. Thus, a topological map of the surface can be constructed. One of the great advantages of the atomic force microscope, particularly with respect to the imaging of biological specimens, is that it can work in fluid, so that experiments can be performed under near-physiological conditions (although in the experiments described

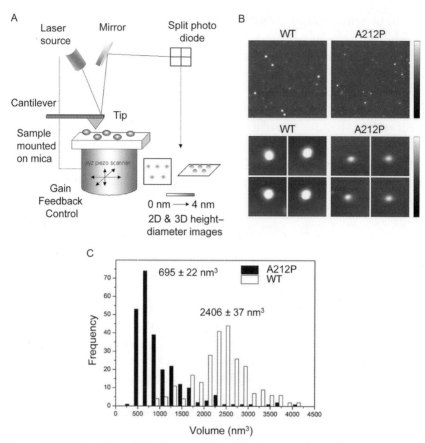

Figure 9.3 AFM analysis of wild-type and the A212P mutant form of seipin. (A) Schematic illustration of AFM analysis. (B) tsA 201 cells were transfected with the long form of wild-type seipin with an N-terminal triple-FLAG tag and a C-terminal Myc tag, or the A212P pathogenic mutant form of this protein that causes lipodystrophy. Representative AFM images of isolated proteins are shown in the upper panels. Scale bar, 200 nm; color-height scale, 0–5 nm. Lower panels show galleries of zoomed images of the wild-type and A212P mutant form of seipin. Scale bar, 20 nm; color-height scale, 0–5 nm. (C) Frequency distributions of molecular volumes of wild-type (white bars) and A212P (black bars) seipin. The means of the distributions ($\pm$SEM) are indicated. (See the color plate.)

here protein samples were imaged in air). The atomic force microscope may be operated with the tip either touching the sample (contact mode) or oscillating rapidly above the sample (tapping mode). This latter mode, which works both in air and under fluid, produces less lateral force on the substrate, a useful feature when investigating relatively delicate and loosely attached biological specimens. In the example below, tapping-mode AFM is used to image complexes of human wild-type seipin and A212-seipin and assess whether these proteins form homo-oligomers.

2.3.1 Procedure

tsA 201 cells (a subclone of human embryonic kidney-293 cells stably expressing the SV40 large T-antigen) were grown in Dulbecco's modified Eagle's medium supplemented with 10% (v/v) fetal calf serum, 100 units/ml penicillin, and 100 μg/ml streptomycin, in an atmosphere of 5% CO_2/air. A total of 250 μg of DNA was used to transiently transfect cells in 5×162 cm^2 culture flasks using the CalPhosTM mammalian transfection kit (Clontech, Mountain View, CA, USA), according to the manufacturer's instructions.

Cells were transfected with N-terminally FLAG-tagged wild-type human seipin or seipin in which the pathogenic A212P single amino acid substitution (A212P-seipin) had been introduced by site-directed mutagenesis. Transfected cells were solubilized in 1% Triton X-100, 25 mM Tris, pH 7.5, 150 mM NaCl, and 10 mM EDTA, containing protease inhibitors, for 1 h, before centrifugation at $61,740 \times g$ to remove insoluble material. Solubilized protein was incubated with anti-FLAG-agarose beads for 3 h. Proteins were eluted with triple-FLAG peptide (0.15 mg/ml). All purification steps were carried out at 4 °C.

Isolated proteins were diluted with an appropriate volume of the above solubilization buffer (without protease inhibitors) to achieve a suitable spread of particles upon AFM imaging. (Typically, samples eluted from the immunobeads are diluted 300–500-fold.) Forty-five microliters of the sample was allowed to adsorb onto freshly cleaved, poly-L-lysine-coated mica disks. After a 5-min incubation, the sample was washed with Biotechnology Performance Certified-grade water (Sigma) and dried in a stream of nitrogen gas. Imaging was performed with a Veeco Digital Instruments Multimode atomic force microscope controlled by a Nanoscope IIIa controller. Samples were imaged in air, using tapping mode. The silicon cantilevers used had a drive frequency ~300 kHz and a specified spring constant of 40 N/m (Olympus, Southend on Sea, UK). The applied imaging force was kept as low as possible ($A_S/A_0 \sim 0.85$). Figure 9.3B shows low-magnification images of the seipin constructs (upper panels). As can be seen, wild-type seipin appeared as large particles of uniform size. By contrast, the A212P mutant appeared noticeably smaller and less homogenous. The galleries of zoomed images of particles shown in the lower panels confirm the difference in size between wild-type and A212P-seipin.

The molecular volumes of wild-type and A212P-seipin particles were determined from particle dimensions based on AFM images. After adsorption of the proteins onto the mica support, the particles adopted the shape of a spherical cap. As has been described previously (Schneider, Larmer, Henderson, & Oberleithner, 1998) the heights and radii were measured from multiple cross-sections of the same particle, and the molecular volume was calculated using Eq. (9.1)

$$V_m = (\pi h/6)(3r^2 + h^2) \quad (9.1)$$

where h is the particle height and r is the radius. Molecular volume based on molecular mass was calculated using the equation

$$V_C = (M_0/N_0)(V_1 + dV_2) \quad (9.2)$$

where M_0 is the molecular mass, N_0 is Avogadro's number, V_1 and V_2 are the partial specific volumes of particle (0.74 cm^3/g) and water (1 cm^3/g), respectively, and d is the extent of protein hydration (taken as 0.4 g water/g protein). Note that it has been shown previously (Schneider et al., 1998) that the molecular volumes of proteins measured by imaging in air are similar to the values obtained by imaging under fluid; hence, the process of drying does not significantly affect the measured molecular volume. It has also been shown by us (Neaves, Huppert, Henderson, & Edwardson, 2009) and by others (Schneider et al., 1998) that there is a close correspondence between the measured and predicted molecular volumes for various proteins over a wide range of molecular masses; hence, molecular volume is measured reasonably accurately by AFM imaging.

Data were separated into appropriate bin widths and histograms plotted. Gaussian curves were fitted and the mean value of the peaks calculated. All errors quoted are standard errors of the mean (SEM).

Figure 9.3C shows that representative frequency distributions of volumes of the particles, calculated using Eq. (9.1), had peaks at 2406 ± 37 (SEM) nm^3 ($n=250$) for wild-type seipin (white bars) and 695 ± 22 nm^3 ($n=250$) for the A212P mutant (black bars).

When isolated from transfected tsA 201 cells, FLAG-tagged wild-type A212P-seipin migrated on polyacrylamide gels at an apparent molecular mass of 80 kDa. According to Eq. (9.2), the expected molecular volume for a protein of this size is 152 nm^3. It has been shown previously that after solubilization in Triton X-100, the detergent used here, the apparent molecular mass of the seipin complex is increased by about 28% because of the presence of bound detergent; this would increase the expected monomer size to 194 nm^3. Dividing the observed molecular volume (2406 nm^3 for wild-type seipin) by 194 nm^3 gives a value of ~12, suggesting that wild-type seipin assembles as a dodecamer. According to this analysis, the A212P-seipin mutant complexes are on average about one-third of this size.

3. DISCUSSION

Proteomic analysis of the yeast seipin ortholog, fld1p, indicated that it may not associate with other proteins (Binns et al., 2010). This may be the case in simple organisms; however, our own proteomic analyses using human seipin suggest that multiple potential binding partners exist. Our preliminary data also indicate that different proteins may associate with seipin at different stages of adipogenesis. We speculate that developmental stage- and cell type-dependent differences in the complement of proteins associated with seipin may explain some of the inconsistencies and apparent paradoxes observed in the effects of seipin disruption or overexpression in different cell types and model organisms.

The methods described above have proven valuable in our initial studies of seipin. Evidently, future analyses would benefit from the use of other fluorescent techniques such as FRET to examine acute changes in the interaction between seipin and potential binding proteins under varying conditions. In addition, viral transduction of adipocytes would provide more efficient expression for such studies, while inducible expression and knockdown systems will prove particularly useful in examining the potentially different effects of altering seipin expression in differentiating preadipocytes at varying stages of their maturation into adipocytes. In a similar way, more sophisticated conditional, *in vivo* models of seipin disruption are likely to give important mechanistic insights regarding the physiological relevance of such molecular studies.

Finally, we are currently extending our use of AFM imaging for the analysis of seipin and the assessment of potential alterations in the subunit structure of pathogenic or artificial mutants of seipin (Sim, Dennis, et al., 2013; Sim, Talukder, et al., 2013). AFM also allows us to visualize the architecture of seipin oligomers in complex with candidate binding proteins permitting the definition of the stoichiometry of such interactions.

Overall, we hypothesize that a major role of seipin in preadipocytes, adipocytes, and other cell types may be to associate with and organize other proteins. This behavior may operate alongside, or have evolved from, a structural role, altering membrane curvature, budding or fluidity. We expect that future studies will uncover multiple seipin-interacting partners, some of which may further explain the importance of seipin in adipocyte development, and others that may report the roles of seipin in mature adipocytes and other cell types.

REFERENCES

Binns, D., Lee, S., Hilton, C. L., Jiang, Q. X., & Goodman, J. M. (2010). Seipin is a discrete homooligomer. *Biochemistry*, *49*(50), 10747–10755.

Chen, W., Chang, B., Saha, P., Hartig, S. M., Li, L., Reddy, V. T., et al. (2012). Berardinelli-seip congenital lipodystrophy 2/seipin is a cell-autonomous regulator of lipolysis essential for adipocyte differentiation. *Molecular and Cellular Biology*, *32*(6), 1099–1111.

Chen, W., Yechoor, V. K., Chang, B. H., Li, M. V., March, K. L., & Chan, L. (2009). The human lipodystrophy gene product Berardinelli-Seip congenital lipodystrophy 2/seipin plays a key role in adipocyte differentiation. *Endocrinology*, *150*(10), 4552–4561.

Cui, X., Wang, Y., Meng, L., Fei, W., Deng, J., Xu, G., et al. (2012). Overexpression of a short human seipin/BSCL2 isoform in mouse adipose tissue results in mild lipodystrophy. *American Journal of Physiology Endocrinology and Metabolism*, *302*(6), E705–E713.

Fei, W., Li, H., Shui, G., Kapterian, T. S., Bielby, C., Du, X., et al. (2011). Molecular characterization of seipin and its mutants: Implications for seipin in triacylglycerol synthesis. *Journal of Lipid Research*, *52*(12), 2136–2147.

Fei, W., Shui, G., Gaeta, B., Du, X., Kuerschner, L., Li, P., et al. (2008). Fld1p, a functional homologue of human seipin, regulates the size of lipid droplets in yeast. *The Journal of Cell Biology*, *180*(3), 473–482.

Fei, W., Shui, G., Zhang, Y., Krahmer, N., Ferguson, C., Kapterian, T. S., et al. (2011). A role for phosphatidic Acid in the formation of "supersized" lipid droplets. *PLoS Genetics*, *7*(7), e1002201.

Garfield, A. S., Chan, W. S., Dennis, R. J., Ito, D., Heisler, L. K., & Rochford, J. J. (2012). Neuroanatomical characterisation of the expression of the lipodystrophy and motor-neuropathy gene Bscl2 in adult mouse brain. *PLoS One*, *7*(9), e45790.

Ito, D., Fujisawa, T., Iida, H., & Suzuki, N. (2008). Characterization of seipin/BSCL2, a protein associated with spastic paraplegia 17. *Neurobiology of Disease*, *31*(2), 266–277.

Kerppola, T. K. (2006). Design and implementation of bimolecular fluorescence complementation (BiFC) assays for the visualization of protein interactions in living cells. *Nature Protocols*, *1*(3), 1278–1286.

Kerppola, T. K. (2008). Bimolecular fluorescence complementation (BiFC) analysis as a probe of protein interactions in living cells. *Annual Review of Biophysics*, *37*, 465–487.

Lundin, C., Nordstrom, R., Wagner, K., Windpassinger, C., Andersson, H., von Heijne, G., et al. (2006). Membrane topology of the human seipin protein. *FEBS Letters*, *580*(9), 2281–2284.

Magre, J., Delepine, M., Khallouf, E., Gedde-Dahl, T., Jr., Van Maldergem, L., Sobel, E., et al. (2001). Identification of the gene altered in Berardinelli-Seip congenital lipodystrophy on chromosome 11q13. *Nature Genetics*, *28*(4), 365–370.

Neaves, K. J., Huppert, J. L., Henderson, R. M., & Edwardson, J. M. (2009). Direct visualization of G-quadruplexes in DNA using atomic force microscopy. *Nucleic Acids Research*, *37*, 6269–6275.

Payne, V. A., Grimsey, N., Tuthill, A., Virtue, S., Gray, S. L., Dalla Nora, E., et al. (2008). The human lipodystrophy gene BSCL2/seipin may be essential for normal adipocyte differentiation. *Diabetes*, *57*(8), 2055–2060.

Reue, K. (2009). The lipin family: Mutations and metabolism. *Current Opinion in Lipidology*, *20*(3), 165–170.

Rochford, J. J. (2010). Molecular mechanisms controlling human adipose tissue development: Insights from monogenic lipodystrophies. *Expert Reviews in Molecular Medicine*, *12*, e24.

Schneider, S. W., Larmer, J., Henderson, R. M., & Oberleithner, H. (1998). Molecular weights of individual proteins correlate with molecular volumes measured by atomic force microscopy. *Pflügers Archiv*, *435*(3), 362–367.

Sim, M. F. M., Dennis, R. J., Aubry, E. M., Ramanathan, N., Sembongi, H., Saudek, V., et al. (2013). The human lipodystrophy protein seipin is an ER membrane adaptor for the adipogenic PA phosphatase lipin 1. *Molecular Metabolism*, *2*(1), 38–46.

Sim, M. F. M., Talukder, M. M. U., Dennis, R. J., O'Rahilly, S., Edwardson, J. M., & Rochford, J. J. (2013). Analysis of naturally occurring mutations in the human lipodystrophy protein seipin reveals multiple potential pathogenic mechanisms. *Diabetologia*, *56*(11), 2498–2506.

Szymanski, K. M., Binns, D., Bartz, R., Grishin, N. V., Li, W. P., Agarwal, A. K., et al. (2007). The lipodystrophy protein seipin is found at endoplasmic reticulum lipid droplet junctions and is important for droplet morphology. *Proceedings of the National Academy of Sciences of the United States of America*, *104*(52), 20890–20895.

Wolinski, H., Kolb, D., Hermann, S., Koning, R. I., & Kohlwein, S. D. (2011). A role for seipin in lipid droplet dynamics and inheritance in yeast. *Journal of Cell Science*, *124*(Pt. 22), 3894–3904.

CHAPTER TEN

Differentiation of Human Pluripotent Stem Cells into Highly Functional Classical Brown Adipocytes

Miwako Nishio, Kumiko Saeki[1]

Department of Disease Control, Research Institute, National Center for Global Health and Medicine, Tokyo, Japan
[1]Corresponding author: e-mail address: saeki@ri.ncgm.go.jp

Contents

1. Introduction	178
2. Experimental Components and Considerations	179
2.1 Materials	179
2.2 Methods	184
3. Notes with Troubleshooting	195
References	197

Abstract

We describe a detailed method for directed differentiation of human pluripotent stem cells, including human embryonic stem cells (hESCs) and human induced pluripotent stem cells (hiPSCs), into functional classical brown adipocytes (BAs) under serum-free and feeder-free conditions. It is a two-tiered culture system, based on very simple techniques, a floating culture and a subsequent adherent culture. It does not require gene transfer. The entire process can be carried out in about 10 days. The key point is the usage of our special hematopoietic cytokine cocktail. Almost all the differentiated cells express uncoupling protein 1, a BA-selective marker, as determined by immunostaining. The differentiated cells show characteristics of classical BA as assessed by morphology and gene/protein expression. Moreover, the expression of myoblast marker genes is transiently induced during the floating culture step. hESC/hiPSC-derived BAs show significantly higher oxygen consumption rates (OCRs) than white adipocytes generated from human mesenchymal stem cell. They also show responsiveness to adrenergic stimuli, with about twofold upregulation in OCR by β-adrenergic receptor (β-AR) agonist treatments. hESC/hiPSC-derived BAs exert *in vivo* calorigenic activities in response to β-AR agonist treatments as assessed by thermography. Finally, lipid and glucose metabolisms are significantly improved in hESC/hiPSC-derived BA-transplanted mice. Our system provides a highly feasible way to produce functional classical BA bearing metabolism-improving capacities from hESC/hiPSC under a feeder-free and serum-free condition without gene transfer.

1. INTRODUCTION

Brown adipose tissue (BAT) is a calorigenic fat tissue involved in nonshivering thermogenesis. It is also involved in preventing aging-associated obesity. BAT is located in specific parts of the body, for example, interscapular, cervical, and axillary regions in mice and supraclavicular and paravertebral areas in humans. It contains dense capillary networks. Brown adipocytes (BAs) show characteristic morphologies: abundant large mitochondria with ladder-shaped dense cristae and multilocular lipid droplets of similar sizes, some of which are located in close vicinity to mitochondria.

Uncoupling protein 1 is essential for heat production: it reduces the proton gradient across inner mitochondrial membrane, as created by the electron transport chain, and thus, converts oxidization-derived chemical energies into thermal energies at the expense of ATP synthesis. Sympathetic innervation plays an indispensable role in the activation and maintenance of BAT. Although the majority of classical BAT disappears after the neonatal period in large size mammals, including humans, small portions of it remain and function through adulthood. ^{18}Fluorodeoxyglucose-positron emission tomography in combination with computed tomography (^{18}F-FDG-PET/CT) along with histological and gene expression studies demonstrated the presence of functional BAT in adult humans. Subsequent clinical studies have shown an inverse correlation between BAT activities and metabolic disorders in humans, suggesting the involvement of BAT in metabolic improvement. However, its direct evidence remains undemonstrated in human cases although a recent study has shown the proof in murine cases via transplantation experiments (Stanford et al., 2013). In addition, human BAT reportedly consumes as little as 20 calories/day, suggesting that there might be still unidentified mechanisms for BAT-mediated metabolic improvement (Muzik et al., 2013). Furthermore, the whole picture of human BAT development remains incompletely understood due to lack of appropriate research tools. To provide answers for those questions and further promote human BAT research, it is of great use to establish a method for a directed differentiation of human pluripotent stem cells into functional classical BA bearing responsiveness to β-adrenergic receptor stimuli.

During our researches on hematopoietic differentiation of human embryonic stem cells (hESCs) and human induced pluripotent stem cells (hiPSCs), we found that hematopoietic areas were exclusively surrounded by multilocular lipid droplet-containing adipocytes. After a process of trial and error, we finally established a method for a high-efficiency differentiation of hESC/hiPSC into functional classical BA via myoblastic differentiation using a special hematopoietic cytokine cocktail (Nishio et al., 2012). Our system provides an excellent tool for an advanced understanding of human BAT.

2. EXPERIMENTAL COMPONENTS AND CONSIDERATIONS

2.1. Materials

2.1.1 A culture medium for mouse embryonic fibroblasts

1. Low-glucose Dulbecco's modified Eagle's medium (DMEM) with L-glutamine (Cat 041-29775, WAKO Pure Chemical Industries, Osaka, Japan); store at 4 °C.
2. 100× Penicillin/streptomycin solution (Cat P0781, Sigma–Aldrich Corporation, St. Louis, MO, USA); store at −20 °C.
3. Fetal bovine serum (FBS) (PAA Laboratories GmbH, Pasching, Austria); store at −20 °C.

2.1.2 Preparation of MEFs

1. A pregnant female mouse with pups at embryonic day 12.5 (ICR strain, for example)
2. Absorbent cotton
3. Sevofrane (WAKO Pure Chemical Industries); store at room temperature
4. A stereo microscope (Model LG-PS2, Olympus Optical Co. Ltd., Tokyo, Japan)
5. Sterilized phosphate buffer saline (PBS); store at 4 °C
6. Sterilized ordinary tweezers and scissors
7. Sterilized ophthalmic tweezers and scissors
8. 60-mm Culture dishes
9. 100-mm Culture dishes
10. 50-ml Conical tube

11. 2.5-ml Syringe
12. 18G Needle
13. 0.25% Trypsin–EDTA (Sigma–Aldrich Corp.); store at −20 °C
14. Bambanker™ (Lymphotec Inc., Tokyo, Japan); store at −20 °C
15. 1.8-ml Innercapped cryotubes (Nunc A/S, Roskilde, Denmark)
16. An X-ray irradiation Device (Model MBR-1520R-3, Hitachi Medical Corporation, Tokyo, Japan) (see Note 1)

2.1.3 hESC and hiPSC

A hESC line, khES-3, was established by Dr. Suemori at Kyoto University in Japan (Suemori et al., 2006). hiPSCs were established from commercially available human umbilical vein endothelial cells, which can be purchased from Lonza Group Ltd. (Basel, Switzerland), by using a Sendai vector system, CytoTune™-iPS ver.1.0 (DNAVEC Corp., Ibaraki, Japan).

2.1.4 Maintenance of immature hESC/hiPSC on MEF layers

1. DMEM: nutrient mixture F-12 Ham 1:1 (WAKO Pure Chemical Industries, Cat 048-29785); store at 4 °C.
2. KNOCKOUT™ Serum Replacement (KSR™) (Life Technologies, Inc., Carlsbad, CA, USA, Cat 10828-028); store at −20 °C.
3. Nonessential Amino Acid (NEAA) Solution (100×) (Sigma–Aldrich Corp., M7145); store at 4 °C.
4. Gelatin type A (Sigma–Aldrich Corp., G2625); store at room temperature.
5. 2-Mercaptoethanol (Sigma–Aldrich Corp., M7522); store at room temperature.
6. Fibroblast growth factor (FGF)-2 (Pepro Tech Inc., Rocky Hill, NJ, USA); store at −20 °C.

2.1.5 Dissociation liquid for hESC/hiPSC

1. KSR™ (Life Technologies, Inc., Cat 10828-028); store at −20 °C.
2. 100 mM $CaCl_2$; store at 4 °C.
3. Sterilized PBS; store at 4 °C.
4. 2.5% Trypsin (Life Technologies, Inc., Cat 15090-046); store at −20 °C.
5. Type IV collagenase (Life Technologies, Inc., Cat 17104-019); store at −20 °C.

2.1.6 Freezing of hESC/hiPSC

1. Freezing medium for hESC/hiPSC (ReproCELL Inc., Tokyo, Japan, Cat RCHEFM001); store at −80 °C.

2.1.7 BA differentiation media (I and II) and culture vessels

1. Iscove's modified Dulbecco's medium (IMDM) (I3390, Sigma–Aldrich Corp.); store at 4 °C.
2. Ham's F12 medium (087-08335, WAKO Pure Chemical Industries); store at 4 °C.
3. Bovine serum albumin (A802, Sigma–Aldrich Corp.); store at 4 °C.
4. 1:100 Synthetic lipids (Gibco # 11905-031, Life Technologies, Inc.); store at 4 °C.
5. α-Monothioglycerol (207-09232, WAKO Pure Chemical Industries); store at 4 °C.
6. 1:100 Insulin–transferrin–selenium (ITS-A, Life Technologies, Inc.); store at 4 °C.
7. 2 mM Glutamax II (Gibco #35050-061, Life Technologies, Inc.); store at −20 °C.
8. Protein-free hybridoma mix (PFHMII, Gibco #12040-077, Life Technologies, Inc.); store at 4 °C.
9. Ascorbic acid-2-phosphate (Sigma–Aldrich Corp., A-8960); store at 4 °C.
10. Vascular endothelial growth factor (VEGFA) (Pepro Tech Inc.); store at −80 °C.
11. Bone morphogenetic protein 4 (BMP4) (R&D Systems Inc., Minneapolis, MN, USA); store at −80 °C.
12. Stem cell factor (SCF) (Pepro Tech Inc.); store at −80 °C.
13. FMS-related tyrosine kinase-3 ligand (Flt3-L) (Pepro Tech Inc.); store at −80 °C.
14. Interleukin 6 (IL6) (Pepro Tech Inc.); store at −80 °C.
15. BMP7 (Pepro Tech Inc.); store at −20 °C.
16. Insulin-like growth factor II (IGF-II) (Pepro Tech Inc.); store at −80 °C.
17. Six-well ultra-low attachment Hydro cell® (CellSeed Inc., Tokyo, Japan) or six-well 2-methacryloyloxyethyl phosphorylcholine (MPC)-coated low attachment plates (Nunc A/S, Roskilde, Denmark).
18. Porcine gelatin (G2500, Sigma–Aldrich Corp.); store at room temperature.

2.1.8 Oil red O staining
1. Isopropanol (WAKO Pure Chemical Industries)
2. Oil red O solution (Muto Pure Chemicals Co., Tokyo, Japan)
3. PBS

2.1.9 Oxygen consumption analyses
1. XF96 Extracellular Flux Analyzer (Seahorse Bioscience Inc., Billerica, MA, USA).
2. XF96 Cell Culture Microplate (Seahorse Bioscience Inc., Billerica, MA, USA); store at room temperature.
3. XF96 Extracellular Flux Assay Kit (Seahorse Bioscience Inc., Billerica, MA, USA); store at room temperature.
4. XF Calibrant, pH 7.4 (100840-000, Seahorse Bioscience Inc., Billerica, MA, USA); store at 4 °C.
5. Sodium bicarbonate-free IMDM (I7633, Sigma–Aldrich Corp.); store at 4 °C.
6. Sodium bicarbonate-free Ham's F12 (21700-075, Life Technologies, Inc.); store at 4 °C.
7. Bovine serum albumin (A802, Sigma–Aldrich Corp.); store at 4 °C.
8. 1:100 Synthetic lipids (Gibco # 11905-031, Life Technologies, Inc.); store at 4 °C.
9. α-Monothioglycerol (207-09232, WAKO Pure Chemical Industries); store at 4 °C.
10. 1:100 Insulin–transferrin–selenium (ITS-A, Life Technologies, Inc.); store at 4 °C.
11. 2 mM Glutamax II (Gibco #35050-061, Life Technologies, Inc.); store at −20 °C.
12. Protein-free hybridoma mix (PFHMII, Gibco #12040-077, Life Technologies, Inc.); store at 4 °C.
13. Ascorbic acid-2-phosphate (Sigma–Aldrich Corp., A-8960); store at 4 °C.
14. Vascular endothelial growth factor (VEGFA) (Pepro Tech Inc.); store at −80 °C.
15. SCF (Pepro Tech Inc.); store at −80 °C.
16. Flt3-L (Pepro Tech Inc.); store at −80 °C.
17. IL6 (Pepro Tech Inc.); store at −80 °C.
18. BMP7 (Pepro Tech Inc.); store at −20 °C.
19. IGF-II (Pepro Tech Inc.); store at −80 °C.

20. Porcine gelatin (G2500, Sigma–Aldrich Corp.); store at room temperature.
21. CL316,243 (C5976, Sigma–Aldrich Corp.).
22. Isoproterenol (12760, Sigma–Aldrich Corp.).

2.1.10 Calorigenic analyses
1. Mice, for example, 5-week-old male ICR mice
2. Hair remover cream, for example, Veet® Hair Removal Cream (Reckitt Benckiser Japan Ltd., Tokyo, Japan)
3. TrypLE™ Express (12604-013, Life Technologies, Inc.)
4. Isoproterenol
5. Pentobarbital sodium (Merck & Co., Inc., Whitehouse Station, NJ, USA)
6. Infrared camera with a high temperature resolution (less than 0.1 °C), for example, Thermography Thermo GEAR G100 (NEC Avio Infrared Technologies Co., Ltd, Tokyo, Japan)

2.1.11 Oral glucose tolerance tests
1. Mice, for example, 6–10-week-old male ICR mice
2. A cylinder-type retaining appliance ($\varphi 25 \times L95$ mm) (#KN-325-C-1) (Natsume Seisakusho Co., Ltd.)
3. Sondes ($\varphi 0.9 \times L50$ mm) (#KN-348) (Natsume Seisakusho Co., Ltd., Tokyo, Japan)
4. Glucose (041-00595, Wako Pure Chemical Industries, Ltd., Osaka, Japan)
5. Isoproterenol
6. Accutrend® Plus (F. Hoffmann-La Roche, Ltd., Basel, Switzerland)
7. Mouse insulin ELISA kit (Morinaga Institute of Biological Science, Inc., Yokohama, Japan)

2.1.12 Oral fat tolerance tests
1. Mice, for example, 6–10-week-old male ICR mice
2. A cylinder-type retaining appliance ($\varphi 25 \times 95$ mm) (#KN-325-C-1) (Natsume Seisakusho Co., Ltd.)
3. Olive oil, for example, purchased from Ajinomoto Co., Inc., Tokyo, Japan
4. Isoproterenol
5. Accutrend® Plus

2.2. Methods

Our feeder-free and serum-free differentiation method enables the production of high-purity (>90%) classical BA from hESC/hiPSC without gene transfer. The differentiation medium is supplemented with a special hematopoictic cytokine cocktail composed of SCF, Flt3-L, VEGF, and IL6. Depletion of any one of these deteriorates the quality of BA (Nishio et al., 2012).

As reported in the protocol for differentiation of hESCs into pancreatic beta cells (Osafune et al., 2008), efficiency of BA differentiation varies significantly depending on line of hESC/hiPSC. There seems to be a tendency that hESC/hiPSC lines with higher hematopoietic differentiation potentials show larger BA differentiation capacities; however, there are exceptions. Thus, we recommend selecting and testing several lines of hESC/hiPSC that bear high hematopoietic differentiation potentials if information about differentiation propensity is available. Moreover, BA differentiation efficiency may differ depending on the condition of immature hESC/hiPSC maintenance, for example, on-feeder maintenance versus feeder-free maintenance. Thus, determining the optical maintenance condition for each selected hESC/iPSC line is another key to successful BA differentiation. In this chapter, we describe a method to maintain immature hESC/iPSC on mouse embryonic fibroblasts (MEFs) layers because it is feasible and inexpensive. However, depending on lines of hESC/iPSC, a feeder-free maintenance is required for an effective BA differentiation because contamination of hESC/hiPSC by MEF, which transmits signals to maintain hESC/hiPSC in undifferentiated states, substantially hampers differentiations. Whether a line of hESC/hiPSC requires a feeder-free maintenance culture or not depends, at least in part, on the detachability of MEF from hESC/hiPSC during a dissociation liquid treatment. In the following section, we illustrate a simple but effective technique to remove MEF from immature hESC/hiPSC clots; however, this procedure may not work in some lines of hESC/hiPSC. We recommend trying feeder-free maintenance in case MEF-cocultured hESC/hiPSC does not provide a satisfactory outcome. Four or more successive passages under feeder-free conditions will minimize the contamination by MEF. There are several commercially available feeder-free hESC/hiPSC maintenance systems such as Essential 8™ Medium (Cat A14666SA, Life Technologies, Inc., Grand Island, NY, USA) and ReproFF2 (Cat RCHEMD006, ReproCELL Inc., Tokyo, Japan). At the same time, we have to emphasize that hESC/hiPSC maintained under feeder-free conditions might show resistance to sphere formation in some cases. A trial-and-error process may be required to determine the optimal condition for sphere formation by checking parameters such as cell density, composition of culture medium, culture plate-coating materials, and composition of dissociation liquid.

Regarding hiPSC, how they are established may influence the quality of BA differentiation. There are various systems to establish hiPSCs such as lentiviral vector-based, retroviral vector-based, Sendai virus vector-based, transposon-based, plasmid-based, episomal vector-based, and protein-based system, among which we recommend a Sendai virus vector system. Sendai virus vectors do not integrate into the host genome because it is an RNA virus vector. Moreover, the vectors can easily be ablated from the establishing hiPSC colonies via a simple dilution during passages.

There are several additional crucial technical points. We will make detailed descriptions on those points in the following sections.

2.2.1 Culture medium for mouse embryonic fibroblasts

Mix the following materials: Low-glucose DMEM with L-glutamine; 445 ml, 100× penicillin/streptomycin solution; 5 ml, heat-inactivated (56 °C, 30 min) FBS; 50 ml.

2.2.2 Preparation of MEFs

Administer an anesthetic to a 12.5 days pregnant female mouse by putting a 50-ml conical tube containing sevofrane-moistened absorbent cotton to her mouth. Euthanize the mouse by cervical dislocation or in compliance with your local animal protection guidelines or laws. Put the body in a sterilized tray on the clean bench and wash it with 70% ethanol. Take out the uteri through abdominal incision and put them into 5 ml of sterile PBS contained in a 60-mm culture dish. Cut the walls of uteri and transfer them into 5 ml of sterile PBS contained in a 60-mm culture dish. Remove the fetuses by cutting amniotic membranes and transfer them into 5 ml of sterile PBS contained in a 60-mm culture dish. Remove the heads and entire internal organs including lungs, heart, liver, spleen, guts, and kidneys using sterilized ophthalmic tweezers and scissors under a stereo microscope. Transfer the ablated fetuses into 5 ml of sterile PBS contained in a 60-mm culture dish. Wash the fetuses with sterile PBS three times. Remove the plunger from 2.5-ml syringes that were fixed with 18G needles. Transfer each fetus into the syringe, reinsert the plunger, push out the fetuses into a 50-ml conical tube containing 2 ml of medium (i.e., DMEM supplemented with 10% FBS).

This fragmented fetus suspension was further homogenized by expelling and reloading the medium two or three times through a 18G needle (see Note 2). Then, transfer the fetal tissue slurry into a 100-mm culture dish containing 8 ml medium and incubate it in a 100% humidified 5% CO_2 incubator at 37 °C. After reaching confluence, collect the cells by 0.25% trypsin–EDTA treatment and seed them on new culture dishes at the density of 7.5×10^5/100-mm dish. At confluence, collect the cells by trypsin–EDTA treatment; suspend the cells in Bambanker solution at the density of 1×10^6 cells/ml; put 1 ml aliquots into cryotubes and keep them at −80 °C (Passage 1 stock). Thaw the stocks when an expansion of MEFs is required. Subculture the cells at the density of 7.5×10^5/100 mm and make frozen stocks at Passage 3 ($1–3 \times 10^6$ cells/cryotube).

2.2.3 Culture medium for hESC/hiPSC
Mix the following materials: DMEM-F12; 500 ml, NEAA solution; 5 ml, 200 mM L-glutamine; 6.25 ml, KSR; 125 ml, 2-mercaptoethanol; 5 μl, FGF-2; 5 ng/ml (final concentration). This medium is usable for 2 weeks when stored at 4 °C.

2.2.4 Dissociation liquid for hESC/hiPSC
Mix the following materials: 2.5% trypsin; 10 ml, 10 mg/ml Type IV collagenase; 10 ml, KSR; 20 ml, 100 mM $CaCl_2$; 1 ml, 1 × PBS; 59 ml. Prepare aliquots and store at −20 °C. After thawing, the solution should be stored at 4 °C and used within 1 week.

2.2.5 Preparation of frozen stocks of hESC/hiPSC
Prepare a confluent culture of hESC/hiPSC on a 60-mm dish. After washing with PBS, incubate the cells with 1 ml of dissociation liquid at room temperature. Observing the cells under microscope, add 5 ml of hESC/hiPSC culture medium at the time point when more than half of hESC/hiPSC colonies are beginning to detach from their margins. Transfer the cells *en bloc* into a 15-ml conical tube. Avoid cell-dissociating procedures such as pipetting and shocking of dishes (see Note 3). Add 5 ml of hESC/hiPSC culture medium to the original dish, transfer the residual cells into the same conical tube. After centrifugation at $170 \times g$ for 5 min at 4 °C, suspend the cell pellets with 200 μl of freezing medium and transfer it into a cryotube. Soak the cryotube in liquid nitrogen as soon as possible, at least within 15 s (see Note 4). After soaking in liquid nitrogen for 1 min (see Note 5), set the cryotube into a liquid nitrogen tank.

2.2.6 Maintenance of hESC/hiPSC on MEF layers

2.2.6.1 Preparation of gelatin-coated dishes
Incubate a 60-mm culture dish with 1.5–2 ml of 0.1% gelatin solution for 10 min at room temperature; aspirate the gelatin solution and wash with sterile PBS.

2.2.6.2 Preparation of MEF layers
Thaw the frozen stock of MEFs (Passage 3) as immediately as possible with mild shaking in a 37 °C water bath. Transfer the thawed MEF suspension into 10 ml of MEF culture medium contained in a 15-ml conical tube. After centrifugation, suspend the MEFs in 5 ml of MEF culture medium and seed them onto a 60-mm gelatin-coated dish above prepared at the density of $1.8–2.4 \times 10^4$ cell/60-mm dish. After overnight culture in a 100% humidified 5% CO_2 incubator at 37 °C, irradiate the cells at 46 Gy. These MEF-layered dishes should be used within 4 days.

2.2.6.3 Thawing of hESC/hiPSC
Add 1 ml of 37 °C-prewarmed hESC/hiPSC culture medium to a frozen hESC/hiPSC stock tube. With pipetting, thaw hESC/hiPSC as immediately as possible. Transfer the thawed cells into a 15-ml conical tube containing 10 ml of 37 °C-prewarmed hESC/hiPSC culture medium. After centrifuge at $170 \times g$ for 5 min, aspirate the supernatant, resuspend the cells with 5 ml of hESC/hiPSC culture medium, and seed them on the MEF layer. Culture the hESC/hiPSC in a 100% humidified 3–5% CO_2 incubator at 37 °C (see Note 6). Change the culture medium every day.

2.2.6.4 Passage of hESC/hiPSC
Aspirate the culture medium; add 1 ml of dissociation liquid that is equilibrated to room temperature; incubate the cells at 37 °C in a CO_2 incubator for 5 min; add 2 ml of hESC/hiPSC culture medium; suspend the cells by using Gilson P-1000 Pipetman® several times to fragment hESC/hiPSC clots into 100-cell clumps; centrifuge at $170 \times g$ for 5 min, aspirate the supernatant as much as possible (see Note 7), mildly resuspend the cells in 2 or 3 ml of hESC/hiPSC culture medium; pour 1 ml aliquot into 4 ml of hESC/hiPSC culture medium contained in a 60-mm MEF-layered dish; add 5 ng/ml of FGF-2 into each dish; culture the diluted hESC/hiPSC culture in a 100% humidified 3–5% CO_2 incubator at 37 °C. Change the culture medium every day. hESC/hiPSC should be maintained by regular passages twice a week (Fig. 10.2, left).

2.2.7 BA differentiation medium I
2.2.7.1 Preparation of basal medium
Mix the following materials: IMDM; 100 ml, Ham's F12; 100 ml, synthetic lipids; 2 ml, 450 mM α-monothioglycerol; 200 μl, insulin–transferrin–selenium; 2 ml, 200 mM Glutamax II, 2 ml, protein-free hybridoma mix; 10 ml, 50 mg/ml ascorbic acid-2-phosphate; 200 μl. After adding 1 g bovine serum albumin, filter the medium using a 0.22-μm pore sterile filter.

2.2.7.2 Supplements
Add the following cytokines to 50 ml basal medium: 20 μg/ml BMP4; 50 μl, 20 μg/ml VEGFA; 12.5 μl, 20 μg/ml KITLG; 50 μl, 10 μg/ml FLT3LG; 12.5 μl, 10 μg/ml IL6; 12.5 μl and 50 μg/ml IGF-II; 5 μl.

This medium is usable for 2 weeks when stored at 4 °C.

2.2.8 BA differentiation medium II
2.2.8.1 Preparation of basal medium
Mix the following materials: IMDM; 100 ml, Ham's F12; 100 ml, synthetic lipids; 2 ml, 450 mM α-monothioglycerol; 200 μl, insulin–transferrin–selenium; 2 ml, 200 mM Glutamax II, 2 ml, protein-free hybridoma mix; 10 ml, 50 mg/ml ascorbic acid-2-phosphate; 200 μl. After adding 1 g bovine serum albumin, filter the medium using a 0.22-μm pore sterile filter.

2.2.8.2 Supplements
Add the following cytokines to 50 ml basal medium: 20 μg/ml BMP7; 25 μl, 20 μg/ml VEGFA; 12.5 μl, 20 μg/ml KITLG; 50 μl, 10 μg/ml FLT3LG; 12.5 μl, 10 μg/ml IL6; 12.5 μl and 50 μg/ml IGF-II; 5 μl for the adherent culture.

This medium is usable for 2 weeks when stored at 4 °C.

2.2.9 Modified BA differentiation medium II
2.2.9.1 Preparation of basal medium
Mix the following materials: sodium bicarbonate-free IMDM; 100 ml, sodium bicarbonate-free Ham's F12; 100 ml, synthetic lipids; 2 ml, 450 mM α-monothioglycerol; 200 μl, insulin–transferrin–selenium; 2 ml, 200 mM Glutamax II, 2 ml, protein-free hybridoma mix; 10 ml, 50 mg/ml ascorbic acid-2-phosphate; 200 μl. After adding 1 g bovine serum albumin, filter the medium using a 0.22-μm pore sterile filter.

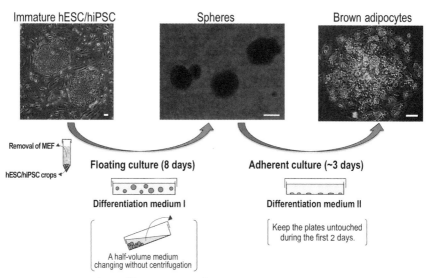

Figure 10.1 Procedure of hESC/hiPSC differentiation into classical BA. Immature hESC/hiPSC maintained on MEF layers were subjected to BA differentiation after removal of MEF. Scale bars indicate 100 μm.

2.2.9.2 Supplements

Add the following cytokines to 50 ml basal medium: 20 μg/ml BMP7; 25 μl, 20 μg/ml VEGFA; 12.5 μl, 20 μg/ml KITLG; 50 μl, 10 μg/ml FLT3LG; 12.5 μl, 10 μg/ml IL6; 12.5 μl and 50 μg/ml IGF-II; 5 μl for the adherent culture.

This medium is usable for 2 weeks when stored at 4 °C.

2.2.10 BA differentiation of hESC/hiPSC

2.2.10.1 Sphere formation

Prepare a 60-mm dish of MEF-cocultured hESC/hiPSC culture (Fig. 10.1, left); aspirate the culture medium; add 1 ml of dissociation liquid prewarmed at room temperature; incubate the cells in a CO_2 incubator for more than 10 min; transfer the detached cell suspension into a 15-ml conical tube; allow about 1 min for the hESC/hiPSC to settle to the bottom; aspirate the supernatant to remove MEFs; add 4 ml immature hESC/hiPSC maintenance medium and mildly resuspend hESC/hiPSC clots; stand the tube still for about 1 min until clots of hESC/hiPSC have sunk at the bottom; aspirate the supernatant to remove the residual MEFs; add 5 ml BA differentiation medium I and mildly resuspend hESC/hiPSC clots; transfer the hESC/hiPSC suspension into 60-mm low attachment dish such as HydroCell® (CellSeed Inc.) and MPC-coated plate (Nalge Nunc International) (see Note 8); culture the cells in a 100% humidified 5% CO_2 incubator at 37 °C; change a half volume of the medium every 3 days as follows: tilt the plate about 30° and keep that position for 1 min to collect the hESC/hiPSC clots at one portion of the bottom of the well, remove a 2 ml of supernatant, adding the same volume of a fresh differentiation medium I, mildly shake the plate to evenly distribute the hESC/hiPSC clots in the well (see Note 9). After a total of 8 days, spheres will be obtained (Fig. 10.1, middle).

2.2.10.2 Adherent culture

Transfer the suspension of hESC/hiPSC-derived spheres into a 15-ml conical tube; stand the tube still for 30 s to 1 min to sink spheres to the bottom; remove the supernatant; add 3 ml of BA differentiation medium II and mildly resuspend the spheres; transfer the suspension into a 15-ml conical tube and centrifuge it at $750 \times g$ for 5 min; resuspend the spheres with the supernatant and seed them on a 0.1% porcine gelatin-coated 60-mm culture dish; incubate the cells in a 100% humidified 5% CO_2 incubator at 37 °C (see Note 10). After 2 or 3 days, high-purity BA-differentiated samples will be obtained (Fig. 10.1, right). The cells contain small even-sized multilocular lipid droplets (Fig. 10.2, left) and abundant mitochondria (Fig. 10.2, right). Change the culture medium every 3 days.

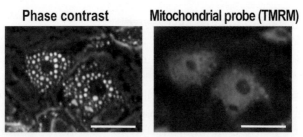

Figure 10.2 Mitochondrial staining of hESC/hiPSC-derived BA. hESC-derived BAs (left) were stained by tetramethyl rhodamine methyl ester (TMRM) (Life Technologies, Inc.), which detects mitochondria of living cells (right). Similar results were obtained from hiPSC-derived BAs (data not shown). Scale bars indicate 50 µm. (See the color plate.)

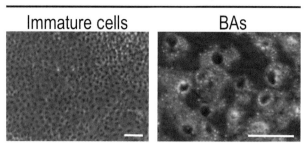

Figure 10.3 Oil red O staining. Immature hESCs (left) or hESC-derived BAs (right) were subjected to oil red O staining. Similar results were obtained from hiPSC-derived BA (Nishio et al., 2012). Scale bars indicate 50 μm. (See the color plate.)

2.2.11 Oil red O staining

Prepare hESC/hiPSC-derived BA as instructed in 2.2.10.2. Wash the cells with PBS twice, fix them by 10% formalin neutral buffer solution for 10 min, wash them with deionized water twice and rinse them with 60% isopropanol for 3 min. Stain the samples with freshly prepared 60% Oil red O solution (Muto pure Chemicals Co., Tokyo, Japan) for 15 min at room temperature. Wash the plates with deionized water to remove unbound dye. Red-colored small multilocular lipid droplets can be detected under microscopy (Fig. 10.3, right).

2.2.12 Oxygen consumption analyses

Prepare hESC/hiPSC-derived spheres as instructed in 2.2.10.1. Seed them onto a XF96 Cell Culture Microplate, which was precoated by 0.1% porcine, at the density of 30 spheres/well in 200 μl differentiation medium II. Culture the spheres at 37 °C in a 5% CO_2 incubator without opening or closing of the incubator door or other disturbances. After 2 days, replace the supernatant by a freshly prepared modified differentiation medium II. Keep the plate in a 37 °C incubator for 1 h for calibration. Then, measure the basal and maximum oxygen consumption rates (OCRs) by using XF96 Extracellular Flux Analyzer according to the guidance of the manufacture (Fig. 10.4A). For evaluation of responsiveness to adrenergic stimuli, add 100 nM of CL316,243, 100 μM of isoproterenol or empty medium (i.e., a modified differentiation medium II) to the corresponding wells and, after another 4-h incubation at 37 °C, measure OCRs (Fig. 10.4B).

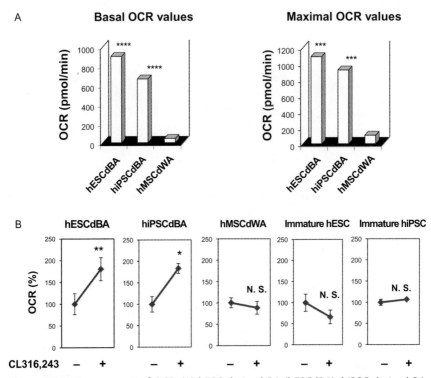

Figure 10.4 Measurement of OCR. (A) hESC-derived BA (hESCdBA), hiPSC-derived BA (hiPSCdBA), and human mesenchymal stem cell-derived white adipocytes (hMSCdBA) (Nishio et al., 2012) were subjected to OCR analyses by using XF96 Extracellular Flux Analyzer (Seahorse Bioscience Inc.). Basal and maximal OCR values were calculated via Mito Stress Test according to the manufacturer's guidance. ****$P<0.00001$, ***$P<0.0001$. (B) hESCdBA, hiPSCdBA, immature hESC, immature ihPSC, and hMSCdBA were treated with CL316,243 and, after 4 h, OCR values were calculated. The horizontal axis indicates the percentage of OCR value of CL316,243-treated cells to untreated cells. **$P<0.01$, *$P<0.05$. Figures were reproduced from Nishio et al., 2012 with minor modifications.

Alternatively, add CL316,243, isoproterenol, or empty media to the corresponding wells via an injection apparatus equipped with XF96 Extracellular Flux Analyzer and measure OCRs over time.

2.2.13 Calorigenic assays

Prepare mice, for example, 5–6-week-old male ICR strain mice. Remove the hair of the mice in broad areas around hip using hair removal cream 3 days before transplantation. Detach hESC/hiPSC-derived BA culture by TrypLE™ Express treatment at 37 °C for 5 min. Transfer cell suspension

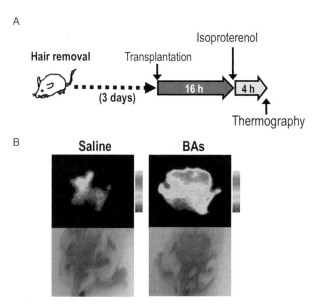

Figure 10.5 Calorigenic assays *in vivo*. (A) The experimental procedure. (B) Saline (left) or hESC-derived BAs (right) were subcutaneously injected to ICR mice and thermographic assessment was performed. Similar calorinogenic activity was detected in hiPSC-derived BA but not in immature hESC or immature hiPSC (Nishio et al., 2012). (See the color plate.)

into 15-ml tube, centrifuge the tube (600 × g, 5 min), remove the supernatant and resuspend the cells by sterile 4 °C PBS at the density of 1×10^6/100 μl. Inject a 100 μl cell suspension subcutaneously into a hip region by using 26G needle. After 24 h, inject 30 μmol/kg of isoproterenol into the tail vein. After another 4 h, anesthetize the mice by intraperitoneally injecting 300 μl/mouse of a 13-fold-diluted pentobarbital sodium with sterile PBS (Fig. 10.5A). Measure dermal temperature by using Thermo GEAR G100 (NEC Avio Infrared Technologies Co., Ltd, Tokyo, Japan). Upregulation of the dermal temperature over 38 °C is observed in BA-transplanted regions (Fig. 10.5B).

2.2.14 Oral glucose tolerance tests

Prepare mice, for example, 6–10-week-old male ICR stain mice. Before starting experiments, get sufficiently friendly with the mice by going to see them every day, putting them on your palms and stroking their backs with a caress (see Note 11). Continue to express tenderness until the mice

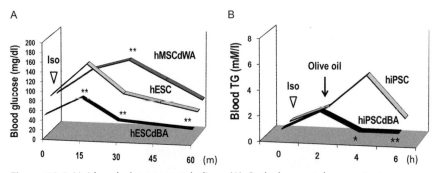

Figure 10.6 Lipid and glucose metabolism. (A) Oral glucose tolerance test was performed in mice transplanted with immature hESC, hESCdBA, and hMSCdWA. Similar blood glucose-lowering effects were observed in hiPSCdBA-transplanted mice (data not shown). (B) Oral olive oil tolerance test was performed in mice transplanted with immature ihPSC or hiPSCdBA. Iso indicates the timing of isoproterenol administration. **$P<0.01$, *$P<0.05$. *Figures were reproduced from Nishio et al., 2012 with minor modifications.*

show no sign of guardedness. Then, subcutaneously transplant ESC/hiPSC-derived BA (1×10^6 cells suspension in 100 μl PBS) as described in Section 2.2.13. Alternatively, put 1×10^6 ESC/hiPSC-derived BA pellet into a small subcutaneously incised part of the hip region by using 1000 μl pipette tip. After transplantation, fast mice for 16 h. Then, inject isoproterenol (30 μmol/kg) into the tail vein. After 4 h, administer 2 g/kg of glucose orally via gavage using a sonde under a cylinder-type retaining appliance. After 0, 15, 30, and 60 min, remove a small volume of blood from the lateral tarsal vein with a 26G needle. Measure glucose concentrations by using Accutrend® Plus (Fig. 10.6A) and insulin concentrations by using mouse insulin ELISA kit. For longer-term analyses, NOG mice (Ito et al., 2002) (CLEA Japan, Inc., Tokyo, Japan), which lack whole lymphocytes including T, B, and NK cells, should be used. Fasting blood glucose level-lowering effects are detectable up to 21 days (see Note 12).

2.2.15 Oral fat tolerance tests

Prepare mice, for example, 6-week-old male ICR stain mice. Then, subcutaneously transplant ESC/hiPSC-derived BA (1×10^6 cells suspension in 100 μl PBS per mouse) as described in Section 2.2.14. Keep the mice without food for 16 h after transplantation. (1) Examinations of fasting blood triglyceride (TG) levels. Inject isoproterenol (30 μmol/kg) via the tail vein. Take a small volume of blood 2 h after isoproterenol administration and measure blood TG levels by using Accutrend® Plus. (2) Blood TG clearance tests. Inject isoproterenol (15 μmol/kg) via the tail vein. After 2 h, administer 200 μl olive oil orally via gavage as described in Section 2.2.14. Take a small volume of blood sample every 2 h and measure blood TG levels by using Accutrend® Plus (Fig. 10.6B).

3. NOTES WITH TROUBLESHOOTING

1. Treatment with mitomycin C (MMC) is an alternative measure to block the proliferation of MEFs if an X-ray irradiation device is not available. There may be a risk that a trace amount of MMC, a genotoxic agent, is released to hESC/hiPSC; nevertheless, we have not yet detected chromosomal abnormalities in MMC-treated MEF-cocultured hESC/hiPSC. A procedure of MMC treatment is as follows: Prepare a 2 mg/ml MMC stock solution; add 10 μg/ml of MMC on the confluent culture of MEFs in a 100-mm dish; mix gently; incubate the cells in 100% humidified 5% CO_2 incubator at 37 °C for 90–120 min. Wash the cells with PBS twice; add 10 ml of MEF culture medium; incubate the cells in a 100% humidified 5% CO_2 incubator at 37 °C for 3 h to overnight.
2. Too much homogenization reduces the viability of MEFs. Thus, stop homogenization when small fragments of tissues are still visible to the naked eye.
3. It is not necessary to dissociate or fragment hESC/hiPSC clots at this step because they are automatically fragmented during the cell thaw process.
4. To minimize cytotoxic effects of the freezing medium, hESC/hiPSC suspension should be soaked in liquid nitrogen as fast as possible, within 15 s at longest.
5. See to it that cell suspension is fully frozen including inner portions.
6. Because of an acidification-prone nature of the hESC/iPSC maintenance medium, it is recommended that the CO_2 concentration of the incubator is set at 2–3%. Alternatively, pH of the medium can be adjusted by adding NaOH.
7. Collagenase and trypsin, which are supplemented to dissociation liquid, cannot be inactivated by an addition of culture medium. Thus, it is recommended that dissociation liquid should be fully aspirated. Alternatively, washing the cell pellets with culture medium will be advantageous to completely eliminate these enzymes.

8. For the best performance, HydroCell® is recommended because it guarantees high-efficiency sphere formation without an attachment of hESC/hiPSC on its surface. The MPC-coated dish is also usable although it occasionally permits hESC/hiPSC to attach on its surface at the margins. Other low-attachment culture dishes are not recommended.
9. Do not transfer hESC/hiPSC clots into tubes nor centrifuge them when performing a half-volume medium change. Pipetting and centrifugation at this step substantially lower the viability of hESC/hiPSC, and as a result, spheres will be fragmented into small pieces, hampering BA differentiation.
10. The culture plates should be kept still in a CO_2 incubator during the first 2 days of the adherence culture. Opening and closing of the incubator door should be avoided because even minor shaking can block the spheres from adhering onto gelatin-coated plates. Coating the plates with fibronectin or Matrigel™ Basement Membrane Matrix (BD Biosciences) is not recommended because those extracellular matrix components substantially lower the efficiency of BA differentiation.
11. Emotional and physical stress significantly affect blood glucose values. Thus, it is particularly important that all the experimental procedures should be performed without adding stress to the mice. To reduce emotional stress, an operator should earn the complete trust of the mice before starting experiments. To minimize physiological stress, all the procedure should be performed quickly under a retaining appliance. Usage of anesthetic agents is not recommended because they might affect blood glucose values.
12. We examined longer-term effects of hESC-derived BAs on glucose metabolism using NOG mice (Ito et al., 2002) by measuring fasting blood glucose values at Day 10, Day 21, and Day 30. Although statistically significant reductions of fasting blood glucose values were observed at Day 10 and Day 21, we could detect a reduced fasting blood glucose value in only one mouse out of three at Day 30. Thus, subcutaneous transplantation method is not suited for long-term (>1 month) transplantation experiments. Similar results were recently reported in a paper on murine BAT transplantation studies, where long-term (12–16 weeks) glucose metabolism-improving effects were detected in mice with peritoneal BAT transplantation but not in mice with subcutaneous transplantation (Stanford et al., 2013). Although mechanisms for the differential physiological effects between subcutaneous and

intraperitoneal transplantation remain elusive, capacities to induce innervation may be a key because sympathetic nerve system plays an indispensable role in maintenance and activation of BAT. Future studies will elucidate the reason for the differential outcome between the two transplantation methods.

REFERENCES

Ito, M., Hiramatsu, H., Kobayashi, K., Suzue, K., Kawahata, M., Hioki, K., et al. (2002). NOD/SCID/gamma(c)(null) mouse: An excellent recipient mouse model for engraftment of human cells. *Blood*, *100*(9), 3175–3182.

Muzik, O., Mangner, T. J., Leonard, W. R., Kumar, A., Janisse, J., & Granneman, J. G. (2013). 15O PET measurement of blood flow and oxygen consumption in cold-activated human brown fat. *Journal of Nuclear Medicine*, *54*(4), 523–531.

Nishio, M., Yoneshiro, T., Nakahara, M., Suzuki, S., Saeki, K., Hasegawa, M., et al. (2012). Production of functional classical brown adipocytes from human pluripotent stem cells using specific hemopoietin cocktail without gene transfer. *Cell Metabolism*, *16*(3), 394–406.

Osafune, K., Caron, L., Borowiak, M., Martinez, R. J., Fitz-Gerald, C. S., Sato, Y., et al. (2008). Marked differences in differentiation propensity among human embryonic stem cell lines. *Nature Biotechnology*, *26*(3), 313–315.

Stanford, K. I., Middelbeek, R. J. W., Townsend, K. L., An, D., Nygaard, E. B., Hitchcox, K. M., et al. (2013). Brown adipose tissue regulates glucose homeostasis and insulin sensitivity. *The Journal of Clinical Investigation*, *123*(1), 215–223.

Suemori, H., Yasuchika, K., Hasegawa, K., Fujioka, T., Tsuneyoshi, N., & Nakatsuji, N. (2006). Efficient establishment of human embryonic stem cell lines and long-term maintenance with stable karyotype by enzymatic bulk passage. *Biochemical and Biophysical Research Communications*, *345*(3), 926–932.

CHAPTER ELEVEN

Analysis and Measurement of the Sympathetic and Sensory Innervation of White and Brown Adipose Tissue

Cheryl H. Vaughan[*], Eleen Zarebidaki[*], J. Christopher Ehlen[*,†], Timothy J. Bartness[*,1]

[*]Department of Biology, Neuroscience Institute and Center for Obesity Reversal, Georgia State University, Atlanta, Georgia, USA
[†]Department of Neurobiology, Neuroscience Institute, Morehouse School of Medicine, Atlanta, Georgia, USA
[1]Corresponding author: e-mail address: bartness@gsu.edu

Contents

1. Introduction 200
2. Surgical Denervation of WAT and BAT Nerves 202
   2.1 Identification of nerves innervating WAT and BAT for surgical denervation 203
   2.2 Required materials for all denervation protocols 204
   2.3 Surgical denervation protocol 204
3. Chemical Denervation of Adipose Tissue Using 6-Hydroxy-Dopamine 207
   3.1 Chemical denervation protocol 208
4. Local Sensory Denervation of Adipose Tissue Using Capsaicin 209
   4.1 Sensory denervation protocol 210
5. Assessment of Sympathetic Denervation Using NETO or Content as Measured by HPLC-EC 211
   5.1 Alpha-methyl-*para*-tyrosine method to measure sympathetic drive (NETO) 211
   5.2 Preparation of samples for HPLC-EC (NETO and NE content) 214
   5.3 HPLC-EC protocol (and NE content) 217
6. Assessment of Sensory Denervation Using CGRP ELIA 218
   6.1 Required materials 218
   6.2 Preparation of samples 219
7. Expression of Data 220
Acknowledgments 220
References 220

Abstract

Here, we provide a detailed account of how to denervate white and brown adipose tissue (WAT and BAT) and how to measure sympathetic nervous system (SNS) activity to these and other tissues neurochemically. The brain controls many of the functions of WAT and BAT via the SNS innervation of the tissues, especially lipolysis and thermogenesis, respectively. There is no clearly demonstrated parasympathetic innervation of WAT or the major interscapular BAT (IBAT) depot. WAT and BAT communicate with the brain neurally via sensory nerves. We detail the surgical denervation (eliminating both innervations) of several WAT pads and IBAT. We also detail more selective chemical denervation of the SNS innervation via intra-WAT/IBAT 6-hydroxy-dopamine (a catecholaminergic neurotoxin) injections and selective chemical sensory denervation via intra-WAT/IBAT capsaicin (a sensory nerve neurotoxin) injections. Verifications of the denervations are provided (HPLC-EC detection for SNS, ELIA for calcitonin gene-related peptide (proven sensory nerve marker)). Finally, assessment of the SNS drive to WAT/BAT or other tissues is described using the alpha-methyl-*para*-tyrosine method combined with HPLC-EC, a direct neurochemical measure of SNS activity. These methods have proven useful for us and for other investigators interested in innervation of adipose tissues. The chemical denervation approach has been extended to nonadipose tissues as well.

1. INTRODUCTION

Because of the epidemic status of obesity, increased interest in the control of the functions of adipose tissues, both white and brown (WAT and BAT, respectively), has grown considerably in the last 20 years. Although there has long been an appreciation of the control of BAT through the activity of its sympathetic nervous system (SNS) (for reviews, see Bartness, Vaughan, & Song, 2010; Cannon & Nedergaard, 2004; Himms-Hagen, 1991), appreciation of the importance of the SNS innervation of WAT has lagged behind. This is possibly due to early misconceptions of the importance of adrenal medullary epinephrine (EPI) in lipolysis (e.g., Havel & Goldfien, 1959; Wool, Goldstein, Ramey, & Levine, 1954). We now know that in the absence of the adrenal medulla, the sole source of circulating EPI, regulated lipolysis occurs in response to food deprivation and cold exposure (factors that increase WAT lipid mobilization, e.g., Paschoalini & Migliorini, 1990; Teixeira, Antunes-Rodrigues, & Migliorini, 1973). We and others have defined the central origins of the SNS outflow from brain to WAT (Adler, Hollis, Clarke, Grattan, & Oldfield, 2012; Bamshad, Aoki, Adkison, Warren, & Bartness, 1998; Shi & Bartness, 2001; Song & Bartness, 2001; Song, Jackson, Harris, Richard, & Bartness, 2005). Destruction of the SNS innervation of WAT blocks lipolysis due to food deprivation, cold, and other stimuli (e.g., Beznak & Hasch, 1937; Cantu & Goodman, 1967; Lazzarini & Wade, 1991; Youngstrom & Bartness, 1998) demonstrating the necessity of the innervation for the control of lipolysis. We and others also have defined the central origins of the SNS outflow from brain to BAT (Bamshad, Song, & Bartness, 1999; Oldfield et al., 2002; Song et al., 2008; Zhang et al., 2011) and destruction of the SNS innervation of BAT blocks or severely diminishes BAT responses to all thermogenic challenges tested to date (for review, see Bartness et al., 2010). Finally, we have demonstrated the sensory inflow to the brain and the central sensory circuits from both WAT (Song, Schwartz, & Bartness, 2009) and BAT (Vaughan & Bartness, 2012) to brain. Our initial studies of the roles of the sensory innervations of WAT and of BAT for physiological responses of these tissues indicate their necessity for normal functioning of the tissues and suggest possible modalities of these neural feedbacks (Foster & Bartness, 2006; Shi & Bartness, 2005; Shi, Song, Giordano, Cinti, & Bartness, 2005; Song & Bartness, 2007; Vaughan & Bartness, 2012). Therefore, testing the necessity of the SNS and sensory innervation of WAT and BAT, as well as other tissues important for energy metabolism (e.g., liver), is apparent.

Because the sympathetic and sensory nerves innervating both WAT and BAT pads travel together in bundles indistinguishable from each other visually by eye, we initially used surgical denervation to test for the contribution of these nerves for normal WAT and BAT physiological responses (Bartness & Wade, 1984; Foster & Bartness, 2006; Rooks et al., 2005; Youngstrom & Bartness, 1998). We recommend surgical denervation as an initial approach because, if done correctly, virtually all innervations are removed.

Previously, chemical denervation of sympathetic (e.g., Johnson & O'Brien, 1976; Levin & Sullivan, 1984; Levin, Triscari, Marquet, & Sullivan, 1984; Mory, Ricquier, Nechad, & Hemon, 1982; Seydoux, Mory, & Girardier, 1981) or sensory (e.g., Cui & Himms-Hagen, 1992a, 1992b; Cui, Zaror-Behrens, & Himms-Hagen, 1990; Kawada, Hagihara, & Iwai, 1986) nerves was performed via systemic injection of sympathetic and sensory neurotoxins resulting in global denervations that are difficult to ascribe specific functional involvement to either WAT or BAT. Therefore, we developed local, selective and effective chemical denervation of the sympathetic and sensory nerve provisions to WAT and BAT pads (Demas & Bartness, 2001a, 2001b; Foster & Bartness, 2006; Rooks et al., 2005; Shi & Bartness, 2005; Shi et al., 2005), the details of which are delineated below.

Thus, the purpose of this chapter is to detail the methods we have used to test the necessity of SNS and sensory innervations of WAT and BAT for their normal functioning. In addition, we describe a direct neurochemical method to assess the sympathetic drive to WAT, BAT or any sympathetically innervated tissue rather than the many inadequate surrogates used by researchers to assess such activity such as histological or static measures of neurochemicals found in the sympathetic or sensory nerves. Note that when possible, we give instructions for how to accomplish the methods in laboratory rats and mice, in addition to Siberian hamsters, the latter being our species of choice. These techniques were first applied to Siberians because of their photoperiod induced, naturally occurring obesity and obesity reversal (for review, see Bartness, Demas, & Song, 2002; Bartness & Wade, 1985).

2. SURGICAL DENERVATION OF WAT AND BAT NERVES

Sympathectomy has been used widely as a means to eliminate postganglionic input to the tissue of choice. We typically have waited ~12 weeks to ensure that consequences of the denervation (axonal degeneration and cell death a.k.a. Wallerian degeneration) are complete (Bartness et al., 1986; Bowers et al., 2004; Foster & Bartness, 2006; Hamilton, Bartness, & Wade, 1989; Shi & Bartness, 2005; Shi et al., 2005; Youngstrom & Bartness, 1998). However, we have observed significant cell denervation-induced increases in white adipocyte proliferation in inguinal WAT (IWAT) as early as 10 days post denervation (Bartness et al., 1986; Bowers et al., 2004; Foster & Bartness, 2006; Hamilton et al., 1989; Shi & Bartness, 2005; Shi et al., 2005; Youngstrom & Bartness, 1998). Therefore, one should conduct time course studies to document the destruction of the sympathetic/sensory innervations (see Sections 5 and 6 for verification methods) for each tissue tested to determine the optimal time to conduct any post-axotomy measures.

Surgical denervation removes both sympathetic output and sensory input from adipose tissues; therefore, to verify successful surgical denervation it is necessary to measure norepinephrine (NE) content (typically by HPLC-EC as current enzyme-linked immunosorbent assay (ELIA) methods, even the currently available ultra-sensitive ELIAs, are not sensitive enough), the principal neurochemical released by sympathetic nerves (described in Section 5), or by ELIA/Western blot for tyrosine hydroxylase (TH) or immunohistochemical assessment of TH-immunoreactivity (Foster & Bartness, 2006). It needs to be emphasized that none of those measures are an assessment of sympathetic activity. Sympathetic activity can only be assessed directly either electrophysiologically, or by norepinephrine turnover (NETO), the latter described in Section 5. As for verification of the extent and specificity of the sensory innervation of WAT and BAT, a marker of sensory nerves such as calcitonin gene-related peptide (CGRP) can be measured via ELIA (described in Section 6), or immunohistochemically. To our knowledge, the only direct method to measure the activity of sensory nerves innervating adipose tissues is electrophysiologically, which we have done for WAT in response to glucoprivation (Song et al., 2009) and we (Murphy et al., 2013) and others (Niijima, 1998, 1999; Shi et al., 2012; Xiong et al., 2012) have done for direct intra-WAT leptin injection.

2.1. Identification of nerves innervating WAT and BAT for surgical denervation

We have performed unilateral and bilateral nerve cuts of inguinal IWAT, epididymal WAT (EWAT), retroperitoneal WAT (RWAT), and interscapular BAT (IBAT) to test for depot-specific effects on anatomy and physiology (Bartness et al., 1986; Bowers et al., 2004; Foster & Bartness, 2006; Hamilton et al., 1989; Shi & Bartness, 2005; Shi et al., 2005; Youngstrom & Bartness, 1998).

The nerves innervating BAT are more easily discernible than those innervating WAT. For IBAT, there are five intercostal nerves that unilaterally innervate each IBAT lobe in laboratory rats (Foster, Depocas, & Zaror-Behrens, 1982; Foster, Depocas, & Zuker, 1982), laboratory mice (Sidman & Fawcett, 1954), and Siberian hamsters (Klingenspor, Meywirth, Stohr, & Heldmaier, 1994). These can be seen and severed without the aid of a dissecting microscope, although visualization and axotomy is improved with magnification.

Because of the difficulty in visualizing the nerves innervating WAT, we frequently use one to two drops of 1% toluidine blue directly onto tissue (Foster & Bartness, 2006; Shi & Bartness, 2005; Shi et al., 2005) and sever any nerves that appear to enter the fat pad. Use of a dissecting microscope is mandatory for visualization and axotomy of WAT innervation.

EWAT nerves are encased in a neurovascular pedicle and pass through the inguinal canal (Correll, 1963). Nerves innervating EWAT are best visualized after testis and associated EWAT are pulled intact from the abdominal cavity through a ventral incision (described in Shi et al., 2005; Song et al., 2009). Any inguinal nerves that innervate IWAT in addition to accessory nerves present nearby that appear to be connected to skin can be visualized after a ventral to dorsal semicircular incision in the skin near the flank.

2.2. Required materials for all denervation protocols
- 95% ethanol
- Sterile gauze
- Povidone-iodine (topical antiseptic, e.g., Betadine)
- At least one of the following to handle skin and/or fat. All available from Roboz Surgical Instrument Co., Gaithersburg, MD
  - Tissue forceps (Graefe: jar width 4.5 mm, length 41/4″; Cat #: RS-8248)
  - Curved serrated forceps (full curve; Cat #: RS-5137)
- Sharp surgical scissors (e.g., Cat #: RS-6702; Roboz Surgical Instrument Co., Gaithersburg, MD)
- Nitrofurozone (antibacterial powder to minimize wound infection, e.g., nfz Puffer; Hess & Clark, Lexington, KY)
- 9 mm wound clips (Stoelting, Wood Dale, IL)
- Wound clip applicator to close skin (EZ Clip ™ Applier; Stoelting, Wood Dale, IL)
- Wound clip remover (Cat #: RS-9263; Roboz Surgical Instrument Co., Gaithersburg, MD)
- Vicryl suture to close musculature (Ethicon, Somerville, NJ)
- Razor or depilatory cream (e.g., Nair®) to clear hair from target area
- Analgesic and/or anti-inflammatory agent (e.g., ketoprofen, buprenorphine)

2.3. Surgical denervation protocol
2.3.1 Required materials
- Materials listed in Section 2.2
- Dissecting microscope
- At least one of the following microdissecting forceps. All available from Roboz Surgical Instrument Co., Gaithersburg, MD
  - Microdissecting angled fine sharp points forceps, length 4″ (Cat #: RS-5095)
  - McPherson: straight smooth tying platform forceps (Cat #: RS-5068)
  - Graefe: serrated, straight forceps (Cat #: RS-5110)
- At least one of the following microdissecting scissors. All available from Roboz Surgical Instrument Co., Gaithersburg, MD
  - McPherson-Vannas: microdissecting straight, sharp spring scissors (Cat #: RS-5600)
  - Vannas: scissors angled on edge (Cat #: RS-5618)
- *Note*: For help with identifying adipose depots, see Cinti (1999)

Adipose Innervation and Measurement

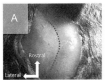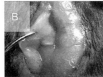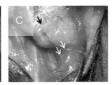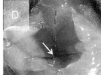

Figure 11.1 IBAT nerve identification. The orientation of all pictures is the dorsal, interscapular surface. (A) Picture of both IBAT lobes. Dotted line delineates left from right IBAT. (B) Forceps revealing the ventral surface of the left IBAT lobe. (C) Picture of right IBAT and associated intercostal nerves supplying sympathetic and sensory innervation (black arrow, 3 nerves; white arrow, 1 nerve each). D. Sulzer's vein draining both BAT lobes (white arrow). Dotted line depicts midline and medial borders of both BAT lobes. (See the color plate.)

1. Anesthetize animal and remove fur from target area. Wipe area with 95% ethanol-soaked sterile gauze followed by additional application of Betadine to the area.
2. *For all fat pads*, use the following descriptions to locate nerves and if difficult use dye to visualize and then sever the nerves.

    For IBAT, make a midline incision in the skin along the upper dorsal surface to expose both IBAT pads (Fig. 11.1A). Gently expose the medial, ventral surface of both pads to visualize nerves beneath pad (Fig. 11.1B). Nerves may appear in bundles of two to three. If seen under the microscope, the bundles can be gently dissociated to view individual nerves (Fig. 11.1C; 5 nerves/BAT pad). Care should be taken not to cut the large Sulzer's vein that is located medially to both pads (Fig. 11.1D).

    For IWAT, make an incision in skin dorsally on the flank from a point near the tail and lateral to the spinal column. Continue the incision rostrally along the dorsum adjacent to the spinal column to a point just rostral to the hind limb, then laterally and ventrally to a point ~2 cm from the ventral midline (Fig. 11.2A and B). Separate the IWAT pad from the abdominal wall and overlying skin by blunt dissection, keeping intact the major blood vessels leading into or through the pad (Fig. 11.2C–F).

    For EWAT, make a single abdominal midline incision in the skin. Through this incision, EWAT pads can be accessed and gently lifted or pushed to visualize nerves (Fig. 11.3A and B). For better visualization, the testis and associated EWAT can be pulled from the peritoneal cavity and laid on 0.9% NaCl-soaked sterile gauze placed on the animal's ventrum for better visualization. Some of the nerves innervating EWAT appear to run along the main artery providing the testis with blood. Do not damage either of these arteries because both are integral to normal

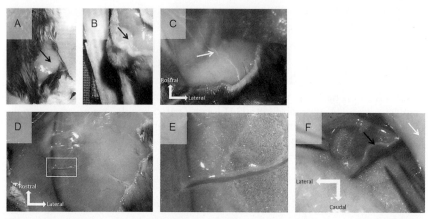

Figure 11.2 IWAT nerve identification. Dorsal (A) and ventral (B) views of animal. Black arrow denotes IWAT pad location. (C) Dorsolateral view of nerves innervating (white arrow) fat and skin. (D) Medial border of the right IWAT (midline to the left of picture). (E) Magnification of white box in D depicting nerves on either side of blood vessel. (F) Ventral surface showing nerve that bifurcates sending one branch to leg and one to IWAT. Do not cut this nerve (black arrow). Peritoneal cavity to the right of this picture (white arrow). (See the color plate.)

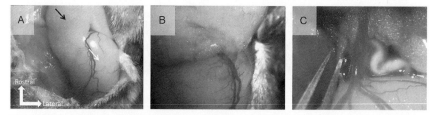

Figure 11.3 EWAT nerve identification. (A) The right testis (white arrow) and right EWAT (black arrow). (B) Toluidine blue dye highlighting nerve innervating EWAT. (C) Forceps holding nerve to EWAT that should not be cut. If cut, loss of testis function occurs. (See the color plate.)

testicular function (Fig. 11.3C). After surgery, close the peritoneal cavity with absorbable sutures (Vicryl).

For RWAT, make an incision in the skin on the ventral surface close to the midline then make an incision into the peritoneal cavity to expose the area medial to the kidneys. RWAT extends longitudinally down the spine and between vertebrae. It is separated from the perirenal depot by a peritoneal fold (Frontini & Cinti, 2010). Gently lift the kidney and cut the three nerve bundles that innervate RWAT (Bowers et al., 2004).

After surgery, close the peritoneal cavity with absorbable sutures (Vicryl).
3. For all initial incisions to access the fat pads, care should be taken with the depth of the incision to avoid damaging the underlying blood vessels and musculature. Nerves identified at ×4 magnification that terminate in the pad should be cut in two or more locations and if possible, remove the nerve sections between the cuts to assure denervation and the unlikely possibility of nerve regrowth.
4. Throughout the surgery, keep the target fat pad moist with 0.9% NaCl-soaked sterile gauze.
5. If using a contralateral pad for a within-animal control (unilateral denervation model; Bartness & Bamshad, 1998; Bartness & Song, 2007), the sham procedure should include fat pad manipulation and nerve identification with the exception that the nerves are left intact. Because of laterality of the innervations, alternate the side for denervation or sham surgery (or chemical denervation, see Sections 3 and 4).
6. After surgery the sham or denervated pad is replaced (if pulled from the peritoneal cavity or moved to the side) and rinsed with sterile 0.9% NaCl.
7. The incision in the skin is closed with sterile wound clips, and nitrofurozone powder is applied to the wound surface to minimize the risk of bacterial infection.

3. CHEMICAL DENERVATION OF ADIPOSE TISSUE USING 6-HYDROXY-DOPAMINE

We previously used intra-fat injections of guanethidine, which is toxic to sympathetic nerves at high concentrations (Burnstock, Evans, Gannon, Heath, & James, 1971). Although we had excellent, nearly axotomy-like complete SNS denervations, as evidenced by nearly or nondetectable NE content (Demas & Bartness, 2001a, 2001b), newer preparations are less reliable; therefore, we now use intra-fat injection of the catecholaminergic neurotoxin, 6-hydroxy-dopamine (6-OHDA; Knyhar, Ristovsky, Kalman, & Csillik, 1969; Ungerstedt, 1968). 6-OHDA has been used by others for chemical sympathectomy in a variety of preparations (Benarroch, Schmelzer, Ward, Nelson, & Low, 1990; Depocas, Foster, Zaror-Behrens, Lacelle, & Nadeau, 1984; Dobbins, Szczepaniak, Zhang, & McGarry, 2003; Joost & Quentin, 1984). Using 6-OHDA for local SNS denervation was inspired by the sympathetic denervation of the testes via direct injection of the toxin into the gonads (Mayerhofer, Amador, Steger, & Bartke, 1990). Therefore, we tried and have successfully sympathetically denervated WAT with direct injections of 6-OHDA (Foster & Bartness, 2006; Giordano, Morroni, Santone, Marchesi, & Cintiet, 2006; Rooks et al., 2005). Previously, 6-OHDA has been injected subcutaneously resulting in a reduction in BAT NE content (denervation) (Depocas et al., 1984; Thureson-Klein, Lagercrantz, & Barnard, 1976), but BAT denervation in light of the global sympathetic denervation that occurs by this type of systemic approach makes interpretation of results exceedingly difficult. Thus, direct intra-WAT or BAT injections of 6-OHDA are both restrictive and selective.

3.1. Chemical denervation protocol
3.1.1 Required materials
- Materials listed in Section 2.2
- 6-OHDA (6-hydroxy-dopamine hydrochloride; Cat #: H4381; Sigma-Aldrich, St. Louis, MO)
- L-Ascorbic acid (Cat #: A5960; Sigma-Aldrich, St. Louis, MO)
- Sodium chloride
- N_2 gas tank
- Hamilton syringe, 10 µl model (Hamilton Company, Reno, NV)

1. To prepare 6-OHDA, first make a solution of 0.15 M NaCl containing 1% ascorbic acid (AA). This AA/saline solution should be gassed in a light-tight container with N_2 for 10 min before adding 6-OHDA. Based on previous work (Foster & Bartness, 2006; Giordano et al., 2006; Rooks et al., 2005), we have found that different fat pads and other organs require slightly different concentrations of 6-OHDA (see directly below).
    a. 6-OHDA is light- and temperature-sensitive so keep solution on ice and covered during surgeries and replace every 2–3 h
2. Anesthetize animal and remove fur from target area. Wipe area with 95% ethanol-soaked gauze followed by application of Betadine.
3. *For all fat pads*, make incisions using descriptions in Section 2.3. With all incisions, care should be taken with the depth of the incision so as not to damage the underlying fat pad and vasculature. Once pad is gently separated from skin, use a 10 µl microsyringe to inject the following across the respective pads:
    - *EWAT*: *Siberian hamsters*: 20 loci with 2 µl injections of 9 mg/ml 6-OHDA solution or vehicle. *Mice*: 10 loci across the pad with

2 µl injections of 8 mg/ml 6-OHDA in 0.01 M PBS containing 1% AA.
- *IWAT: Siberian hamsters*: 40 loci with 2 µl injections of 4 mg 6-OHDA in 100 µl vehicle or vehicle. *Mice*: 12 loci with 2 µl injections of 9 mg/ml 6-OHDA in PBS containing 1% AA (Harris, 2012; Rooks et al., 2005).
- *RWAT: Rats*: 10 loci with 2 µl injections of either 9 mg/ml 6-OHDA in 0.01 M PBS containing 1% AA or vehicle.

4. Once the needle is inserted into the adipose tissue, fluid should be expelled slowly over a ~30 s period for each injection. The tip of the needle should be held in place for ~30–60 s to avoid reflux.
5. The incision is closed with sterile wound clips and nitrofurozone powder is applied to minimize sepsis.
6. Administer an analgesic and monitor animal appropriately.
7. Determine the postsurgical time in pilot studies for significant reductions in sympathetic and sensory nerve markers. Generally, allow anywhere from 10 days to 4 weeks post injection for testing the denervations (Foster & Bartness, 2006; Giordano et al., 2006; Harris, 2012; Rooks et al., 2005).

4. LOCAL SENSORY DENERVATION OF ADIPOSE TISSUE USING CAPSAICIN

Capsaicin, the pungent part of red chili peppers, selectively destroys small, unmyelinated mostly C-fiber sensory nerves (Jansco, Kiraly, & Jansco-Gabor, 1980; Jansco, Kiraly, Joo, Such, & Nagy, 1985). We have shown that injections of capsaicin into WAT selectively destroys sensory nerves, as documented immunohistochemically by significant decreases in, CGRP, but not TH (Foster & Bartness, 2006), the rate-limiting enzyme in NE synthesis and a proven sympathetic nerve marker.

Other investigators have tested the effects of IBAT sensory denervation via repeated, large doses of systemic capsaicin in neonatal or adult rats resulting in denervation of all peripheral tissues (Cui & Himms-Hagen, 1992a, 1992b; Cui et al., 1990; Himms-Hagen, Cui, & Sigurdson, 1990; Melnyk & Himms-Hagen, 1994). This approach results in capsaicin-desensitization, as the animals no longer respond to the hypothermic effects of an acute capsaicin injection (e.g., Jancso-Gabor, Szolcsanyi, & Jancso, 1970). Acute intraperitoneal (ip) injections of capsaicin also create lesions in some central structures that have a weak blood–brain barrier, such as the nucleus of the solitary tract (Castonguay & Bellinger, 1987). Due to these unclear interpretations regarding whole animal capsaicin-desensitization, we choose to use local injections into fat as above for SNS denervations using 6-OHDA. This approach was inspired by the direct application of capsaicin to the vagus nerve to promote vagal afferent denervation (Raybould, Holzer, Reddy, Yang, & Tache, 1990; Yoneda & Raybould, 1990).

4.1. Sensory denervation protocol
4.1.1 Required materials
- Materials listed in Section 2.2
- Hamilton syringe, 10 µl model (Hamilton Company, Reno, NV)
- Hot block (Isotemp Heat block, Fisher Scientific, Pittsburgh, PA)
- Heating pad and pump (Gaymar T Pump & Multi-T pad, Braintree Scientific, Inc., Braintree, MA)
  - *Note*: surgery takes about ≥ 1 h
- Olive oil (Cat #: O1514; Sigma-Aldrich, St. Louis, MO)
- Capsaicin (Cat #: M2028; Sigma-Aldrich, St. Louis, MO). *Note*: All of the capsaicin is used to make the stock solution and is not measured out on a balance due to risk of skin, eye, and/or respiratory irritation
- Capsaicin stock solution: Add 1.25 ml of 100% ethanol to 250 mg bottle of capsaicin to make the final concentration of 200 µg/µl and keep in dark at 4 °C
- Capsaicin working solution: Dilute stock solution in olive oil at 1:10 (stock:olive oil) to yield a final concentration of 20 µg/µl

1. Make capsaicin stock solution.
2. Before surgery, make fresh working solution under a fume hood. Warm solution to 37 °C using hot block or water bath; make sure solution is covered during warming. Change out for fresh solution if crystals begin to form in container with working solution. Anesthetize and prepare animal by removing hair in target area. Wipe area with 95% ethanol-soaked sterile gauze followed by application of Betadine.
3. Expose fat pads and keep them moist. Be careful when moistening tissue after injecting capsaicin as it can spread to unwanted areas.
4. Change gloves between handling capsaicin and handling the next animal.
    a. If at any point, there is a possibility there may be capsaicin solution on gloves, change gloves.
    b. Wear a mask at all times when handling capsaicin.

5. *For all adipose tissues*, make incisions using descriptions in Section 2.3. For all fat pads, inject across 20–30 s and wait for ∼30 s with tip of needle in the adipose tissue and then move to the next injection site in that pad. *Test that fluid is being expelled often.* Occasionally, olive oil solution will clog the needle; therefore, use 100% ethanol to flush needle and syringe if necessary between injections. Once pads are gently separated from skin, inject the following volumes across the respective pads:
    - IBAT: *Siberian hamsters*: inject with 30 microinjections (1 μl/injection) of 20 μg/μl of capsaicin or vehicle (1:10, ethanol: olive oil).
    - IWAT: *Siberian hamsters and mice*: inject with 40 microinjections (1 μl per injection) of 20 μg/μl of capsaicin or vehicle (1:10, ethanol: olive oil).
    - EWAT: *Siberian hamsters*: inject with 40 microinjections (1 μl per injection) of 20 μg/μl of capsaicin or vehicle (1:10, ethanol:olive oil).
6. During the injections minimize the handling of the adipose tissue.
7. Determine the postsurgical time in pilot studies for significant reductions in sympathetic and sensory nerve markers. We typically allow 21 days before conducting tests in Siberian hamsters.

5. ASSESSMENT OF SYMPATHETIC DENERVATION USING NETO OR CONTENT AS MEASURED BY HPLC-EC

NE content is measured to verify the effectiveness of the surgical or chemical sympathetic denervations. We use reverse-phase HPLC-EC following our modifications (Bowers et al., 2004) to the method of Mefford (1981). NETO is used as a direct neurochemical measure of sympathetic drive; as noted above, there is no surrogate for this method of assessment except for direct measures of sympathetic nerve activity electrophysiologically. The benefit of NETO, however, is that there is no restriction on the number of tissues where NETO can be measured from the same animal (e.g., various WAT pads, IBAT, muscle, liver, etc.).

5.1. Alpha-methyl-*para*-tyrosine method to measure sympathetic drive (NETO)

NETO is measured using the alpha methyl-*para*-tyrosine (AMPT) method (Cooper, Bloom, & Roth, 1982). AMPT is a competitive inhibitor of TH. Without available TH and thus no new NE, the endogenous tissue levels of NE decline at a rate proportional to initial NE concentrations (Cooper et al., 1982). NETO is measured over the last 2–4 h of a test, regardless of the stimulus applied (e.g., central or peripheral drugs/hormones/neurochemicals, cold exposure, food deprivation, etc.). For WAT or BAT, which are less heavily sympathetically innervated than some tissues (e.g., heart), the 4 h time period seems best. Because AMPT is a severe behavioral depressant that is expected to markedly affect many physiological/biochemical responses, only NETO should be measured in these animals, with a parallel set of animals used for non-NETO measures.

It is important to emphasize that handling and other manipulations (previous surgery, intraventicular injection, peripheral injection) can increase sympathetic drive to WAT or BAT. Therefore, based on our experience, animals should be handled daily for ~1–3 weeks for 5 min to adapt them to the handling associated with the AMPT injection procedure and thereby minimize stress-induced increases in sympathetic drive (Brito, Brito, Baro, Song & Bartness, 2007; Brito, Brito, & Bartness, 2008).

The assumption for NETO is that the decline across the NETO test time is linear; therefore, this should be demonstrated first by using three time points (e.g., 0, 2, and 4 h; see Fig. 11.4). Once established in the laboratory for tissues of interest in pilot experiments, we have found that the 0 and the longest time point (typically 4 h) are sufficient. Half the animals are killed by decapitation without anesthesia. This is necessary because anesthesia can cause NE release from sympathetic terminals in the tissue, but rapid

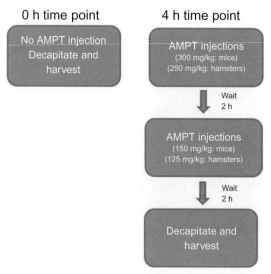

Figure 11.4 Timeline of AMPT injections for NETO measurement. (See the color plate.)

decapitation without anesthesia does not (Depocas & Behrens, 1977; Popper, Chiueh, & Kopin, 1977). Animals should be divided into two time point groups on measures relevant to the study. One group (time 0 h) will be killed with no injection of AMPT, whereas the second group will be killed later (typically time 4 h) after initial AMPT injection.

5.1.1 Required materials

- Syringes and needles for AMPT injections (appropriate needles for ip injections and syringes for calculated volume)
- Glacial acetic acid (Fisher Scientific, Pittsburgh, PA)
- AMPT methyl ester hydrochloride (Cat #: M3281-1G; Sigma-Aldrich, St. Louis, MO)
    - Solution: Add 1 μl of glacial acetic acid per 1 mg of AMPT to make the initial AMPT dose. Then, for the subsequent doses dilute the initial dose with $0.15\ M$ NaCl.
- Balance to weigh animals day before test to calculate volume for AMPT injections.
- Microbalance for weighing tissue on test day.
- Dissection tools
    - For example, tissue forceps (Graefe: jar width 4.5 mm, length 4¼"; Cat #: RS-8248), curved serrated forceps (full curve; Cat #: RS-5137) or sharp surgical scissors (e.g., Cat #: RS-6702). All available from Roboz Surgical Instrument Co., Gaithersburg, MD.
- Weighing papers (Cat #: 09-898-12A; Fisher Scientific, Pittsburgh, PA) or aluminum foil for weighing tissue
    - Use adipose tissue weight to calculate mg tissue/ml NETO (see below for comment).
- Sharp scissors for rapid decapitation (Fiskars scissors, small animals) or guillotine (rats).

1. One group will be a 0 h time point group that receives no AMPT ip and are immediately decapitated for tissue harvest (see Fig. 11.4). The second set of animals will receive AMPT (hamsters: 250 mg AMPT/kg; 25 mg/ml; mice: 300 mg AMPT/kg; rats 250 mg AMPT/kg; Levin, 1995) injected ip.
2. Two hours later (or the halfway time point), the animals will be injected with a supplemental dose of AMPT (hamsters: 125 mg AMPT/kg; 25 mg/ml; mice: 150 mg AMPT/kg; rats: 125 mg AMPT/kg; Levin, 1995) to assure the maintenance of TH inhibition.

3. Two hours later (or at the last time point, e.g., 6 h later for brain, given the short half-life of NE in that tissue; Levin, 1995), the animals are decapitated and tissues are harvested.
4. During harvest, it is critical to do this rapidly and to keep them cold. Therefore, the headless carcass should be packed in crushed ice and the tissues of interest should be bathed frequently with ice cold 0.15 M NaCl during dissection. Tissues are weighed, placed in a suitable container (e.g., aluminum foil) labeled with animal number and tissue, and snap frozen in liquid nitrogen for subsequent storage at $-80\ °C$ until assayed for NE content to determine NETO.

5.2. Preparation of samples for HPLC-EC (NETO and NE content)

CATs are thermal and light labial and will therefore degrade easily. Handling them on ice and shielding from light throughout the preparation of samples ensures longevity of CATs. After extraction (see below), the liquid samples should be stored at $-80\ °C$ in acidic conditions (perchloric acid and ascorbic acid (PCA/AA); recipe below). While processing the samples, especially during bullet blending (see below), make certain that the sample is kept cold. To reduce electrochemical "noise" during subsequent HPLC analysis, HPLC-grade water should be used for all solutions.

5.2.1 Required materials
- Microbalance for weighing tissue
- Dissection and mincing tools (see Section 5.1)
- 70% PCA (Cat #: AC42403; Fisher Scientific, Pittsburgh, PA)
- L-Ascorbic acid (Cat #: A5960; Sigma-Aldrich, St. Louis, MO)
- Dihydroxybenzylamine (DHBA) (Sigma-Aldrich, St. Louis, MO)
- Pipets and appropriate tips
- Tris (Trizma base; Cat #:T1503-1 kg; Sigma-Aldrich, St. Louis, MO)
- Aluminum oxide (Al_2O_3; Sigma-Aldrich, St. Louis, MO)
- Multiflex round tips (1–200 µl; Sorenson, Salt Lake City, UT)
- Eppendorf® Safe-Lock tubes (Fisher Scientific, Pittsburgh, PA).
- 2 ml conical screw cap tubes (Fisher Scientific, Pittsburgh, PA)
- Bullet Blender (Next Advance, Averill Park, NY) and accessories below:
  - 50 µl scoop for Bullet Blender beads (Next Advance, Averill Park, NY)
  - Zirconium oxide beads (0.5 mm) for Bullet Blender (Next Advance, Averill Park, NY)
- Parafilm (Fisher Scientific, Pittsburgh, PA)
- Degree Vertical Multi-Function Rotator (Grant-Bio, Grants Pass, OR)

1. Make PCA/AA. Add 0.3 mg of L-ascorbic acid (AA) for every milliliter of 0.2 M PCA. Solution is light-sensitive and can be kept at room temperature.
2. Weigh out 0.015–0.250 g of frozen tissue and add to new 2 ml Eppendorf® Safe-Lock tubes containing 790 µl of PCA/AA and one scoop (~0.15 g) of Bullet Blender beads.
3. Make stock solution of DHBA (1,000,000 ng/ml), aliquot and store at $-80\ °C$.
4. Add 10 µl of the internal standard, DHBA in the appropriate concentration to each sample as followed (values based on hamster/mice tissues, but should be empirically determined for the animals and tissues used):
   - WAT, liver, skeletal muscle—10 ng of DHBA/10 µl (1 µg/ml)
   - BAT, pancreas—50 ng of DHBA/10 µl (5 µg/ml)
   - Heart—100 ng of DHBA/10 µl (10 µg/ml)
   - Adrenal—2.0 µg of DHBA/10 µl (200 µg/ml)
   - Plasma—2 ng of DHBA/10 µl (200 ng/ml)
5. Thoroughly mince with surgical scissors if tissue is >0.03 g.
6. Tightly wrap the cap of the tubes with a small piece of parafilm. Only Eppendorf Safe-Lock tubes should be used with the Bullet Blender. Do not wrap the entire tube. Place tubes in Bullet Blender. It is unnecessary to balance tube placement.
7. Using a Bullet Blender at speed setting #8, disrupt tissue at 1 min intervals, letting samples sit on ice for 2–3 min in between each run. To ensure the tubes are held tightly into place, lay a few napkins on top of the samples to seal. This should be repeated until tissue is fully homogenized, seemingly opaque with no visible large pieces remaining. Sonication is not a desirable method for homogenization of these tissues due to the thermolability of catecholamines by sonication-induced heat generation. For samples not minced in step 4, bullet blend 1–2 min longer. Maximum total bullet blending time is usually 5–6 min and a minimum of 3 min. For samples that are not completely homogenized at the end of 5–6 min, place back in Bullet Blender and bullet blend separately for 30 s, holding tube into place with hands. BAT will be more difficult to homogenize and will require more blending time.
8. Remove parafilm and centrifuge at $8000 \times g$ for 20 min at $4\ °C$. For BAT, centrifuge at $10,000 \times g$ for 20 min at $4\ °C$.
9. Carefully remove the homogenate under the fat layer and add to sterile 2 ml conical screw cap microcentrifuge tubes containing ~150 g of

ACTIVATED alumina (Al_2O_3). Use a 1000 µl pipette and appropriate tip to remove ~750 µl, then a 200 µl pipette with a gel loading tip (multiflex round tips 1–200 µl) to aspirate out the remaining fluid, allowing the Bullet Blender beads to act as a filter.

Note: Alumina can be purchased already activated from Sigma-Aldrich (Cat #: 199974). Alumina can be activated for catecholamine extraction using the following steps:
- Add 100 g of AL_2O_3 to a 1 l beaker. Add 500 ml of 2 N HCl. Heat continuously and shake for 45 min at 90–100 °C.
- Decant HCl to remove supernatant and small (thin) particles (after ±5 min off heat).
- Wash the precipitate (the remaining Al_2O_3) with 250 ml of 2 N HCl for 10 min at 70 °C. Decant HCl to remove supernatant and small (thin) particles (after ±5 min off heat).
- Wash with 500 ml 2 N HCl for 10 min at 50 °C. Decant HCl.
- Wash ±10 times with pure, Milli-Q water using ~200 ml each time until pH 3–4. Decant water.
- Heat at 120 °C for 1 h.
- Heat at 200 °C for 2 h.
- To cool, put on vacuum desiccator at 37 °C.

10. Add 1 ml of 0.5 *M* Tris. pH to 8.6. Adjust with 70% PCA.
11. Mix samples well using Degree Vertical Multi-Function Rotator for 20 min at 4 °C on speed 7, then vortex for 20 s to ensure the binding of CATs to the alumina.
12. Centrifuge at $8000 \times g$ for 2 min at 4 °C. For BAT centrifuge at $10,000 \times g$ for 2 min at 4 °C.
13. Remove and discard the supernatant. To avoid alumina aspiration, use a 1000 µl pipette to remove and discard ~850 µl of supernatant followed by using a 200 µl pipette using redi-tip 200 µl pipette tip for the remaining 150 µl.
14. Add 1 ml of HPLC-grade water to each tube to wash the alumina. Invert 10 times by hand then vortex for 5 s.
15. Centrifuge at $8000 \times g$ for 1 min at 4 °C. For BAT, centrifuge at $10,000 \times g$ for 1 min at 4 °C.
16. Repeat steps 12–15.
17. Add 200 µl of PCA/AA to elute the CATs. Allow samples to spin on Degree Vertical Multi-Function Rotator for 10 min in 4 °C fridge then vortex for 20 s to ensure desorption.
18. Centrifuge at $8000 \times g$ for 1 min at 4 °C.

19. Remove 180 μl of supernatant and store in −80 °C until run on the HPLC-EC. Be careful not to transfer alumina with the supernatant in order to prevent clogging of HPLC column during HPLC analysis.

5.3. HPLC-EC protocol (and NE content)

Settings and suggestions will vary with the HPLC-EC system used. The following is for the Coulochem II (ESA/DIONEX) system. Separation is performed by injecting 50 μl of sample onto a C-18 reverse-phase column (model HR-80; DIONEX/Thermo Fisher Scientific, Rockford, IL). Mobile phase is Cat-A-Phase II (Product: 45-0216; DIONEX/Thermo Fisher Scientific, Rockford, IL) at a flow rate of 1 ml/min. In-line filters are recommended to prevent accumulation of alumina in the analytical column. Coulometric electrochemical detection (Coulochem II; DIONEX/Thermo Fisher Scientific, Rockford, IL) settings are as follows: *guard cell* (DIONEX model 5021; Thermo Fisher Scientific, Rockford, IL) +350 mV, *analytical cell* (DIONEX model 5011; Thermo Fisher Scientific, Rockford, IL) *cell 1* +10 mV, and *cell 2* −300 mV. A three-point external standard curve is used for quantitation. Standards containing known concentrations of NE, E, and DA are injected every 10 unknowns. A single-point internally standardized method using dihydroxybenzylamine (DHBA; added during sample preparation) is used to control for extraction efficiency. Typical retention times for NE, E, DHBA, and DA on this system with settings/processing as above are 4, 4.5, 7, and 9.5 min, respectively.

NE concentrations are determined for the pre- and post-AMPT treatment groups. The total tissue weight of the adipose depot, the amount of adipose tissue used for catecholamine extraction, and the NE content obtained from HPLC-EC quantitation are used to calculate the concentration of NE per gram of adipose for both baseline and AMPT-injected animals (e.g., time=4 h; Brodie, Costa, Dlabar, Neff, & Smooker, 1966). Then, the NE content of the total adipose tissue (calculating NETO at fat pad level) is calculated by multiplying the adipose tissue mass (at time of harvest) and the concentration of NE per gram of adipose tissue for both baseline and AMPT-injected animals (e.g., time=4 h).

We use the following equation based on the method of Brodie et al. (1966): $k=(\log[NE]_0 - \log[NE]_4)/(0.434 * 4)$ and $K=k^*NE_0$. To obtain the rate of NE efflux (k) that is used to obtain NETO, take the log NE content in the total fat tissue mass (log[NE]) and subtract the log of NE content of the total tissue mass at baseline and at "time=4 h" animals

$(\log[NE]_0 - \log[NE]_4/(0.434*4)$. The NETO ($K$) is obtained by multiplying the rate of NE efflux by the initial NE content of baseline animals (Brito et al., 2007, 2008; Shi, Bowers, & Bartness, 2004; Youngstrom & Bartness, 1995).

6. ASSESSMENT OF SENSORY DENERVATION USING CGRP ELIA

Direct injection of capsaicin into WAT and BAT selectively and effectively destroys small unmyelinated C-fiber sensory nerves. This can be tested by assessing levels of the sensory nerve-associated peptides CGRP and/or substance P for effectiveness of the denervation and TH (ELIA) or NE (HPLC-EC) for selectivity. In our experience (unpublished observations), we find equivalent values for the percent depletions of SP and CGRP following intra-WAT or intra-BAT capsaicin injections. Therefore, we usually assess only CGRP. The presence of these sensory peptides has been identified immunohistochemically at the level of both WAT and BAT depots (Foster & Bartness, 2006; Giordano et al., 1998; Giordano, Morroni, Santone, Marchesi, & Cinti, 1996; Norman, Mukherjee, Symons, Jung, & Lever, 1988; Shi et al., 2005).

6.1. Required materials

- Microbalance for weighing tissue
- Dissection tools (see Section 5.1)
- Weighing papers (Fisher Scientific, Pittsburgh, PA; Cat #: 09-898-12A) or aluminum foil for weighing tissue
- Sharp scissors for rapid decapitation (Fiskars scissors, small animals) or guillotine (rats)
- Bullet Blender (Next Advance, Averill Park, NY) and accessories (see Section 5.2)
- 2 M acetic acid
- 2 ml conical screw cap tubes (Fisher Scientific, Pittsburgh, PA)
- Eppendorf® Safe-Lock tubes (Fisher Scientific, Pittsburgh, PA)
- Lyophilizer (We use the FreeZone 61 Shell Freezer model from Labconco, Kansas City, CO)
  - Accessories we use for FreeZone lyophilizer: Fast-freeze flask, Fast-freeze flask top and Lyph-lock flask adapter tube (Labconco, Kansas City, CO). Assemble these items together and use to hold samples for attachment to lyophilizer

- CGRP ELIA kit (Cat # 589101; Cayman Chemical, Ann Arbor, MI)
- Spectrophotometer

6.2. Preparation of samples

We typically use fresh adipose tissue for homogenization before the CGRP ELIA and it is unnecessary to fix tissue. If tissue will not be processed the same day as sacrifice, then weigh tissue, place in aluminum foil, snap freeze, and store at $-80\ °C$. When using tissue that has been frozen, place frozen tissue in Safe-Lock tube with $2\ M$ acetic acid and one scoop of zirconium beads and follow the homogenization steps (see below).

6.2.1 Protocol for homogenizing adipose tissue with Bullet Blender

1. Use 150–300 mg of tissue for analysis and place into a Safe-Lock microcentrifuge tube.
2. Add 750 µl $2\ M$ acetic acid into each tube; for tissue <0.40 g, 400 µl is enough.
3. Thoroughly mince with surgical scissors if tissue is >0.03 g.
4. Add one scoop of zirconium oxide beads (0.5 mm) to each tube.
5. Close the microcentrifuge tubes and place tubes into the Bullet Blender.
6. Using the Bullet Blender at speed setting # 6, disrupt tissue at 1 min intervals. This should be repeated until tissue is fully homogenized, relatively opaque and containing no visible large pieces.
7. Heat at $90\ °C$ for 10 min in a hot block.
8. Centrifuge samples at $4\ °C$ at $13,000 \times g$ for 15 min.
9. Aspirate about 725 µl of the infranatant under the fat cake and add to sterile 2 ml conical screw cap microcentrifuge tube. Use a 200 µl pipette with a gel loading tip (multiflex round tips 1–200 µl) to aspirate out the remainder allowing the Bullet Blender beads to act as a filter. Place conical screw cap tube immediately on dry ice to freeze and store at $-80\ °C$ until ready to lyophilize.

6.2.2 Preparation of samples for ELIA using lyophilizer

1. Remove samples from $-80\ °C$.
2. Lyophilize samples according to manufacturer's directions for operation.
3. If using a FastFreeze flask, unscrew caps about halfway to allow vacuum suction access to the samples. Do not place the flask in freezer (-20 to $-80\ °C$) or dry ice while placing samples in flask (use box grid). Use wet ice while placing samples into flask or refrigerate ($4\ °C$) if loading more than one flask as the flasks will crack if placed in freezer.

4. When samples are all lyophilized, screw on caps tightly and place in −80 °C until assaying CGRP by ELIA.
5. Follow directions in CGRP ELIA kit to test for CGRP content.

7. EXPRESSION OF DATA

There are numerous arguments for how any or all of the above data should be expressed. For verification of sympathetic (NE content) or sensory (CGRP or SP content), the data can be expressed as the substance amount/mg tissue or substance amount/mg protein. As long as one is making comparisons between similar tissues (IBAT controls and IBAT denervated) with relatively the same amount of tissue, the total amount of the substance/pad seems more appropriate because physiology does not work on a relative basis. For example, a small change in a substance found in a large tissue (e.g., a liver enzyme) might be statistically nonsignificant when expressed per mg protein or mg tissue, but because of the size of the tissue (e.g., liver or muscle), physiologically this may be highly significant. Alternatively, the data can be expressed as a % of the control tissue (i.e., % depletion) based on total or relative values. When analyzing NETO, we also take the view that physiology does not work in a relative manner and we therefore express the data per adipose depot (e.g., Brito et al., 2007, 2008).

ACKNOWLEDGMENTS

The authors thank all the people who have contributed to the protocols presented herein, in particular: Drs. Marcia and Nilton Brito, Dr. Haifei Shi, Dr. Michelle Foster, Dr. Yogendra Shrestha, and Dr. Yang Liu. We thank John T. Garretson and Ngoc Ly Nguyen for technical assistance. This work was funded by National Institutes of Health R37 DK35254 to T. J. B. and NRSA F32 DK082143 to C. H. V.

REFERENCES

Adler, E. S., Hollis, J. H., Clarke, I. J., Grattan, D. R., & Oldfield, B. J. (2012). Neurochemical characterization and sexual dimorphism of projections from the brain to abdominal and subcutaneous white adipose tissue in the rat. *Journal of Neuroscience, 32,* 15913–15921.
Bamshad, M., Aoki, V. T., Adkison, M. G., Warren, W. S., & Bartness, T. J. (1998). Central nervous system origins of the sympathetic nervous system outflow to white adipose tissue. *The American Journal of Physiology, 275,* R291–R299.
Bamshad, M., Song, C. K., & Bartness, T. J. (1999). CNS origins of the sympathetic nervous system outflow to brown adipose tissue. *The American Journal of Physiology, 276,* R1569–R1578.
Bartness, T. J., & Bamshad, M. (1998). Innervation of mammalian white adipose tissue: Implications for the regulation of total body fat. *The American Journal of Physiology, 275,* R1399–R1411.

Bartness, T. J., Billington, C. J., Levine, A. S., Morley, J. E., Brown, D. M., & Rowland, N. E. (1986). Insulin and metabolic efficiency in rats. I: Effect of sucrose feeding and BAT axotomy. *The American Journal of Physiology, 251*, R1109–R1117.

Bartness, T. J., Demas, G. E., & Song, C. K. (2002). Seasonal changes in adiposity: The roles of the photoperiod, melatonin and other hormones and the sympathetic nervous system. *Experimental Biology and Medicine, 227*, 363–376.

Bartness, T. J., & Song, C. K. (2007). Sympathetic and sensory innervation of white adipose tissue. *Journal of Lipid Research, 48*, 1655–1672.

Bartness, T. J., Vaughan, C. H., & Song, C. K. (2010). Sympathetic and sensory innervation of brown adipose tissue. *International Journal of Obesity, 34*(Suppl. 1), S36–S42.

Bartness, T. J., & Wade, G. N. (1984). Effects of interscapular brown adipose tissue denervation on estrogen-induced changes in food intake, body weight and energy metabolism. *Behavioral Neuroscience, 98*, 674–685.

Bartness, T. J., & Wade, G. N. (1985). Photoperiodic control of seasonal body weight cycles in hamsters. *Neuroscience and Biobehavioral Reviews, 9*, 599–612.

Benarroch, E. E., Schmelzer, J. D., Ward, K. K., Nelson, D. K., & Low, P. A. (1990). Noradrenergic and neuropeptide Y mechanisms in guanethidine-sympathectomized rats. *The American Journal of Physiology, 259*, R371–R375.

Beznak, A. B. L., & Hasch, Z. (1937). The effect of sympathectomy on the fatty deposit in connective tissue. *Quarterly Journal of Experimental Physiology, 27*, 1–15.

Bowers, R. R., Festuccia, W. T. L., Song, C. K., Shi, H., Migliorini, R. H., & Bartness, T. J. (2004). Sympathetic innervation of white adipose tissue and its regulation of fat cell number. *The American Journal of Physiology, 286*, R1167–R1175.

Brito, M. N., Brito, N. A., Baro, D. J., Song, C. K., & Bartness, T. J. (2007). Differential activation of the sympathetic innervation of adipose tissues by melanocortin receptor stimulation. *Endocrinology, 148*, 5339–53347.

Brito, N. A., Brito, M. N., & Bartness, T. J. (2008). Differential sympathetic drive to adipose tissues after food deprivation, cold exposure or glucoprivation. *American Journal of Physiology. Regulatory, Integrative and Comparative Physiology, 294*, R1445–R1452.

Brodie, B. B., Costa, E., Dlabar, A., Neff, H., & Smooker, H. H. (1966). Application of steady state kinetics to the estimation of synthesis rate and turnover time of tissue catecholamines. *Journal of Pharmacology and Experimental Therapeutics, 154*, 494–498.

Burnstock, G., Evans, B., Gannon, B. J., Heath, J. W., & James, V. (1971). A new method of destroying adrenergic nerves in adult animals using guanethidine. *British Journal of Pharmacology, 43*(2), 295–301.

Cannon, B., & Nedergaard, J. (2004). Brown adipose tissue: Function and physiological significance. *Physiological Reviews, 84*, 277–359.

Cantu, R. C., & Goodman, H. M. (1967). Effects of denervation and fasting on white adipose tissue. *The American Journal of Physiology, 212*, 207–212.

Castonguay, T. W., & Bellinger, L. L. (1987). Capsaicin and its effects upon meal patterns, and glucagon and epinephrine suppression of food intake. *Physiology and Behavior, 40*, 337–342.

Cinti, S. (1999). *The adipose organ*. Milano: Editrice Kurtis.

Cooper, J. R., Bloom, F. E., & Roth, R. H. (1982). *The biochemical basis of neuropharmacology*. New York: Oxford University Press.

Correll, J. W. (1963). Adipose tissue: Ability to respond to nerve stimulation in vitro. *Science, 140*, 387–388.

Cui, J., & Himms-Hagen, J. (1992a). Long-term decrease in body fat and in brown adipose tissue in capsaicin-desensitized rats. *The American Journal of Physiology, 262*, R568–R573.

Cui, J., & Himms-Hagen, J. (1992b). Rapid but transient atrophy of brown adipose tissue in capsaicin-desensitized rats. *The American Journal of Physiology, 262*, R562–R567.

Cui, J., Zaror-Behrens, G., & Himms-Hagen, J. (1990). Capsaicin desensitization induces atrophy of brown adipose tissue in rats. *The American Journal of Physiology, 259*, R324–R332.
Demas, G. E., & Bartness, T. J. (2001a). Direct innervation of white fat and adrenal medullary catecholamines mediate photoperiodic changes in body fat. *The American Journal of Physiology, 281*, R1499–R1505.
Demas, G. E., & Bartness, T. J. (2001b). Novel method for localized, functional sympathetic nervous system denervation of peripheral tissue using guanethidine. *Journal of Neuroscience Methods, 112*, 21–28.
Depocas, F., & Behrens, W. A. (1977). Effects of handling, decapitation, anesthesia, and surgery on plasma noradrenaline levels in the white rat. *Canadian Journal of Physiology and Pharmacology, 55*, 212–219.
Depocas, F., Foster, D. O., Zaror-Behrens, G., Lacelle, S., & Nadeau, B. (1984a). Recovery of function in sympathetic nerves of interscapular brown adipose tissue of rats treated with 6-hydroxydopamine. *Canadian Journal of Physiology and Pharmacology, 62*, 1327–1332.
Dobbins, R. L., Szczepaniak, L. S., Zhang, W., & McGarry, J. D. (2003). Chemical sympathectomy alters regulation of body weight during prolonged ICV leptin infusion. *American Journal of Physiology. Endocrinology and Metabolism, 284*, E778–E787.
Foster, M. T., & Bartness, T. J. (2006). Sympathetic but not sensory denervation stimulates white adipocyte proliferation. *American Journal of Physiology. Regulatory, Integrative and Comparative Physiology, 291*, R1630–R1637.
Foster, D. O., Depocas, F., & Zaror-Behrens, G. (1982). Unilaterality of the sympathetic innervation of each pad of interscapular brown adipose tissue. *Canadian Journal of Physiology and Pharmacology, 60*, 107–113.
Foster, D. O., Depocas, F., & Zuker, M. (1982). Heterogeneity of the sympathetic innervation of rat interscapular brown adipose tissue via intercostal nerves. *Canadian Journal of Physiology and Pharmacology, 60*, 747–754.
Frontini, A., & Cinti, S. (2010). Distribution and development of brown adipocytes in the murine and human adipose organ. *Cell Metabolism, 11*, 253–256.
Giordano, A., Morroni, M., Carle, F., Gesuita, R., Marchesi, G. F., & Cinti, S. (1998). Sensory nerves affect the recruitment and differentiation of rat periovarian brown adipocytes during cold acclimation. *Journal of Cell Science, 111*, 2587–2594.
Giordano, A., Morroni, M., Santone, G., Marchesi, G. F., & Cinti, S. (1996). Tyrosine hydroxylase, neuropeptide Y, substance P, calcitonin gene-related peptide and vasoactive intestinal peptide in nerves of rat periovarian adipose tissue: An immunohistochemical and ultrastructural investigation. *Journal of Neurocytology, 25*, 125–136.
Giordano, A., Song, C. K., Bowers, R. R., Ehlen, J. C., Frontini, A., Cinti, S., et al. (2006). White adipose tissue lacks significant vagal innervation and immunohistochemical evidence of parasympathetic innervation. *The American Journal of Physiology, 291*, R1243–R1255.
Hamilton, J. M., Bartness, T. J., & Wade, G. N. (1989). Effects of norepinephrine and denervation on brown adipose tissue in Syrian hamsters. *The American Journal of Physiology, 257*, R396–R404.
Harris, R. B. (2012). Sympathetic denervation of one white fat depot changes norepinephrine content and turnover in intact white and brown fat depots. *Obesity (Silver Spring), 20*, 1355–1364.
Havel, R. J., & Goldfien, A. (1959). The role of the sympathetic nervous system in the metabolism of free fatty acids. *Journal of Lipid Research, 1*, 102–108.
Himms-Hagen, J. (1991). Neural control of brown adipose tissue thermogenesis, hypertrophy, and atrophy. *Frontiers in Neuroendocrinology, 12*, 38–93.

Himms-Hagen, J., Cui, J., & Sigurdson, S. L. (1990). Sympathetic and sensory nerves in control of growth of brown adipose tissue: Effects of denervation and of capsaicin. *Neurochemistry International, 17*, 271–279.

Jansco, G., Kiraly, E., & Jansco-Gabor, A. (1980). Direct evidence for an axonal site of action of capsaicin. *Nauyn-Schmiedeberg's Archives of Pharmacology, 31*, 91–94.

Jansco, G., Kiraly, E., Joo, F., Such, G., & Nagy, A. (1985). Selective degeneration by capsaicin of a subpopulation of primary sensory neurons in the adult rat. *Neuroscience Letters, 59*, 209–214.

Jancso-Gabor, A., Szolcsanyi, J., & Jancso, N. (1970). Stimulation and desensitization of the hypothalamic heat-sensitive structures by capsaicin in rats. *The Journal of Physiology, 208*, 449–459.

Johnson, J. E., Jr., & O'Brien, F. (1976). Evaluation of the permanent sympathectomy produced by the administration of guanethidine to adult rats. *Journal of Pharmacology and Experimental Therapeutics, 196*, 53–61.

Joost, H. G., & Quentin, S. H. (1984). Effects of chemical sympathectomy on insulin receptors and insulin action in isolated rat adipocytes. *Journal of Pharmacology and Experimental Therapeutics, 229*, 839–844.

Kawada, T., Hagihara, K. I., & Iwai, K. (1986). Effects of capsaicin on lipid metabolism in rats fed a high fat diet. *Journal of Nutrition, 116*, 1272–1278.

Klingenspor, M., Meywirth, A., Stohr, S., & Heldmaier, G. (1994). Effect of unilateral surgical denervation of brown adipose tissue on uncoupling protein mRNA level and cytochrom-c-oxidase activity in the Djungarian hamster. *Journal of Comparative Physiology. B, 163*, 664–670.

Knyhar, E., Ristovsky, K., Kalman, G., & Csillik, B. (1969). Chemical sympathectomy: Histochemical and submicroscopical consequences of 6-hydroxy-dopamine treatment in the rat iris. *Experientia 25*(5), 518–520.

Lazzarini, S. J., & Wade, G. N. (1991). Role of sympathetic nerves in effects of estradiol on rat white adipose tissue. *The American Journal of Physiology, 260*, R47–R51.

Levin, B. E. (1995). Reduced norepinephrine turnover in organs and brains of obesity-prone rats. *The American Journal of Physiology, 268*, R389–R394.

Levin, B. E., & Sullivan, A. C. (1984). Dietary obesity and neonatal sympathectomy. II. Thermoregulation and brown adipose metabolism. *The American Journal of Physiology, 247*, R988–R994.

Levin, B. E., Triscari, J., Marquet, E., & Sullivan, A. C. (1984). Dietary obesity and neonatal sympathectomy. I. Effects on body composition and brown adipose. *The American Journal of Physiology, 247*, R979–R987.

Mayerhofer, A., Amador, A. G., Steger, R. W., & Bartke, A. (1990). Testicular function after local injection of 6-hydroxydopamine or norepinephrine in the golden hamster (Mesocricetus auratus). *Journal of Andrology, 11*, 301–311.

Mefford, I. N. (1981). Application of high performance liquid chromatography with electrochemical detection to neurochemical analysis: Measurement of catecholamines, serotonin and metabolites in rat brain. *Journal of Neuroscience Methods, 3*, 207–224.

Melnyk, A., & Himms-Hagen, J. (1994). Leanness in capsaicin-desensitized rats one year after treatment. *International Journal of Obesity, 18*(Suppl. 2), 130, Ref Type: Abstract.

Mory, G., Ricquier, D., Nechad, M., & Hemon, P. (1982). Impairment of trophic response of brown fat to cold in guanethidine-treated rats. *The American Journal of Physiology, 242*, C159–C165.

Murphy, K. T., Schwartz, G. J., Nguyen, N. L., Mendez, J. M., Ryu, V., & Bartness, T. J. (2013). Leptin sensitive sensory nerves innervate white fat. *American Journal of Physiology. Endocrinology and Metabolism, 304*, E1338–E1347.

Niijima, A. (1998). Afferent signals from leptin sensors in the white adipose tissue of the epididymis, and their reflex effect in the rat. *Journal of the Autonomic Nervous System, 73,* 19–25.

Niijima, A. (1999). Reflex effects from leptin sensors in the white adipose tissue of the epididymis to the efferent activity of the sympathetic and vagus nerve in the rat. *Neuroscience Letters, 262,* 125–128.

Norman, D., Mukherjee, S., Symons, D., Jung, R. T., & Lever, J. D. (1988). Neuropeptides in interscapular and perirenal brown adipose tissue in the rat: A plurality of innervation. *Journal of Neurocytology, 17,* 305–311.

Oldfield, B. J., Giles, M. E., Watson, A., Anderson, C., Colvill, L. M., & McKinley, M. J. (2002). The neurochemical characterisation of hypothalamic pathways projecting polysynaptically to brown adipose tissue in the rat. *Neuroscience, 110,* 515–526.

Paschoalini, M. A., & Migliorini, R. H. (1990). Participation of the CNS in the control of FFA mobilization during fasting in rabbits. *Physiology and Behavior, 47,* 461–465.

Popper, C. W., Chiueh, C. C., & Kopin, I. J. (1977). Plasma catecholamine concentrations in unanesthetized rats during sleep, wakefulness, immobilization and after decapitation. *Journal of Pharmacology and Experimental Therapeutics, 202,* 144–148.

Raybould, H. E., Holzer, P., Reddy, S. N., Yang, H., & Tache, Y. (1990). Capsaicin-sensitive vagal afferents contribute to gastric acid and vascular responses to intracisternal TRH analog. *Peptides, 11,* 789–795.

Rooks, C. R., Penn, D. M., Kelso, E., Bowers, R. R., Bartness, T. J., & Harris, R. B. (2005). Sympathetic denervation does not prevent a reduction in fat pad size of rats or mice treated with peripherally administered leptin. *American Journal of Physiology. Regulatory, Integrative and Comparative Physiology, 289,* R92–R102.

Seydoux, J., Mory, G., & Girardier, L. (1981). Short-lived denervation of brown adipose tissue of the rat induced by chemical sympathetic denervation. *Journal of Physiology, 77,* 1017–1022.

Shi, H., & Bartness, T. J. (2001). Neurochemical phenotype of sympathetic nervous system outflow from brain to white fat. *Brain Research Bulletin, 54,* 375–385.

Shi, H., & Bartness, T. J. (2005). White adipose tissue sensory nerve denervation mimics lipectomy-induced compensatory increases in adiposity. *The American Journal of Physiology, 289,* R514–R520.

Shi, H., Bowers, R. R., & Bartness, T. J. (2004). Norepinephrine turnover in brown and white adipose tissue after partial lipectomy. *Physiology & Behavior, 81,* 535–543.

Shi, Z., Chen, W. W., Xiong, X. Q., Han, Y., Zhou, Y. B., Zhang, F., et al. (2012). Sympathetic activation by chemical stimulation of white adipose tissues in rats. *Journal of Applied Physiology, 112,* 1008–1014.

Shi, H., Song, C. K., Giordano, A., Cinti, S., & Bartness, T. J. (2005). Sensory or sympathetic white adipose tissue denervation differentially affects depot growth and cellularity. *The American Journal of Physiology, 288,* R1028–R1037.

Sidman, R. L., & Fawcett, D. W. (1954). The effect of peripheral nerve section on some metabolic responses of brown adipose tissue in mice. *Anatomical Record, 118,* 487–501.

Song, C. K., & Bartness, T. J. (2001). CNS sympathetic outflow neurons to white fat that express melatonin receptors may mediate seasonal adiposity. *The American Journal of Physiology, 281,* R666–R672.

Song, C. K., & Bartness, T. J. (2007). Central projections of the sensory nerves innervating brown adipose tissue. In *Neuroscience meeting planner,* San Diego, CA: Society for Neuroscience, Online.

Song, C. K., Jackson, R. M., Harris, R. B., Richard, D., & Bartness, T. J. (2005). Melanocortin-4 receptor mRNA is expressed in sympathetic nervous system outflow neurons to white adipose tissue. *American Journal of Physiology. Regulatory, Integrative and Comparative Physiology, 289,* R1467–R1476.

Song, C. K., Schwartz, G. J., & Bartness, T. J. (2009). Anterograde transneuronal viral tract tracing reveals central sensory circuits from white adipose tissue. *American Journal of Physiology. Regulatory, Integrative and Comparative Physiology, 296*, R501–R511.

Song, C. K., Vaughan, C. H., Keen-Rhinehart, E., Harris, R. B., Richard, D., & Bartness, T. J. (2008). Melanocòrtin-4 receptor mRNA expressed in sympathetic outflow neurons to brown adipose tissue: Neuroanatomical and functional evidence. *The American Journal of Physiology, 295*, R417–R428.

Teixeira, V. L., Antunes-Rodrigues, J., & Migliorini, R. H. (1973). Evidence for centers in the central nervous system that selectively regulate fat mobilization in the rat. *Journal of Lipid Research, 14*, 672–677.

Thureson-Klein, A., Lagercrantz, H., & Barnard, T. (1976). Chemical sympathectomy of interscapular brown adipose tissue. *Acta Physiologica Scandinavica, 98*, 8–18.

Ungerstedt, U. (1968). 6-Hydroxy-dopamine induced degeneration of central monoamine neurons. *European Journal of Pharmacology, 5*(1), 107–110.

Vaughan, C. H., & Bartness, T. J. (2012). Anterograde transneuronal viral tract tracing reveals central sensory circuits from brown fat and sensory denervation alters its thermogenic responses. *The American Journal of Physiology, 302*, R1049–R1058.

Wool, I. B., Goldstein, M. S., Ramey, E. R., & Levine, B. (1954). Role of epinephrine in the physiology of fat mobilization. *The American Journal of Physiology, 178*, 427–432.

Xiong, X. Q., Chen, W. W., Han, Y., Zhou, Y. B., Zhang, F., Gao, X. Y., et al. (2012). Enhanced adipose afferent reflex contributes to sympathetic activation in diet-induced obesity hypertension. *Hypertension, 60*, 1280–1286.

Yoneda, M., & Raybould, H. E. (1990). Capsaicin-sensitive vagal afferent fibers do not contribute to histamine H_2 receptor agonist-induced gastric acid secretion in anesthetized rats. *European Journal of Pharmacology, 186*, 349–352.

Youngstrom, T. G., & Bartness, T. J. (1995). Catecholaminergic innervation of white adipose tissue in the Siberian hamster. *The American Journal of Physiology, 268*, R744–R751.

Youngstrom, T. G., & Bartness, T. J. (1998). White adipose tissue sympathetic nervous system denervation increases fat pad mass and fat cell number. *The American Journal of Physiology, 275*, R1488–R1493.

Zhang, Y., Kerman, I. A., Laque, A., Nguyen, P., Faouzi, M., Louis, G. W., et al. (2011). Leptin-receptor-expressing neurons in the dorsomedial hypothalamus and median preoptic area regulate sympathetic brown adipose tissue circuits. *The Journal of Neuroscience, 31*, 1873–1884.

CHAPTER TWELVE

Measurement and Manipulation of Human Adipose Tissue Blood Flow Using Xenon Washout Technique and Adipose Tissue Microinfusion

Richard Sotornik[1], Jean-Luc Ardilouze
Division of Endocrinology, Department of Medicine, Université de Sherbrooke, Quebec, Canada
[1]Corresponding author: e-mail address: richard.sotornik@usherbrooke.ca

Contents

1. Introduction	228
2. AT ^{133}Xenon Washout—Principle of the Method	229
3. AT Microinfusion—Principle of the Method	230
4. Materials	231
4.1 Technical equipments	231
4.2 Consumable materials	232
5. Procedure	232
5.1 ^{133}Xe preparation	232
5.2 Preparation of agents	232
5.3 Subjects preparation	233
5.4 Microinfusion protocol	234
5.5 Modifications to the method and drawbacks	236
6. Calculations	236
6.1 ATBF values	236
6.2 Baseline ATBF and fasting ATBF under the influence of a vasoactive agent	236
6.3 Stimulated (postglucose) ATBF	237
7. Other Techniques Used in ATBF Measurement	237
7.1 The microdialysis technique	237
7.2 Laser doppler flowmetry	238
7.3 Positron emission tomography	238
7.4 Real-time contrast-enhanced ultrasound	239
8. Conclusion	239
Acknowledgments	239
References	240

Abstract

Adipose tissue (AT) is a very active organ, both metabolically and hormonally. These important functions depend on adequate blood flow (BF). Metabolic, hormonal, and vascular processes within AT are highly interconnected and any disruption will invariably impact the others. Therefore, any alteration of ATBF with obesity and/or insulin resistance will impact metabolic and hormonal AT functions. Similarly, metabolic or hormonal changes in AT will lead to ATBF disturbance. Thus, it is plausible that insufficient ATBF alters AT metabolic processes and response to regulatory signals, and may even aggravate the negative impacts of dysfunction in AT. The role of BF in AT metabolism can be evaluated by several techniques, but the xenon washout method is considered the "gold" standard. This technique can be combined with local microinfusion protocols, and the combination allows for precise assessment of mechanisms implicated in ATBF regulation.

1. INTRODUCTION

Adipose tissue (AT) has complex metabolic and hormonal roles. Therefore, the regulation of blood flow (BF) within the AT is highly regulated. Various techniques have been developed to measure ATBF in physiological and pathophysiological situations. In healthy people, AT is a very active organ with fasting ATBF around 3–5 ml (100 g/min) (Ardilouze, Fielding, Currie, Frayn, & Karpe, 2004; Sotornik et al., 2012), whereas BF oscillates around 1.5–3 ml (100 g/min) in resting skeletal muscle (Elia & Kurpad, 1993; Tobin, Simonsen, & Bulow, 2010). ATBF increases two to four times in response to a meal or physical activity (Bulow & Madsen, 1976; Frayn, 2010). Mental stress also increases ATBF (Linde, Hjemdahl, Freyschuss, & Juhlin-Dannfelt, 1989). However, both fasting- and meal-stimulated ATBF are significantly reduced in obese, insulin-resistant individuals (Andersson et al., 2010; Karpe, Fielding, Ilic, Macdonald, et al., 2002; Summers, Samra, Humphreys, Morris, & Frayn, 1996) and in subjects with type 2 diabetes (Dimitriadis et al., 2007).

Measurement of metabolic fluxes within AT is based on Fick's equation: $Ex = BF \times (A - V)$, where any metabolic or hormonal exchange (Ex) depends on the product of BF and arteriovenous difference $(A - V)$. Thus, using arterialized, venous, or interstitial concentrations, it is possible to calculate the local kinetics of glucose, lactate, triglycerides (Samra et al., 1996), glycerol, or fatty acids (Jansson, Larsson, Smith, & Lonnroth, 1992; McQuaid et al., 2011). ATBF measurement techniques also allow the assessment of ATBF regulatory mechanisms. For example, using

Table 12.1 List of factors known to influence ATBF

Basal ATBF	Postprandial ATBF	ATBF in physical activity
NO (+)	β2-Adrenergic effect (+)	α,β-Adrenergic effects
α-Adrenergic effect (−)	Insulin (+; indirect effect)	ANP (+)
ANGII (−)		Adenosine (+)

ANGII, angiotensin II; ANP, atrial natriuretic peptide; ATBF, adipose tissue blood flow; NO, nitric oxide; (+), stimulation; (−), inhibition.

systemic (intravenous) (Quisth et al., 2005) and local (microdialysis or AT microinfusion) administration (Martin et al., 2011), it was found that basal ATBF is controlled by nitric oxide (NO), with a vasodilatatory effect (Ardilouze, Fielding, et al., 2004), and angiotensin II and α-adrenergic system signaling, both with inhibitory effects (Ardilouze, Fielding, et al., 2004; Goossens et al., 2006). Whereas the postprandial ATBF surge is driven by stimulation of β-adrenergic receptors (Karpe, Fielding, Ardilouze, et al., 2002; Simonsen, Bulow, Astrup, Madsen, & Christensen, 1990), during physical activity, atrial natriuretic peptide (Galitzky et al., 2001), β-adrenergic agonists, and locally produced adenosine (Heinonen et al., 2012) are the principal regulatory factors (Table 12.1).

2. AT ^{133}Xenon WASHOUT—PRINCIPLE OF THE METHOD

In general, tissue blood flow (TBF) is measured quantitatively by analyzing the clearance of a local radioisotope depot (Lassen, Lindbjerg, & Munck, 1964) injected directly into a tissue. This is based on the assumption that the faster the disappearance of the isotope, the higher the BF. TBF is then calculated as the product of the exponential rate constant per second (k, in ln counts/s) (calculated from the semilog plot of disappearance of counts vs. time) and the tissue/blood coefficient of partition (λ, in ml/g) of the isotope: $\text{TBF} = k \times \lambda \times 100(g) \times 60(s)$. The results are expressed in milliliter (100 g of tissue/min). The theory relies on the assumption that a diffusion coefficient equilibrium is maintained between tissue-dissolved and blood isotope throughout the study: $C = \lambda \times C_{\text{blood}}$, where C is the amount of isotope per gram of tissue and C_{blood} is the amount of isotope per milliliter of blood (Larsen, Lassen, & Quaade, 1966).

In AT, ^{133}xenon (^{133}Xe), a γ-emitting isotope, is most often employed. The ATBF measurement technique, called xenon washout, was introduced in 1966 (Larsen et al., 1966). The inert and lipophilic nature of ^{133}Xe makes it ideal for ATBF studies. ^{133}Xe has a high partition coefficient (λ) between plasma and fat tissue, which means a considerable fat solubility and longer retention in the tissue. Moreover, xenon is quantitatively exhaled during passage through the lungs, so recirculation is not a problem. A prerequisite for ATBF calculation is knowledge of the value of the partition coefficient for xenon. λ has been estimated from the chemical composition of AT (Bulow, Jelnes, Astrup, Madsen, & Vilmann, 1987), skinfold thickness (Jansson & Lonnroth, 1995), and other indirect calculations (Blaak et al., 1995). The best, but complicated method, is based on the biochemical analysis of biopsies (Bulow, 2001). With this method, the average λ for ^{133}Xe was 8.2 ± 1.2 ml/g in the subcutaneous tissue of the abdomen and the thigh in a cohort of lean subjects (Bulow et al., 1987). A difference between lean and obese subjects was found when λ was estimated from the abdominal fat cell size (needle biopsies were taken from subcutaneous AT). λ was 8.6 ml/g in lean subjects and 9.9 ml/g in obese subjects (Jansson & Lonnroth, 1995). Nevertheless, because of the difficulty of determining λ in humans, an average value of 10 ml/g is usually used for calculation of ATBF (Martin et al., 2011).

3. AT MICROINFUSION—PRINCIPLE OF THE METHOD

To test the ATBF response to a potential vasoactive substance (i.e., to study the underlying mechanisms of ATBF regulation) several approaches have been used (see Section 7). One possibility is the combination of xenon washout technique with systemic administration of potential vasoactive agents (Blaak et al., 1995). However, exposure of the whole body to the drug may interfere with the interpretation of direct results in AT due to indirect effects of the drug elsewhere (e.g., metabolic, cardiovascular, neural). The microdialysis technique in combination with the ethanol escape method can be seen as another solution, although the substance use and delivery is limited by the properties of the microdialysis membrane (Lafontan & Arner, 1996). Moreover, the ethanol escape method has several methodological limitations (Martin et al., 2011). The best choice is adipose tissue microinfusion, which was introduced in 2002 (Karpe, Fielding, Ardilouze, et al., 2002), and which overcomes the drawbacks of systemic drug administration and the limitations of microdialysis. In principle, the infusion of a vasoactive agent is made directly into the exact location that ^{133}Xe has been deposited, allowing the recording of ^{133}Xe disappearance in relation to changes in local tissue environment (Karpe, Fielding, Ardilouze, et al., 2002).

4. MATERIALS
4.1. Technical equipments
4.1.1 Scintillation detectors

The first detectors used to measure the residual radioactivity were of the NaI crystal photomultiplier tube type, and were mounted on a fixed frame at a set distance from the injected pool. A lightweight detector with a cesium crystal became available in 1995 (Oakfield Instruments, Eynsham, UK) (Fig. 12.1) which can be directly strapped onto the abdominal wall (Samra, Frayn, Giddings, Clark, & Macdonald, 1995). The γ-counter probes are connected to the Mediscint device (Oakfield Instruments, Eynsham, UK) to monitor radioactivity disappearance by continuous readings summed over 20 s blocks.

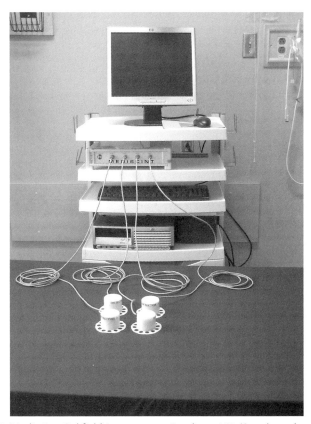

Figure 12.1 Mediscint, Oakfield Instruments, Eynsham, UK. (See the color plate.)

4.1.2 Microinfusion pumps

A commonly used precision pump is the CMA100 pump (Chromatography Sciences Company, St-Laurent, Canada).

4.2. Consumable materials

4.2.1 Syringes

Syringes for ^{133}Xe dilution, drug injections (Tuberculin syringe, 1 ml, 25G5/8) and ^{133}Xe retrieval (Insulin Syringe, U-100, 27G × 16 mm, or U-100, 29G × 12.7 mm) are from BD-Canada (Oakville, Ontario, Canada).

4.2.2 Catheters

Two types of catheters can be used: the Quick-set Infusion Sets (Minimed, Medtronic of Canada Ltd., Mississauga, Canada) (Fig. 12.2) and the Inset® II Infusion Set (Animas Corporation, LifeScan Canada Ltd., Burnaby, British Columbia, Canada) (Fig. 12.3).

The sets are comprised of a teflon catheter (length: 60, 80, or 110 mm) with detachable tubing from the head (or connector) at the infusion site. The head is incorporated into an adhesive pad and contains a soft cannula 6 or 9 mm long. The priming volume of the chamber of the head and cannula is 10 μl for the Quick-set catheter and 4 μl for the Inset II.

4.2.3 ^{133}Xenon

^{133}Xe is supplied as a gas (one glass vial, 10 mCi) by Lantheus Medical Imaging Inc., Montreal, Canada, or other local providers.

5. PROCEDURE

5.1. ^{133}Xe preparation

- 1–1.5 ml of the 0.9% saline is injected into the vial containing 10 mCi of ^{133}Xe.
- The vial is placed in a refrigerator (4 °C) for at least 1 h before an experiment (low temperature assures higher proportion of ^{133}Xe in the saline).
- ^{133}Xe and all the contaminated material must be handled and disposed of according to local safety standards.

5.2. Preparation of agents

- Stock solutions of vasoactive compounds are prepared in advance under sterile conditions by dilution with saline. Solutions are then aliquoted and stored, usually at −20 °C.

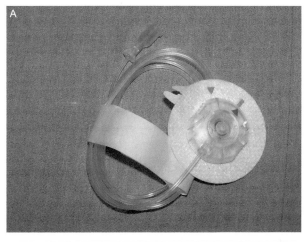
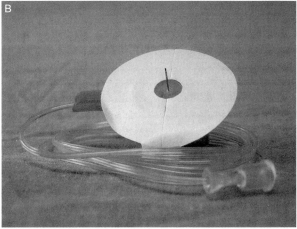

Figure 12.2 Quick-set Infusion Set; Minimed, Medtronic of Canada Ltd., Mississauga, Canada. (See the color plate.)

- Desired concentrations (depending on the type of the agent) are prepared just before the experiments.

5.3. Subjects preparation

- Volunteers are asked to refrain from strenuous exercise, or intake of alcohol or caffeine, for 48 h before the experiment. Testing follows an overnight fast.
- Unless specific research objectives are pursued, subjects are tested in a recumbent position; they remain as motionless as possible throughout

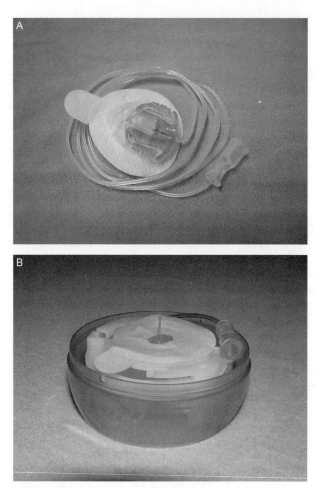

Figure 12.3 Inset II Infusion Set, Animas Corporation, LifeScan Canada Ltd., Burnaby, British Columbia, Canada. (See the color plate.)

the experiment, so that only smooth respiratory movements of the abdominal wall influence ATBF monitoring.

5.4. Microinfusion protocol

- After their respective connection to tuberculin syringes (1 ml, 25G5/8) filled with saline, the microinfusion catheters are flushed until the first drop of saline appears at the end of the cannula.
- After local disinfection, catheters are applied into the abdominal wall 6–8 cm above and below the umbilicus and 6–8 cm from the medial line.

- The insertion needle is removed and the adhesive hub of the catheter is carefully kept in place until removal at the end of the experiment.
- A saline microinfusion is started at 2 µl/min using a microinfusion pump.
- After allowing 30 min for the tissue to recover, ^{133}Xe at a dose 1–2 MBq per site is retrieved from the vial using an insulin syringe and slowly (1–2 min) injected through the hub of the catheter connector; the volume injected should not exceed 100–110 µl (bigger volume could be painful) although it has been shown that volumes 20–500 µl have no influence on BF values (Nielsen, 1991; Summers et al., 1996). The syringe can be withdrawn 30 s later.
  * Syringes matching the Quick-set Infusion Set hub hole and cannula are U-100, 27G × 16-mm insulin syringes.
  * Syringes matching the Inset II Infusion Set hub hole and cannula are U-100, 29G × 12.7-mm insulin syringes.
- Then, γ-counter probes are firmly taped to the skin over the infusion devices, that is, over the ^{133}Xe-labeled AT. Interconnection of all four probes is necessary to minimize movements due to breathing.
- After this, catheters are perfused for a further 60 min to allow ^{133}Xe diffusion/equilibration into the tissue.
- Thereafter, a "Baseline ATBF" is recorded, usually for 40–60 min.
- After this "Baseline ATBF" period, γ-counter probes are temporarily removed and two catheters (one above, one below the umbilicus, chosen at random) are disconnected and replaced by catheters filled with vasoactive agents; then, γ-counter probes are retaped over the infusion devices.
- Before ATBF recording, the chambers and cannulas must be filled by the vasoactive agent solution: at the infusion rate of 2 µl/min, 5 and 2 min are needed for the Quick-set system and the Inset II system, respectively.
- At the two contralateral control sites, parallel saline microinfusions continue.
- Afterward, the fasting ATBF under the influence of potential vasoactive agents is recorded for the next 40–60 min at the same infusion rate ("Fasting ATBF + Agent").
- A 75-g glucose load is given to provoke an endogenous stimulation of ATBF, while recording is maintained for 120 min, since ATBF peak normally occurs 30–90 min postglucose ingestion (Ardilouze, Karpe, Currie, Frayn, & Fielding, 2004).

5.5. Modifications to the method and drawbacks

- Up to four microinfusions can be performed at the same time: right and left sides at the upper and lower level of the abdomen. As there is no fasting or postprandial ATBF difference between the right and the left side (Ardilouze, Karpe, et al., 2004), two agents can be tested in each subject for each experiment to allow direct comparison of the effect of a vasoactive substance on one side with the contralateral saline control side on the other (each subject acts as his own control).
- The number of tested sites may vary according to study objectives and the number of γ-counter probes available. However, ATBF is faster above the umbilicus (Ardilouze, Karpe, et al., 2004).
- Other fat depots can be tested: femoral or gluteal subcutaneous AT (Nielsen et al., 2009; Tan, Goossens, Humphreys, Vidal, & Karpe, 2004).
- The glucose load can be replaced by a standard meal and the postprandial ATBF recording can be prolonged (Perez-Matute, Neville, Tan, Frayn, & Karpe, 2009).
- Injection of ^{133}Xe mixed with atmospheric air is also possible (Enevoldsen et al., 2007; Simonsen, Enevoldsen, & Bulow, 2003).
- The main drawback of xenon washout method dwells in high intra-individual variability (Summers et al., 1996). To minimize this, all the steps of the protocol should be standardized and carefully observed.
- The technique does not permit evaluation of BF in intraabdominal AT.
- Even 6 mm cannula carries a slight risk of an intramuscular administration in slim subjects (Gibney, Arce, Byron, & Hirsch, 2010).

6. CALCULATIONS

6.1. ATBF values

The Mediscint System provides the results as counts per 20 s in Excel format (three values/min). These data are used to calculate ATBF according to the aforementioned equation.

6.2. Baseline ATBF and fasting ATBF under the influence of a vasoactive agent

"Baseline ATBF" is calculated as the mean of the last two or three values of the fasting period. "Fasting ATBF + Agent" is calculated as the mean of the last two or three values of the second fasting period. The effect of an agent on fasting ATBF is calculated as absolute and relative differences between "Fasting ATBF + Agent" minus "Baseline ATBF."

6.3. Stimulated (postglucose) ATBF

"Peak ATBF" is calculated as the mean of three points, the maximum value and the two adjacent points. Postglucose response of ATBF can be evaluated using: (1) the difference "Peak ATBF" minus fasting ATBF, and (2) the ATBF area or the incremental area under the curve (AUC or iAUC).

7. OTHER TECHNIQUES USED IN ATBF MEASUREMENT
7.1. The microdialysis technique

The microdialysis technique in combination with ethanol escape method can be another solution for indirect estimation of local ATBF (Galitzky, Lafontan, Nordenstrom, & Arner, 1993).

The rationale behind the technique lies in the equilibration between the interstitial tissue and the dialysate. It is assumed that the turnover of interstitial fluid is directly proportional to the TBF. Therefore, if a compound (ethanol) is introduced via the dialysate into the interstitial fluid and monitored over time, its removal rate is proportional to the BF.

During the procedure, the microdialysis probes are inserted into subcutaneous AT, connected to a microperfusion pump and perfused, usually with Ringer's solution at a rate of 0.5–2.5 µl/min. Ethanol is added to the perfusion fluid and, after a period of equilibration, 10-min fractions of outgoing perfusate are collected for ethanol analysis. Vasoactive compounds can be introduced in the system and local alterations of BF can be monitored.

Nevertheless, the ethanol escape method has several methodological limitations (Martin et al., 2011). The substance use and delivery is limited by the properties of the microdialysis membrane (Lafontan & Arner, 1996). The technique is a semiquantitative method, ATBF values are expressed in relative terms as a change in dialysate/perfusate ratio. Moreover, pharmacological concentrations of stimulating agents are required to enable detection of a change in BF. When compared to the xenon washout technique, ethanol escape shows a delayed response in ATBF elicited by the ingestion of glucose as well as slow and incomplete subsequent normalization afterward (Karpe, Fielding, Ilic, Humphreys, & Frayn, 2002). This method is therefore either not sufficiently sensitive, or sufficiently rapid (Martin et al., 2011).

7.2. Laser doppler flowmetry

The basic principle of the method is to analyze changes in the spectrum of light reflected from moving blood cells (Doppler shift). The magnitude and frequency distribution of these changes are directly related to the number and velocity of blood cells. The quantity measured by laser Doppler flowmetry (LDF) is referred to as perfusion and expressed in relative units (perfusion units). Rather than BF, it reflects global tissue perfusion in capillaries, arterioles, venules, and shunts (Hoff, Gregersen, & Hatlebakk, 2009).

A beam of monochromatic laser light is carried by a flexible optic fiber (Crandall, Hausman, & Kral, 1997) inserted into the AT and attached to LDF probe. The signal generated by frequency shift is registered by a flow meter.

LDF technique is easily performed and minimally invasive and has excellent time resolution. Nevertheless, there are several inherent limitations of the LDF method relating mainly to multiple Doppler shifts, tissue optical properties, motion artifacts, lack of knowledge regarding the depth of measurement, and the biological zero signal (residual signal from tissues present even if there is no BF) (Rajan, Varghese, van Leeuwen, & Steenbergen, 2009). Unlike in large vessels, the results of LDF cannot be expressed in absolute values and compared between studies.

7.3. Positron emission tomography

Positron emission tomography (PET) is based on detection of paired gamma rays arising from the annihilation of emitted positrons by a specific radioactive isotope. The use of [^{15}O]-labeled water allows direct and regional assessment of BF often together with metabolic studies based on administration of [^{18}F]-labeled fluorodeoxyglucose.

The method requires intravenous injection of a radioactive tracer and simultaneous and continuous withdrawals of arterial blood throughout the study to monitor the content of radioactive water. PET scanners are co-equipped with computed tomography (CT) or more recently with magnetic resonance imaging (MRI) scanners to yield both anatomic and functional information.

PET represents another technique that provides absolute values of ATBF (ml (100 g/min)) comparable with ^{133}Xe washout method (Virtanen et al., 2001). The method has confirmed that the abdominal BF is lower in obese than in the nonobese subjects during an euglycemic–hyperinsulinemic clamp, and that both subcutaneous and visceral AT BF are impaired in obese subjects (Virtanen et al., 2001). PET offers information from deep AT depots and the possibility to monitor simultaneously perfusion in other tissues (e.g., muscles). It seems to be reliable, but is expensive due to the high cost of equipment. Steady-state conditions are required to obtain reliable and constant flux into the tissue and measurements can be done only once, making the method unsuitable for monitoring over several hours. Another concern is that the radiation dose, which is usually around 6 mSv, is several fold higher when PET/CT scans are performed (Brix et al., 2005).

7.4. Real-time contrast-enhanced ultrasound

Contrast-enhanced ultrasound (CEU) is a noninvasive technique originally developed to image myocardial perfusion (Wei et al., 1998) and later adapted to determine capillary recruitment (CR) in muscles (Mulder, van Dijk, Smits, & Tack, 2008) and AT (Tobin et al., 2010). A bolus of a suspension of phospholipid-stabilized microbubles filled with inert gas is administered intravenously. Afterward, the changes in signal intensity from AT are captured by ultrasound scanning.

Relative microvascular blood volume (MBV) is determined as the mean signal intensity in decibels after subtraction of the signal intensity before the contrast injection (Mulder et al., 2008; Tobin et al., 2010). After stimulation (e.g., glucose load), the magnitude of the CR is derived from the changes of MBV. In healthy people, oral glucose load stimulates CR, whereas the increase is abolished in obese type 2 diabetics (Tobin et al., 2010; Tobin, Simonsen, & Bulow, 2011).

8. CONCLUSION

ATBF is an integral component of AT function. The most effective ATBF measurement method seems to be the xenon washout technique because, in comparison to microdialysis, it is more discriminative and assesses ATBF in absolute terms. Unlike PET, it requires relatively simple and affordable equipment, and the patient is exposed to lower levels of radiation during the experiment. In combination with microinfusion, the xenon washout technique overcomes the technical limitations of microdialysis and allows local BF manipulation, without the systemic effects of vasoactive drugs.

ACKNOWLEDGMENTS

We are grateful to our colleagues Pascal Brassard, Maude Gagnon-Auger, Elisabeth Martin, and Julie Ménard who helped to implement these techniques. A special thank to Amélie Ardilouze who edited the manuscript.

REFERENCES

Andersson, J., Sjostrom, L. G., Karlsson, M., Wiklund, U., Hultin, M., Karpe, F., et al. (2010). Dysregulation of subcutaneous adipose tissue blood flow in overweight postmenopausal women. *Menopause, 17,* 365–371.

Ardilouze, J. L., Fielding, B. A., Currie, J. M., Frayn, K. N., & Karpe, F. (2004). Nitric oxide and beta-adrenergic stimulation are major regulators of preprandial and postprandial subcutaneous adipose tissue blood flow in humans. *Circulation, 109,* 47–52.

Ardilouze, J. L., Karpe, F., Currie, J. M., Frayn, K. N., & Fielding, B. A. (2004). Subcutaneous adipose tissue blood flow varies between superior and inferior levels of the anterior abdominal wall. *International Journal of Obesity and Related Metabolic Disorders, 28,* 228–233.

Blaak, E. E., van Baak, M. A., Kemerink, G. J., Pakbiers, M. T., Heidendal, G. A., & Saris, W. H. (1995). Beta-adrenergic stimulation and abdominal subcutaneous fat blood flow in lean, obese, and reduced-obese subjects. *Metabolism, 44,* 183–187.

Brix, G., Lechel, U., Glatting, G., Ziegler, S. I., Munzing, W., Muller, S. P., et al. (2005). Radiation exposure of patients undergoing whole-body dual-modality ^{18}F-FDG PET/CT examinations. *Journal of Nuclear Medicine, 46,* 608–613.

Bulow, J. (2001). Measurement of adipose tissue blood flow. *Methods in Molecular Biology, 155,* 281–293.

Bulow, J., Jelnes, R., Astrup, A., Madsen, J., & Vilmann, P. (1987). Tissue/blood partition coefficients for xenon in various adipose tissue depots in man. *Scandinavian Journal of Clinical and Laboratory Investigation, 47,* 1–3.

Bulow, J., & Madsen, J. (1976). Adipose tissue blood flow during prolonged, heavy exercise. *Pflügers Archiv, 363,* 231–234.

Crandall, D. L., Hausman, G. J., & Kral, J. G. (1997). A review of the microcirculation of adipose tissue: Anatomic, metabolic, and angiogenic perspectives. *Microcirculation, 4,* 211–232.

Dimitriadis, G., Lambadiari, V., Mitrou, P., Maratou, E., Boutati, E., Panagiotakos, D. B., et al. (2007). Impaired postprandial blood flow in adipose tissue may be an early marker of insulin resistance in type 2 diabetes. *Diabetes Care, 30,* 3128–3130.

Elia, M., & Kurpad, A. (1993). What is the blood flow to resting human muscle? *Clinical Science (London), 84,* 559–563.

Enevoldsen, L. H., Polak, J., Simonsen, L., Hammer, T., Macdonald, I., Crampes, F., et al. (2007). Post-exercise abdominal, subcutaneous adipose tissue lipolysis in fasting subjects is inhibited by infusion of the somatostatin analogue octreotide. *Clinical Physiology and Functional Imaging, 27,* 320–326.

Frayn, K. N. (2010). Fat as a fuel: Emerging understanding of the adipose tissue-skeletal muscle axis. *Acta Physiologica (Oxford, England), 199,* 509–518.

Galitzky, J., Lafontan, M., Nordenstrom, J., & Arner, P. (1993). Role of vascular alpha-2 adrenoceptors in regulating lipid mobilization from human adipose tissue. *The Journal of Clinical Investigation, 91,* 1997–2003.

Galitzky, J., Sengenes, C., Thalamas, C., Marques, M. A., Senard, J. M., Lafontan, M., et al. (2001). The lipid-mobilizing effect of atrial natriuretic peptide is unrelated to sympathetic nervous system activation or obesity in young men. *Journal of Lipid Research, 42,* 536–544.

Gibney, M. A., Arce, C. H., Byron, K. J., & Hirsch, L. J. (2010). Skin and subcutaneous adipose layer thickness in adults with diabetes at sites used for insulin injections: Implications for needle length recommendations. *Current Medical Research and Opinion, 26,* 1519–1530.

Goossens, G. H., McQuaid, S. E., Dennis, A. L., van Baak, M. A., Blaak, E. E., Frayn, K. N., et al. (2006). Angiotensin II: A major regulator of subcutaneous adipose tissue blood flow in humans. *The Journal of Physiology, 571,* 451–460.

Heinonen, I., Bucci, M., Kemppainen, J., Knuuti, J., Nuutila, P., Boushel, R., et al. (2012). Regulation of subcutaneous adipose tissue blood flow during exercise in humans. *Journal of Applied Physiology, 112*, 1059–1063.

Hoff, D. A., Gregersen, H., & Hatlebakk, J. G. (2009). Mucosal blood flow measurements using laser Doppler perfusion monitoring. *World Journal of Gastroenterology, 15*, 198–203.

Jansson, P. A., Larsson, A., Smith, U., & Lonnroth, P. (1992). Glycerol production in subcutaneous adipose tissue in lean and obese humans. *The Journal of Clinical Investigation, 89*, 1610–1617.

Jansson, P. A., & Lonnroth, P. (1995). Comparison of two methods to assess the tissue/blood partition coefficient for xenon in subcutaneous adipose tissue in man. *Clinical Physiology, 15*, 47–55.

Karpe, F., Fielding, B. A., Ardilouze, J. L., Ilic, V., Macdonald, I. A., & Frayn, K. N. (2002). Effects of insulin on adipose tissue blood flow in man. *The Journal of Physiology, 540*, 1087–1093.

Karpe, F., Fielding, B. A., Ilic, V., Humphreys, S. M., & Frayn, K. N. (2002). Monitoring adipose tissue blood flow in man: A comparison between the (133)xenon washout method and microdialysis. *International Journal of Obesity and Related Metabolic Disorders, 26*, 1–5.

Karpe, F., Fielding, B. A., Ilic, V., Macdonald, I. A., Summers, L. K., & Frayn, K. N. (2002). Impaired postprandial adipose tissue blood flow response is related to aspects of insulin sensitivity. *Diabetes, 51*, 2467–2473.

Lafontan, M., & Arner, P. (1996). Application of in situ microdialysis to measure metabolic and vascular responses in adipose tissue. *Trends in Pharmacological Sciences, 17*, 309–313.

Larsen, O. A., Lassen, N. A., & Quaade, F. (1966). Blood flow through human adipose tissue determined with radioactive xenon. *Acta Physiologica Scandinavica, 66*, 337–345.

Lassen, N. A., Lindbjerg, J., & Munck, O. (1964). Measurement of blood-flow through skeletal muscle by intramuscular injection of xenon-133. *Lancet, 1*, 686–689.

Linde, B., Hjemdahl, P., Freyschuss, U., & Juhlin-Dannfelt, A. (1989). Adipose tissue and skeletal muscle blood flow during mental stress. *The American Journal of Physiology, 256*, E12–E18.

Martin, E., Brassard, P., Gagnon-Auger, M., Yale, P., Carpentier, A. C., & Ardilouze, J. L. (2011). Subcutaneous adipose tissue metabolism and pharmacology: A new investigative technique. *Canadian Journal of Physiology and Pharmacology, 89*, 383–391.

McQuaid, S. E., Hodson, L., Neville, M. J., Dennis, A. L., Cheeseman, J., Humphreys, S. M., et al. (2011). Downregulation of adipose tissue fatty acid trafficking in obesity: A driver for ectopic fat deposition? *Diabetes, 60*, 47–55.

Mulder, A. H., van Dijk, A. P., Smits, P., & Tack, C. J. (2008). Real-time contrast imaging: A new method to monitor capillary recruitment in human forearm skeletal muscle. *Microcirculation, 15*, 203–213.

Nielsen, S. L. (1991). Sources of variation in 133Xenon washout after subcutaneous injection on the abdomen. *Acta Physiologica Scandinavica. Supplementum, 603*, 93–99.

Nielsen, N. B., Hojbjerre, L., Sonne, M. P., Alibegovic, A. C., Vaag, A., Dela, F., et al. (2009). Interstitial concentrations of adipokines in subcutaneous abdominal and femoral adipose tissue. *Regulatory Peptides, 155*, 39–45.

Perez-Matute, P., Neville, M. J., Tan, G. D., Frayn, K. N., & Karpe, F. (2009). Transcriptional control of human adipose tissue blood flow. *Obesity (Silver Spring), 17*, 681–688.

Quisth, V., Enoksson, S., Blaak, E., Hagstrom-Toft, E., Arner, P., & Bolinder, J. (2005). Major differences in noradrenaline action on lipolysis and blood flow rates in skeletal muscle and adipose tissue in vivo. *Diabetologia, 48*, 946–953.

Rajan, V., Varghese, B., van Leeuwen, T. G., & Steenbergen, W. (2009). Review of methodological developments in laser Doppler flowmetry. *Lasers in Medical Science, 24*, 269–283.

Samra, J. S., Frayn, K. N., Giddings, J. A., Clark, M. L., & Macdonald, I. A. (1995). Modification and validation of a commercially available portable detector for measurement of adipose tissue blood flow. *Clinical Physiology, 15*, 241–248.

Samra, J. S., Simpson, E. J., Clark, M. L., Forster, C. D., Humphreys, S. M., Macdonald, I. A., et al. (1996). Effects of epinephrine infusion on adipose tissue: Interactions between blood flow and lipid metabolism. *The American Journal of Physiology, 271*, E834–E839.

Simonsen, L., Bulow, J., Astrup, A., Madsen, J., & Christensen, N. J. (1990). Diet-induced changes in subcutaneous adipose tissue blood flow in man: Effect of beta-adrenoceptor inhibition. *Acta Physiologica Scandinavica, 139*, 341–346.

Simonsen, L., Enevoldsen, L. H., & Bulow, J. (2003). Determination of adipose tissue blood flow with local 133Xe clearance. Evaluation of a new labelling technique. *Clinical Physiology and Functional Imaging, 23*, 320–323.

Sotornik, R., Brassard, P., Martin, E., Yale, P., Carpentier, A. C., & Ardilouze, J. L. (2012). Update on adipose tissue blood flow regulation. *American Journal of Physiology. Endocrinology and Metabolism, 302*, E1157–E1170.

Summers, L. K., Samra, J. S., Humphreys, S. M., Morris, R. J., & Frayn, K. N. (1996). Subcutaneous abdominal adipose tissue blood flow: Variation within and between subjects and relationship to obesity. *Clinical Science (London), 91*, 679–683.

Tan, G. D., Goossens, G. H., Humphreys, S. M., Vidal, H., & Karpe, F. (2004). Upper and lower body adipose tissue function: A direct comparison of fat mobilization in humans. *Obesity Research, 12*, 114–118.

Tobin, L., Simonsen, L., & Bulow, J. (2010). Real-time contrast-enhanced ultrasound determination of microvascular blood volume in abdominal subcutaneous adipose tissue in man. Evidence for adipose tissue capillary recruitment. *Clinical Physiology and Functional Imaging, 30*, 447–452.

Tobin, L., Simonsen, L., & Bulow, J. (2011). The dynamics of the microcirculation in the subcutaneous adipose tissue is impaired in the postprandial state in type 2 diabetes. *Clinical Physiology and Functional Imaging, 31*, 458–463.

Virtanen, K. A., Peltoniemi, P., Marjamaki, P., Asola, M., Strindberg, L., Parkkola, R., et al. (2001). Human adipose tissue glucose uptake determined using [(18)F]-fluoro-deoxyglucose ([(18)F]FDG) and PET in combination with microdialysis. *Diabetologia, 44*, 2171–2179.

Wei, K., Jayaweera, A. R., Firoozan, S., Linka, A., Skyba, D. M., & Kaul, S. (1998). Quantification of myocardial blood flow with ultrasound-induced destruction of microbubbles administered as a constant venous infusion. *Circulation, 97*, 473–483.

CHAPTER THIRTEEN

Isolation and Quantitation of Adiponectin Higher Order Complexes

Joseph M. Rutkowski[*], Philipp E. Scherer[*,†,1]

[*]Touchstone Diabetes Center, Department of Internal Medicine, UT Southwestern Medical Center, Dallas, Texas, USA
[†]Department of Cell Biology, UT Southwestern Medical Center, Dallas, Texas, USA
[1]Corresponding author: e-mail address: philipp.scherer@utsouthwestern.edu

Contents

1. Introduction 244
2. Sample Collection and Preparation 246
3. Gel Fractionation by FPLC 247
   3.1 Required materials 248
   3.2 Technical procedure 248
   3.3 Customization and optimization 250
4. Western Blot Analysis 251
   4.1 Required materials 251
   4.2 Technical aspects 252
5. Complex Distribution Quantitation and Presentation 254
   5.1 The art of quantification 254
   5.2 Data presentation 254
6. Comparison to Other Techniques 256
   6.1 SDS-PAGE fraction analysis 256
   6.2 HMW adiponectin enzyme-linked immunosorbent assay analysis 257
7. Concluding Remarks 258
Acknowledgments 258
References 258

Abstract

Adiponectin is a circulating bioactive hormone secreted by adipocytes as oligomers ranging in size from 90 kDa trimers and 180 kDa hexamers to larger high molecular weight oligomers that may reach 18- or 36-mers in size. While total circulating adiponectin levels correlate well with metabolic health, it is the relative distribution of adiponectin complexes that is most clinically relevant to glucose sensitivity and inflammation. High molecular weight adiponectin best mirrors insulin sensitivity, while trimeric adiponectin dominates with insulin resistance and adipose tissue inflammation. Experimental animal and *in vitro* models have also linked the relative fraction of high molecular weight adiponectin to its positive effects. Quantitating adiponectin size distribution thus provides a window into metabolic health and can serve as a surrogate marker for adipose tissue fitness.

Here, we present a detailed protocol for isolating and quantitating adiponectin complexes in serum or plasma that has been extensively utilized for both human clinical samples and numerous animal models under various experimental conditions. Examples are presented of different adiponectin distributions and tips are provided for optimization using available equipment. Comparison of this rigorous approach to other available methods is also discussed. In total, this summary is a blueprint for the expanded quantitation and study of adiponectin complexes.

1. INTRODUCTION

Adipose tissue is at the center of the worldwide obesity epidemic as an important endocrine tissue that plays a literally expanding role in systemic metabolic health with weight gain. While most characterized adipocyte secretory proteins increase with increased adiposity and serve as signals of metabolic dysregulation or adipose tissue inflammation, adiponectin is unique in its positive correlation with metabolic health (Turer & Scherer, 2012). As adipose tissue expands, Body Mass Index (BMI) is increased, and insulin sensitivity is limited, and circulating levels of adiponectin are reduced (Kusminski & Scherer, 2009). Inversely, adiponectin levels are "normal" at lower BMIs and can be increased upon weight loss. These extensive clinical correlations have been confirmed in a host of animal models and *in vitro* studies (Turer & Scherer, 2012). This relationship is not merely correlative, however, as adiponectin is a potent regulator of cellular glucose and lipid metabolism and has demonstrated numerous cytoprotective effects (Kusminski & Scherer, 2009). These findings have made measuring adiponectin levels across a host of pathologies both clinically interesting and experimentally essential. Adiponectin circulates as multimeric complexes; its oligomerization state may reflect the degree of adiponectin's metabolic benefits.

Adiponectin is an approximately 30 kDa protein secreted from adipocytes as a highly stable 90 kDa trimer, two-trimer hexamers (sometimes referred to as low molecular weight, LMW), and multihexamers of 12, 18, or 36 total monomers in size (Pajvani et al., 2003). It is these high molecular weight (HMW) complexes that have demonstrated the most profound examples of adiponectin's prometabolic and anti-inflammatory potential (Pajvani et al., 2004). Measurements capable of differentiating HMW adiponectin, or its fractional distribution (as compared to circulating trimer abundance), have been the most well-correlated to metabolic health and general wellbeing (Baessler et al., 2011; Hamilton et al., 2011; Lo et al., 2011; Matsumoto, Toyomasu, Uchimura, & Ishitake, 2013). Thus, while total levels are important—and indeed reflective of adipose tissue health—the fractional amount of HMW may be the best biomarker of clinical and experimental model relevance.

There have been more than 400 manuscripts in which the researchers have performed distribution measurements and then taken advantage of analyzed differences in adiponectin oligomer size when interpreting their results. Some early examples include Bobbert et al. (2005), Fisher et al. (2005), Kobayashi et al. (2004), Lara-Castro, Luo, Wallace, Klein, and Garvey (2006), and Pajvani et al. (2004); each highlighting the importance and contribution of adiponectin oligomers. The fast protein liquid chromatography (FPLC)-based protocol presented here has been utilized for many years by many users and researchers and has been refined continuously with column performance and sample volumes. This approach was originally established in Schraw, Wang, Halberg, Hawkins, and Scherer (2008). Figure 13.1 is a representative flowchart of the protocol that highlights the customization and optimization of the procedures as the methods are put into place locally and feedback is generated; potential changes to this outline are highlighted in each section.

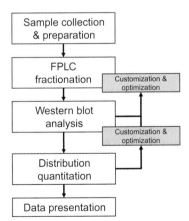

Figure 13.1 Flowchart of adiponectin complex separation, isolation, and quantitation utilizing FPLC and Western blot. At multiple steps during the suggested procedure there are opportunities for customization and optimization of individual equipment or sample needs.

2. SAMPLE COLLECTION AND PREPARATION

Adiponectin complexes are highly stable. Briefly, the structure and stability of adiponectin oligomers is based in the collagenous domain: numerous posttranslational modifications along the collagenous domain are necessary for trimer formation and mutations to this sequence result in adiponectin that is unable to be secreted (Wang, Lam, Yau, & Xu, 2008). This synthesis is a primary reason why recombinant production of full-length adiponectin complexes requires mammalian cells. Higher order adiponectin complex formation requires two trimers forming a hexamer by disulfide bond at cysteine 36 (C39 in mice); hexamer formation is requisite for the formation of subsequent higher molecular weight forms (Pajvani et al., 2003; Wang et al., 2008). The noncovalent bonding of HMW forms is thus only broken under low pH or reducing conditions.

The stability of adiponectin to light, temperature, and serum proteases allows for analysis of adiponectin complexes in samples that have not been optimally maintained (e.g., having experienced multiple freeze–thaw cycles). Adiponectin can be measured in samples that have been stored at -20 or $-80\ °C$ for many years, if not decades, provided they have not been dehydrated over years in storage. It is still recommended, however, that samples be carefully handled and maintained as any serum or plasma sample.

Clinical collections of serum or plasma using lithium heparin or EDTA have been tested and are recommended.

Complexes are, however, sensitive to even modest acidic conditions and become destabilized below pH 7. If samples are shipped on dry ice, care has to be taken to use tightly capped tubes. Alternatively, additional buffer (e.g., 50 mM HEPES, pH 7.5 final concentration) can be added to ensure that the pH stays above 7.0 upon the unintended exposure to CO_2 in shipment.

Samples must be centrifuged before proceeding to fractionation. Even if care was taken in removing serum or plasma from the collected blood, it is essential for best practice that centrifugation at $>10,000 \times g$ for 5 min be utilized to ensure that no debris enters the FPLC system. Filtration of samples is strongly discouraged due to the potential exclusion of the large-sized HMW adiponectin complexes. Samples may then either be loaded directly to the FPLC or diluted in running buffer (Section 3.1) depending on loading method or capabilities of a given system (Table 13.1).

Table 13.1 Suggested sample dilutions based on FPLC loading method and standard loop sizes

Loading method	Volume sample (μL)	Volume buffer (μL)
Micro syringe		
25 μL sample loop	25	10
50 μL sample loop	25	35
200 μL sample loop	25	200
Standard syringe[a]		
200 μL sample loop	40	360
Autosampler	Dead volume-dependent. Sample must be prepared such that actual injection volume contains 20–25 μL sample	

[a]Standard syringes are not recommended for smaller sample loops; this loading volume will affect elution time precision.
Volumes are based on 20 μL of sample reaching the column and no air potentially remaining in the sample loop. Samples low in adiponectin may require increased sample volumes (with reduced buffer volume).

3. GEL FRACTIONATION BY FPLC

To analyze abundance of adiponectin complexes, high-resolution molecular weight separation must be employed. While defined here as fractionation by FPLC, the actual type of pump system used is irrelevant because the end pressure and flow rate considerations are determined by the selected column performance capabilities. All of the work carried out in our laboratory has utilized an ÁKTA FPLC system equipped with a single Superdex 200 10/300 GL column (both GE Healthcare Biosciences; Pittsburgh, PA). One of our units relies on an ÁKTA HPLC pump, but the end settings in operation are essentially identical when the same column is employed.

The Superdex 200 10/300 GL column is ideal for the task of separating adiponectin complexes from serum samples because it offers a large bed volume (50 μL sample volume) and provides excellent separation from 10 to 600 kDa. Another column with similar separation characteristics may be utilized; the following procedure details are based on the Superdex 200 10/300 GL column. Optimization of the fractionation protocol (Section 3.3) may help when using another system or column to achieve comparable separation results to proceed to analysis. None of this analytical procedure or system needs to be performed in or placed into a 4 °C cooler.

3.1. Required materials

3.1.1 Separation system
The flow rates and volumetric settings described in the protocol are based on using the listed gel filtration column.
- Superdex 200 10/300 GL column
- FPLC/HPLC system equipped with
  - 280 nm UV lamp or 214 nm for visualizing protein absorbance
  - an automatic fraction collector for collection into tubes or, ideally, 96-well plates
  - 96-well, nonbinding plates for sample collection

3.1.2 Solutions
All solutions utilized in the FPLC system should be vacuum filtered using a 0.22 μm filter and sufficiently degassed.
- Running buffer: HEPES/Ca^{2+}, pH 7.4
  - 25 mM HEPES
  - 150 mM NaCl
  - 1 mM $CaCl_2$
- ddH_2O for column rinsing
- 20% EtOH (prepared with ddH_2O) for column storage

3.2. Technical procedure

During separation, the 280 nm UV display of the sample permits identification of adiponectin complexes. Each of the three primary complexes elutes before albumin with distinct protein peaks (Fig. 13.2A). Albumin concentrations do not vary across samples so immediately during the separation the relative abundance of each species may be discernable based on each peak's approximate mAU value on the y-axis compared to albumin. Lower sample loading volumes (Fig. 13.2B) may make this unblinded initial assessment more difficult.

Our protocol is designed to separate and collect nearly 100% of the adiponectin in a sample across 24 eluted fractions to maximize the size separation while allowing all of the adiponectin to be detected on one gel (Section 4). A stepwise protocol with flow rates and volumes is summarized in Table 13.2. Collected elution volumes and overall collection time/volume should be adjusted so as not to lose significant amounts of adiponectin to the waste stream (Fig. 13.2C).

With our current FPLC-driven gel fractionation system, we strive to maintain column performance for maximum separation while making

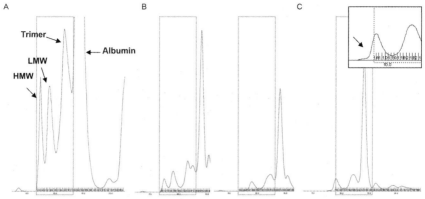

Figure 13.2 Example FPLC gel filtration elution curves assessed at 280 nm detection. Samples are plotted as mAU (y-axis) versus volume (x-axis) with the red dashed box indicating the first 24 collected eluted fractions. (A) HMW, LMW, and trimeric adiponectin each elute as a distinct protein peak ahead of albumin. (B) Lower sample loading volumes result in more unique peaks (left), but too low of plasma volume coupled with low adiponectin makes discerning the peaks difficult (right). Here, this sample may contain limited HMW adiponectin. (C) Improper elution collection setting can result in adiponectin lost to the waste stream. The first 24 collected contain all of the serum albumin and fail to capture nearly half of the HMW adiponectin (inset, arrow). (See the color plate.)

Table 13.2 Suggested operating parameters for FPLC operation, gel filtration, and sample collection based on a GE Superdex 200 10/300 GL column

Procedure step	Flow rate (mL/min)	Approx. pressure (MPa)	Volume (mL)	Column fraction[a]
Equilibration	0.6	0.6–0.7	4.8	0.2
Sample injection	0.3	0.3–0.35	0.75	0.03
Elution before collection	0.3	0.3–0.35	6.24	0.26
Fraction collection	0.3	0.3–0.35	15.6	0.65
Regeneration	0.6	0.6–0.7	21.6	0.9
Water rinse	0.6	0.6	25	1.04
Ethanol storage	0.6	0.6–0.7	25	1.04

[a]Specific to Superdex 200 10/300 GL column with 24 mL packed bed.
Sample system pressures are for a column of moderate age. New columns will operate at 0.1 MPa less; older columns should be cleaned once pressures exceed 0.75 MPa. Volumes and column fractions are given to provide protocol programming guidelines following the Section 3 methodology.

subsequent adiponectin assessment fast and easy. For these reasons, we operate at lower flow rates and operating pressure than necessary for the Superdex 200 10/300 GL column. Our protocol (Table 13.2) also includes an equilibration and regeneration step totaling one column volume to essentially wash the columns between samples. For increased speed of fractionation, these steps may be eliminated or reduced and flow rates may be increased to operate the column at 0.75 MPa (the manufacturer's suggested operating pressure) throughout. Rinsing with ddH$_2$O and storage of the column, even for short periods (i.e., weekends), is suggested.

3.3. Customization and optimization

The numbers listed for this protocol in Table 13.2 are based on normal operation of a routinely maintained system. Even so, as columns age and in-line filters slowly clog, performance decreases. It is recommended that a regular column cleaning and filter replacement routine be utilized to ensure optimal system performance. The flow rates listed in the protocol limit the column pressure to reduce column matrix compaction, but heavy usage will inevitably reduce column separation performance. One can either replace the column or make slight adjustments to the protocol as follows. Every system and column may perform differently so it is necessary to adjust the collection protocol accordingly. Changes and adjustments should be applied to all samples equivalently, as separation performance and comparisons can be negatively affected when calculating distribution (Section 5.1).

Protocol adjustment is based on a feedback of multiple outputs and should be performed with a sample containing measurable HMW and trimeric species. The primary feedback is the 280 nm UV detector curve. As described in the technical procedure (Section 3.2), the start of fraction collection should be adjusted so as to begin collecting once initial protein elution occurs—the first peak is HMW adiponectin. With high protein loads or increased column compaction, this point may drift (Fig. 13.2C). We have designed our protocol to have the first appearance of adiponectin be in collected fraction 1 (plate well A1) and end to be by fraction 24 (plate well H3) (Fig. 13.2A), but a starting protocol could begin collection earlier. A wider collection window ensures that all adiponectin forms are collected, and the 24 fractions analyzed through Western blotting can be selected based on the 280 nm UV curve. The secondary feedback is following membrane imaging (Sections 4 and 5). Once the distribution of adiponectin is quantified, changes can be made such that minimum values are achieved in the first and last collected fractions. To calculate a change, multiply the current fraction volume (260 µL in our case) by the number of fractions adiponectin currently appears in or is believed to stretch into (e.g., 26 fractions). Divide this number by 24 to yield a new fraction collection volume (282 µL in the current example). This will optimize all subsequent runs such that all adiponectin is fractionated into 24 wells for easier Western blot quantitation.

Feedback from Western blot analysis may also suggest that loading more sample is necessary due to the dilution of adiponectin. The volumes provided in Table 13.1 should provide sufficient adiponectin to be analyzed if plasma concentrations are within a normal physiologic range. At times, however, patients or experimental animals may exhibit markedly low adiponectin. While reporting these low levels is important, the purpose of this protocol is to quantitate complex distribution, so sufficient sample must be loaded to later visualize. It is not recommended to exceed the Superdex 200 10/300 GL column's 50 µL loading limit of plasma. Even though adiponectin levels are low in these samples, other plasma proteins are not, so the increased protein load may affect complex peak separation and should be considered later in quantitation.

4. WESTERN BLOT ANALYSIS

Following optimized gel fractionation, the adiponectin complexes are effectively distributed throughout 24 collected elutions from the FPLC system. While the UV 280 nm curve of the system yields an approximation of adiponectin complex distribution to its three most abundant forms, Western blot analysis is necessary to complete quantitation of the collected fractions. This ensures that only adiponectin is included in the protein concentration analysis and provides a linear means by which to quantitate and present adiponectin complex distribution following immunoblotting. From the standard method presented here, it will be clear why quantitation works best and is most precise if the fractionation (Section 3) was optimized for maximal separation of adiponectin across the largest elution volume.

4.1. Required materials

Each laboratory likely has its own procedure for immunoblotting. The procedures for SDS-PAGE and membrane transfer are therefore intentionally short. We suggest a few antibodies for labeling adiponectin and list some helpful technical aspects based on experience and our in-house procedure.

We utilize the LI-COR Odyssey Imaging System (LI-COR Biosciences; Lincoln, NE) for immunoblot imaging and quantification.

4.1.1 SDS-PAGE

We have found that the Bio-Rad Criterion system works well as 26-well gels for running all 24 fractions at once (with two protein ladder bookends) are offered. We have utilized both Criterion XT Bis–Tris and the newer Criterion TGX precast gels with no impact on the method (both Bio-Rad Laboratories; Hercules, CA).

The SDS-PAGE buffers for these two gels are different: MOPS and Tris/glycine/SDS, respectively, according to the manufacturer's instructions (both are available in easy-to-use concentrated forms).

- Bio-Rad Criterion XT Bis–Tris 26-well 4–12% gel (345-0125) or TGX 26-well 4–15% gel (567-1085)
- MOPS running buffer
- 5× Laemmli sample buffer with dithiothreitol[1] (DTT) added just before use

4.1.2 Immunoblot analysis

- Primary antibody against adiponectin. Do bear in mind that the primary antigen is frequently the amino-terminal variable domain, and hence, the antibodies offer limited reactivity across species. Millipore AB3784P and AB269P (both from EMD Millipore; Billerica, MA) have been tested against human and mouse samples, respectively.
- Secondary antibody labeled with an infrared dye emitting at 700 nm (Cy5, Alexa Fluor 680, LI-COR IRDye 700) or 800 nm (Cy7, Alexa Fluor 790, LI-COR IRDye 800).
- LI-COR Odyssey Imaging System.
- Nitrocellulose membrane (0.22 or 0.45 μm pore size) or Bio-Rad Midi Transfer Pack (170-4159).

4.2. Technical aspects

For this method, the gel percentage or gradient do not affect the ability to later identify adiponectin with a reasonably specific primary antibody; because gels may serve other purposes and serum may be run directly for

---

[1] Make sure you achieve maximal reducing power. Human adiponectin, in particular, is occasionally reduced incompletely and may be observed as a 60 kDa band.

native analysis without fractionation (Section 6), 4–12% or 4–15% gels are recommended to both permit migration of unreduced HMW and hexamer adiponectin at the top and sufficient separation of adiponectin monomers from IgG light chains at the bottom. Reduced adiponectin will run on the gel at approximately 29 kDa. Some protein molecular weight markers utilize a standard at 26 (or 29) kDa in size that runs at different speeds depending on the buffer system; check the compatibility so as to not lose adiponectin or mis-cut the immunoblot membrane.

Following fractionation, the adiponectin concentration in each fraction will be low, having been diluted by over 100-fold from that of the original plasma sample. Heat 50 µL of the first 24 fractions (or 24 fractions containing adiponectin based on the 280 nm curve) with 10 µL Laemmli buffer containing fresh $2\,M$ DTT at 95 °C for 5 min. To maximize detection, 15 µL—the full well capacity—of each reduced fraction are then loaded onto the gel for SDS-PAGE separation. If adiponectin concentrations are known to be low and/or the 280 nm fractionation curve has protein peaks of under 5 mAU, a different gel system with larger well capacity may be necessary to load sufficient adiponectin for subsequent immunoblot labeling. Trichloroacetic acid (TCA) precipitation of each 260 µL fraction with direct resuspension in loading buffer may also aid in detection of low adiponectin fractions. End detection may also be limited by primary antibody sensitivity.

Membrane selection and secondary detection should be made based on the laboratory's imaging of choice. We have listed materials for use the with LI-COR Odyssey system that couples infrared detection with straightforward image analysis of the scanned membrane. With infrared detection, polyvinylidene difluoride (PVDF) membrane may yield higher background intensities with the LI-COR scanner, so this procedure uses nitrocellulose. Anecdotally, milk used as a blocking buffer or antibody diluent may also increase infrared background. If using the Bio-Rad Criterion 26-well gels, the Bio-Rad Trans-Blot Turbo packs provide a prepackaged system for protein transfer (midi size; used with the Mixed MW setting) to nitrocellulose. Secondary antibody concentrations should be kept low (1:5000) for best signal to noise. If adiponectin bands are barely detectable, or absent, more reduced sample may need to be loaded requiring a switch to a larger well gel system. Longer antibody incubations can be tried first on the initial membrane, then larger capacity gels, then TCA precipitation of collected fractions, and then finally (if possible) loading a larger amount of plasma to the FPLC for a new fractionation.

5. COMPLEX DISTRIBUTION QUANTITATION AND PRESENTATION

Quantification of adiponectin complex distribution from the Western blot is elementary with image analysis software that yields band intensities. We have utilized the LI-COR Odyssey system and software for several years in which the integrated infrared intensity value is linear with concentration. This system and analysis also permits precise quantification of adiponectin concentrations if a few known standards are run concurrently to create a standard curve covering the infrared intensity range. In the absence of running standards in parallel, all fraction intensities are plotted as measured or compared to each other to obtain a fractional distribution. Using the LI-COR Odyssey software, each integrated band intensity is measured using the rectangular region of interest (ROI) band tool, with the background measurements for the tool set to "All" (using Top and Bottom and Side to Side for normalization). Care should be taken to center the ROI on the band and the same size/shape ROI can be copied and pasted to maintain a consistent and normal area across all bands (as integrated intensity is area dependent).

5.1. The art of quantification

To designate each peak as HMW, hexamer, and trimer, the lowest intensity value between the apices is set as the division. This fraction cannot be necessarily standardized from sample to sample, however, (e.g., fraction 7 is the division between HMW and hexamer) as complete gel filtration will shift with protein concentration and should be noted (Section 3.3). A sample with very little HMW may have a discernable hexamer peak start before that of one enriched in HMW. Also, even with optimized fractionation there is never a zero value achieved between peaks. In the absence of a known adiponectin concentration standard, the intensity values for each fraction can be normalized by subtracting the lowest value from all intensities, thus yielding a fraction intensity curve.

5.2. Data presentation

The resultant data from intensity quantifications can be expressed in several ways, each with its inherent pros and cons. The x-axis of fraction intensity/concentration plots has no units, so a summation of intensities or calculated areas under the curves are equivalent in each option. None of these presentation methods allow data to be misrepresented and are only a personal visual choice.

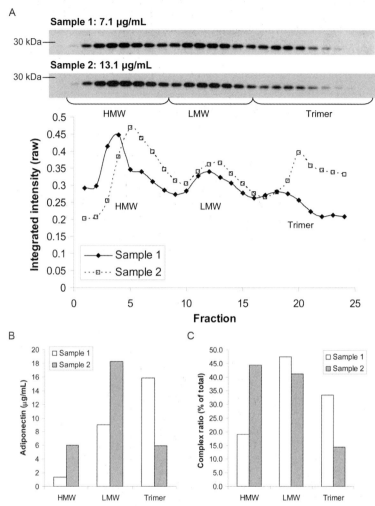

Figure 13.3 Presentation of adiponectin complex quantitative data can differ by choice and purpose. (A) From their immunoblots (top), raw quantified elution curves from two samples (bottom) highlight the peak distribution. The brackets demarcate the fractions identified as HMW, LMW, and trimeric adiponectin. (B) Expression of the summed peak intensities from (A) expressed as adiponectin concentrations based on comparison of fraction band intensities to the sample concentration from ELISA. (C) Expressed as each peaks value divided by the total can highlight changes in complex distribution across groups of experimental samples.

Option 1 (Fig. 13.3A): The data are presented as the raw plots of two sample fraction intensities. *Pros*: the relative peak distribution is clear between groups; best for mutations in which one complex is absent. *Cons*: only represents one sample of each group rather than averages and may be difficult to interpret ratios of each complex to each other if total concentration makes the peaks of one sample much greater.

Option 2 (Fig. 13.3B): Area under the curve calculations of each peak presented as concentrations. *Pros*: provides a relative value of each species; averages of multiple samples can be used for statistical analysis. *Cons*: ratios may be difficult to discern, subtle changes such as a HMW shoulder would disappear, nonzero values between peaks makes precise quantification impossible, and requires knowing the total concentration by another method.

Option 3 (Fig. 13.3C): Percent distribution of each peak/complex using area under the curve. *Pros*: changes are clear on a bar graph and statistics from sample to sample readily applied. *Cons*: Nonzero values between peaks makes precise quantification impossible and subtle changes such as a HMW shoulder disappear.

6. COMPARISON TO OTHER TECHNIQUES

Utilizing an FPLC to fractionate every serum or plasma sample and then identify adiponectin in each fraction by running a full gel for each sample is time consuming for the number of samples processed. It does, however, provide a complete representation of adiponectin complex distribution and thus represents the best method for demonstrating changes in oligomer size. This is particularly powerful for experimental groups in which the changes may be subtle and clinical populations in which a shift toward trimer from LMW may represent a negative metabolic shift. Other less rigorous techniques are possible; we briefly assess these with their respective pros and cons.

6.1. SDS-PAGE fraction analysis

The FPLC fractionation protocol presented here relies on gel filtration/size exclusion chromatography to separate adiponectin complexes by molecular size. Gel electrophoresis provides essentially the same capability without the burden or cost of FPLC equipment and time. Using nonreducing conditions and limited detergents, adiponectin complexes can be reasonably identified by immunoblot following electrophoresis. Complexes can be further stabilized by chemical crosslinking (Scherer, Williams, Fogliano, Baldini, & Lodish, 1995). The Bio-Rad gels described in Section 4 can be used for this alternative method, though care must be taken to run the samples slowly on ice, at low power (setting the amperage low), and with no added reducing agents or detergents. While less labor-intensive than the FPLC-based method, there are two large problems with the technique. First, it is challenging to not have some complexes break down with heat or in using gels that may contain some reducing agents or detergents. Since it is unknown as to which complexes break down, the relative abundance is skewed and it cannot be assumed, for example, that monomers are solely from the trimer pool. Second, FPLC isolation of complexes followed by reduction to monomers allows for antibody detection to be linear with abundance. The band intensities of a native gel immunoblot will not be linear with adiponectin concentration due to the number of potential antigens present within each "fraction." Distribution calculations are, therefore, skewed by the number of immunoglobulins per complex.

Pros: Relatively inexpensive and fast. Multiple samples can be assessed on the sample gel. Small sample volumes needed.

Cons: Difficult to perform without detergents or reduction so challenging to not have some complex breakdown. HMW complexes do not sufficiently enter gel for separation. Antibody detection of nonreduced forms not linear with concentration.

6.2. HMW adiponectin enzyme-linked immunosorbent assay analysis

In recent years, the desire to measure adiponectin complex distribution in a more rapid, high-throughput technique has led to the commercial development of enzyme-linked immunosorbent assays (ELISAs) specific to human HMW adiponectin. Some kits utilize monoclonal antibodies that are claimed to be specific to the HMW oligomer, while others utilize sample pretreatment targeted to eliminate LMW and trimeric adiponectin (Ebinuma et al., 2006; Komura et al., 2008; Sinha et al., 2007). Comparison of these methods to the gel fractionation FPLC protocol in our laboratory suggests that these kits generally yield a good representation of the sample, except perhaps at the extremes of very high and very low HMW levels (Sinha et al., 2007). To date, however, these are limited to human adiponectin analysis.

Pros: Fast and high throughput. Small sample volumes needed.

Cons: Cost per sample when run in triplicate. Must run both full and HMW ELISAs for distribution and still lack LMW or trimeric abundance. Sample collection may alter digestion when necessary. Species use limited to available kits.

7. CONCLUDING REMARKS

Adiponectin is a metabolically active adipokine that is increasingly appreciated as a powerful marker of glucose homeostasis and inflammation. More than 10,000 publications have been published on the topic. An increasing number of clinical studies and correlations have linked total adiponectin and the abundance of its HMW oligomers with a variety of pathological states (Aleksandrova et al., 2012; Baessler et al., 2011; Hamilton et al., 2011; Lo et al., 2011; Matsumoto et al., 2013; Mazaki-Tovi et al., 2009). This has made measuring adiponectin levels across a range of medical and experimental disciplines, from endocrinology and nephrology to reproduction and oncology, a noteworthy biomarker and disease effector. This protocol for quantitating adiponectin complexes presents the most reliable method for analysis of oligomer distribution and has been extensively used for clinical and experimental samples. While other methods to estimate this distribution exist, complex separation by FPLC isolation and fractionation is the most thorough in determining changes in the three most abundant species. Despite its advantages, this is a rather laborious technique, and its use for large-scale clinical studies with hundreds or thousands of samples is rather limited. Further individual customization, optimization and potential automatization of the protocol to maximize costs and time benefits hopefully makes this protocol operational for many laboratories interested in adiponectin.

ACKNOWLEDGMENTS

J. M. R. is supported by the American Heart Association Scientist Development Grant 12SDG12050287. This work was also supported by the National Institutes of Health (Grants R01-DK55758, R01-DK099110 and P01-DK088761 to P. E. S.).

REFERENCES

Aleksandrova, K., Boeing, H., Jenab, M., Bueno-de-Mesquita, H. B., Jansen, E., van Duijnhoven, F. J., et al. (2012). Total and high-molecular weight adiponectin and risk of colorectal cancer: The European Prospective Investigation into Cancer and Nutrition Study. *Carcinogenesis*, *33*(6), 1211–1218.

Baessler, A., Schlossbauer, S., Stark, K., Strack, C., Riegger, G., Schunkert, H., et al. (2011). Adiponectin multimeric forms but not total adiponectin levels are associated with myocardial infarction in non-diabetic men. *Journal of Atherosclerosis and Thrombosis*, *18*(7), 616–627.

Bobbert, T., Rochlitz, H., Wegewitz, U., Akpulat, S., Mai, K., Weickert, M. O., et al. (2005). Changes of adiponectin oligomer composition by moderate weight reduction. *Diabetes*, *54*(9), 2712–2719.

Ebinuma, H., Miyazaki, O., Yago, H., Hara, K., Yamauchi, T., & Kadowaki, T. (2006). A novel ELISA system for selective measurement of human adiponectin multimers by using proteases. *Clinica Chimica Acta, 372*(1–2), 47–53.

Fisher, F. M., Trujillo, M. E., Hanif, W., Barnett, A. H., McTernan, P. G., Scherer, P. E., et al. (2005). Serum high molecular weight complex of adiponectin correlates better with glucose tolerance than total serum adiponectin in Indo-Asian males. *Diabetologia, 48*(6), 1084–1087.

Hamilton, M. P., Gore, M. O., Ayers, C. R., Xinyu, W., McGuire, D. K., & Scherer, P. E. (2011). Adiponectin and cardiovascular risk profile in patients with type 2 diabetes mellitus: Parameters associated with adiponectin complex distribution. *Diabetes & Vascular Disease Research, 8*(3), 190–194.

Kobayashi, H., Ouchi, N., Kihara, S., Walsh, K., Kumada, M., Abe, Y., et al. (2004). Selective suppression of endothelial cell apoptosis by the high molecular weight form of adiponectin. *Circulation Research, 94*(4), e27–e31.

Komura, N., Kihara, S., Sonoda, M., Kumada, M., Fujita, K., Hiuge, A., et al. (2008). Clinical significance of high-molecular weight form of adiponectin in male patients with coronary artery disease. *Circulation Journal, 72*(1), 23–28.

Kusminski, C. M., & Scherer, P. E. (2009). The road from discovery to clinic: Adiponectin as a biomarker of metabolic status. *Clinical Pharmacology and Therapeutics, 86*(6), 592–595.

Lara-Castro, C., Luo, N., Wallace, P., Klein, R. L., & Garvey, W. T. (2006). Adiponectin multimeric complexes and the metabolic syndrome trait cluster. *Diabetes, 55*(1), 249–259.

Lo, M. M., Salisbury, S., Scherer, P. E., Furth, S. L., Warady, B. A., & Mitsnefes, M. M. (2011). Serum adiponectin complexes and cardiovascular risk in children with chronic kidney disease. *Pediatric Nephrology, 26*(11), 2009–2017.

Matsumoto, Y., Toyomasu, K., Uchimura, N., & Ishitake, T. (2013). Low-molecular-weight adiponectin is more closely associated with episodes of asthma than high-molecular-weight adiponectin. *Endocrine Journal, 60*(1), 119–125.

Mazaki-Tovi, S., Romero, R., Vaisbuch, E., Kusanovic, J. P., Erez, O., Gotsch, F., et al. (2009). Maternal serum adiponectin multimers in preeclampsia. *Journal of Perinatal Medicine, 37*(4), 349–363.

Pajvani, U. B., Du, X., Combs, T. P., Berg, A. H., Rajala, M. W., Schulthess, T., et al. (2003). Structure-function studies of the adipocyte-secreted hormone Acrp30/adiponectin. Implications fpr metabolic regulation and bioactivity. *Journal of Biological Chemistry, 278*(11), 9073–9085.

Pajvani, U. B., Hawkins, M., Combs, T. P., Rajala, M. W., Doebber, T., Berger, J. P., et al. (2004). Complex distribution, not absolute amount of adiponectin, correlates with thiazolidinedione-mediated improvement in insulin sensitivity. *Journal of Biological Chemistry, 279*(13), 12152–12162.

Scherer, P. E., Williams, S., Fogliano, M., Baldini, G., & Lodish, H. F. (1995). A novel serum protein similar to C1q, produced exclusively in adipocytes. *Journal of Biological Chemistry, 270*(45), 26746–26749.

Schraw, T., Wang, Z. V., Halberg, N., Hawkins, M., & Scherer, P. E. (2008). Plasma adiponectin complexes have distinct biochemical characteristics. *Endocrinology, 149*(5), 2270–2282.

Sinha, M. K., Songer, T., Xiao, Q., Sloan, J. H., Wang, J., Ji, S., et al. (2007). Analytical validation and biological evaluation of a high molecular-weight adiponectin ELISA. *Clinical Chemistry, 53*(12), 2144–2151.

Turer, A. T., & Scherer, P. E. (2012). Adiponectin: Mechanistic insights and clinical implications. *Diabetologia, 55*(9), 2319–2326.

Wang, Y., Lam, K. S., Yau, M. H., & Xu, A. (2008). Post-translational modifications of adiponectin: Mechanisms and functional implications. *Biochemical Journal, 409*(3), 623–633.

CHAPTER FOURTEEN

Genome-Wide Profiling of Transcription Factor Binding and Epigenetic Marks in Adipocytes by ChIP-seq

Ronni Nielsen, Susanne Mandrup[1]

Department of Biochemistry and Molecular Biology, University of Southern Denmark, Odense, Denmark
[1]Corresponding author: e-mail address: s.mandrup@bmb.sdu.dk

Contents

1. Introduction	261
2. Technical Aspects	262
2.1 How to obtain ChIP-seq grade material from adipocytes	262
2.2 Chromatin immunoprecipitation	263
2.3 Library preparation for Illumina sequencing	269
2.4 Downstream data analyses	276
3. Future Challenges	277
Acknowledgments	277
References	277

Abstract

The recent advances in high-throughput sequencing combined with various other technologies have allowed detailed and genome-wide insight into the transcriptional networks that control adipogenesis. Chromatin immunoprecipitation (ChIP) combined with high-throughput sequencing (ChIP-seq) is one of the most widely used of these technologies. Using these methods, association of transcription factors, cofactors, and epigenetic marks can be mapped to DNA in a genome-wide manner. Here, we provide a detailed protocol for performing ChIP-seq analyses in preadipocytes and adipocytes. We have focused mainly on critical points, limitations of the assay, and quality controls required in order to obtain reproducible ChIP-seq data.

1. INTRODUCTION

The transcriptional network that controls adipogenesis and adipocyte function has been extensively studied during the last two decades. Most of these studies have been carried out in immortalized preadipocyte cell lines such as 3T3-L1 or 3T3-F442A; however, subsequent validation in primary cells and animals models has confirmed that a lot of knowledge obtained from cell lines also applies to primary fat cells. Collectively, data from numerous laboratories have shown that the adipogenic transcription factors (TFs) act in two waves. The first wave, which is induced immediately after stimulation of differentiation includes CCAAT/enhancer-binding protein (C/EBP)β and C/EBPδ. The factors in the first wave induce the expression of the factors involved in the second wave of which peroxisome proliferator-activated receptor γ and C/EBPα are key players (Farmer, 2006; Lefterova & Lazar, 2009; Siersbæk, Nielsen, & Mandrup, 2012).

Until the last decade transcriptional regulation was exclusively studied at a gene-by-gene level; however, the recent advances in deep sequencing combined with various other technologies have allowed investigators to study transcriptional networks at a genome-wide level (Ren et al., 2000). One of the most useful technologies is the combination of chromatin immunoprecipitation (ChIP) (Solomon & Varshavsky, 1985) with deep sequencing (ChIP-seq) (Barski et al., 2007; Robertson et al., 2007). Recently, we as well as others have used ChIP-seq to generate genome-wide maps of the association of TFs, cofactors, and epigenetic marks with DNA in mature adipocytes and throughout adipocyte differentiation (Lefterova et al., 2010; Mikkelsen et al., 2010; Nielsen et al., 2008; Siersbæk et al., 2011; Siersbæk, Loft, et al., 2012; Siersbæk, Nielsen, et al., 2012; Steger et al., 2010). Genome-wide techniques like ChIP-seq have revolutionized the way we think about transcriptional regulation and have greatly increased our understanding of the complexity of transcriptional networks regulating adipogenesis (reviewed by Siersbæk et al., 2010; Siersbæk, Loft, et al., 2012; Siersbæk, Nielsen, et al., 2012; Cristancho & Lazar, 2011).

In the last few years, a number of groups have further developed and refined the ChIP-seq protocols and recently the Encyclopedia of DNA Elements (ENCODE) Consortium has published a comprehensive and useful list of working standards and guidelines for doing ChIP-seq (Landt et al., 2012). However, it is important to note that protocols need to be optimized for each cell type, and that conditions must be adjusted depending on the antibody. Thus, herein, we provide a detailed protocol for performing ChIP-seq in adipocytes.

2. TECHNICAL ASPECTS

2.1. How to obtain ChIP-seq grade material from adipocytes

The first important issue in order to obtain high quality and reproducible ChIP reactions in adipocytes is to ensure that the percentage of adipocytes is high and similar between samples. In case of *in vitro* differentiated adipocytes, care should be taken that the percentage of differentiated cells is as high as possible and similar between experiments.

One of the most widely used model systems for studying adipogenesis is the murine 3T3-L1 preadipocyte cell line (Green & Kehinde, 1974). Using a cocktail of hormonal inducers (Helledie et al., 2002), we routinely obtain >95% differentiation of these cells. Guidelines to obtain this high degree of differentiation include that (1) the 3T3-L1 cells should not grow confluent during expansion of the cells, (2) the culture should be plated so that confluence is reached within 48 h, (3) cells must be confluent for 48 h before hormonal inducers are added, and (4) serum batches need to be screened for their ability to support differentiation.

2.2. Chromatin immunoprecipitation
2.2.1 Preparation of cross-linked chromatin from adipocytes
Cross-linking of protein–DNA interactions in living cells is the first essential step in the ChIP procedure. This is most often done using formaldehyde, which cross-links protein–DNA/RNA and protein–protein in close proximity (~2 Å), thereby generating a snapshot of endogenous chromatin complexes. For studying proteins that are indirectly associated with DNA, such as transcriptional coregulators, protein–protein cross-linking by disuccinimidyl glutarate (DSG) prior to formaldehyde treatment can significantly increase the efficiency of the ChIP analysis (Nowak et al., 2005).

The amount of starting material (i.e., cell number) is critical for the success of the experiment. Depending on the abundance of the target protein and the efficiency of the antibody, typically $0.5–1 \times 10^6$ cells are needed for a single TF ChIP experiment followed by quantitative real-time PCR (qPCR) evaluation (ChIP-PCR). For ChIP-seq experiments one should aim at $10–20 \times 10^6$ cells to obtain enough immunoprecipitated DNA to perform reproducible library preparations. In cases where such amounts of cells are difficult to obtain, as for example, for adipose tissue and primary adipocytes (Siersbæk, Loft, et al., 2012; Siersbæk, Nielsen, et al., 2012), one can adopt the procedure and include DNA amplification after the ChIP (e.g., linear amplification protocols (Shankaranarayanan et al., 2011)) or optimized low-scale library preparation procedures (e.g., nano-ChIP-seq (Adli & Bernstein, 2011)). Very recently, a refined carrier-assisted ChIP-seq procedure has been published. It uses recombinant histones and random human mRNA during immunoprecipitation to significantly increase the ChIP-seq signal from as few as 0.1×10^6 cells (Zwart et al., 2013). The protocol below is optimized for one 15 cm dish of 3T3-L1 adipocytes and can be scaled accordingly.

1. Remove the medium from the plate.
2. *Optional*—Add 25 ml PBS+DSG[1] and cross-link for 20–45 min at room temperature (RT) and remove PBS/DSG.
3. Cross-link cells for 10 min at RT by adding 30 ml PBS+1% formaldehyde (prepare fresh for each experiment).
4. Add 4.3 ml 1 M glycine (final conc. 0.125 M) to quench the formaldehyde cross-linking. Incubate 10 min RT.
5. Wash the cells on the plate two times in cold PBS (~25 ml) and remove as much PBS as possible.
6. Scrape cross-linked cells off the plate in 1 ml of ChIP lysis buffer (see Table 14.1) and transfer to a 15 ml tube. Keep samples on ice from this step forward to avoid undesired decross-linking due to heating of the samples.
7. *Optional.* Isolate nuclei by pelleting the cross-linked cells, add 10 ml nuclei isolation buffer (20 mM Tris, pH 8, 0.5% NP-40, 85 mM KCl), and homogenize using a douncer (e.g., Wheaton).
8. *Optional.* Resuspend crude nuclei at 10×10^6 per milliliter in RIPA buffer (1% NP-40, 0.5% sodium deoxycholate, 0.1% SDS).

Table 14.1 ChIP lysis and wash buffers

ChIP lysis buffer	ChIP wash buffer 1	ChIP wash buffer 2	ChIP wash buffer 3	ChIP wash buffer 4 (10×)
0.1% SDS	0.1% SDS	0.1% SDS	0.25 M LiCl	10 mM EDTA
1% Triton X-100	0.1% NaDOC	0.1% NaDOC	0.5% NaDOC	200 mM Tris, pH 8
0.15 M NaCl	1% Triton X-100	1% Triton X-100	0.5% NP-40	
1 mM EDTA	0.15 M NaCl	0.5 M NaCl	1 mM EDTA	
20 mM Tris, pH 8	1 mM EDTA	1 mM EDTA	20 mM Tris, pH 8	
	20 mM Tris, pH 8	20 mM Tris, pH 8		

---

[1] To prepare PBS-DSG solution: Weigh the required amount of DSG directly in a 1.5 ml tube. Add DMSO (1 ml per 0.16 g DSG = 10 × 15 cm plates) to make a solution with a final concentration of 0.5 M DSG–DMSO. Immediately before use (<30 min), slowly add droplets of 0.5 M DSG–DMSO to PBS (100 μl/25 ml) while swirling. If not careful, the DSG will polymerize (use p100). DSG cross-linking time needs to be optimized for each cell type (45 min for 3T3-L1 cells).

Note: Before sonication of the cells or nuclei it is recommended to measure the amount of DNA in each sample to minimize sample-to-sample variation in sonication and subsequent immunoprecipitation efficiency. This can be done by decross-linking an aliquot of each sample, purifying DNA, and measuring the concentration. Alternatively, the cell numbers can be determined with a cell counter/hemocytometer from parallel noncross-linked cultures.

2.2.2 Fragmentation of chromatin

Chromatin needs to be fragmented to smaller pieces to increase the resolution of the binding site analysis (Fig. 14.1). Fragmentation of chromatin for subsequent immunoprecipitation is most commonly obtained by ultrasound sonication introducing randomly shearing of the chromatin. We routinely use the Bioruptor® Twin (Diagenode), which processes up to 12 samples simultaneously within cooled water baths that reduce thermal decrosslinking of the samples. Other fragmentation instruments include the Covaris AFA (KBiosciences). Sonication must be optimized for each setup, as cell type, cell density, type of plastic tube, and composition of lysis buffer greatly influence the fragmentation efficiency. It is generally best practice to shear chromatin directly from fresh cells, as sonication of frozen cells has been shown to be less efficient (Schoppee Bortz & Wamhoff, 2011). Cells can also be frozen for later shearing of the chromatin, but further optimizing of sonication conditions is then recommended. Sonication optimization has been thoroughly addressed elsewhere ((Schoppee Bortz & Wamhoff, 2011) and Diagenode sonication guidelines[2]). The following fragmentation protocol is optimized for differentiated 3T3-L1 cells. Special attention should be paid to removing the floating lipid from the mature adipocytes after sonication.

9. Sonicate 1.5 ml chromatin (adjusted for similar DNA concentration between samples) in 15 ml tubes with metal probe.
10. Sonicate samples 4×10 bursts of 30 s ON and 30 s OFF at high-level output. Place samples on ice 5 min between each round of 10 bursts. Total sonication time for 12 samples is approximately 60 min.
11. Transfer the chromatin to a 2 ml tube and spin down cell debris ($10,000 \times g$) for 1 min at 4 °C.
12. For adipocyte samples excess lipid after cell lysis and sonication will float on top of the supernatant. This needs to be carefully removed by repeating centrifugation and lipid removal two to three times.

---

[2] http://www.diagenode.com/en/support/protocols.php.

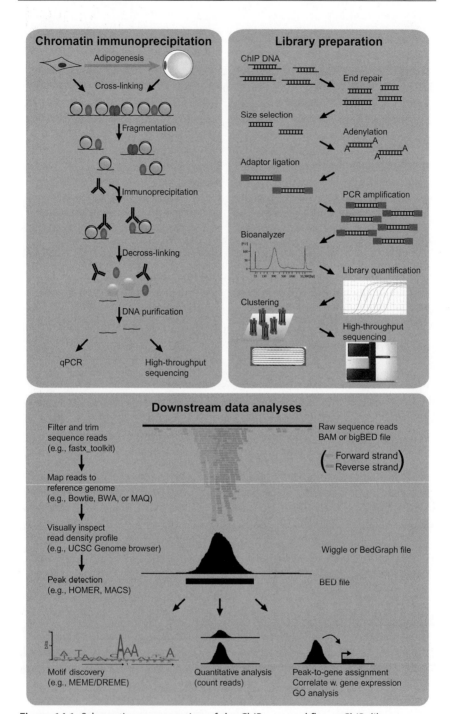

Figure 14.1 Schematic representation of the ChIP-seq workflow—ChIP, library preparation, and downstream data analyses. See text for details. (See the color plate.)

13. Transfer the cleared supernatant into a new tube, pool samples if multiple of same and aliquot the chromatin before freezing the samples at −80 °C. *Note*: Cross-linked and sonicated chromatin can be stored at −80 °C for at least 2 years without loss of immunoprecipitation signal. Avoid repeated freeze–thaw cycles.

2.2.3 Chromatin immunoprecipitation

When the chromatin has been cross-linked and sonicated, the DNA fragments associated with the protein of interest can be immunoprecipitated with a specific antibody. The ChIP procedure greatly relies on availability of highly specific antibodies, and inclusion of positive and negative controls is highly recommended. Even when a good source of antibody has been identified, batch quality may vary, and it is therefore important to keep batches of good antibodies on stock.

ChIP can be performed either manually or in an automated work station (e.g., SX-8G IPstar® from Diagenode). The protocol below describes the manual workflow including preparation of beads, immunoprecipitation, DNA purification, and qPCR validation, which has been used with success for a broad selection of antibodies in our laboratory. Using approximately 1×10^6 cells for each ChIP reaction, 10–15 ChIP experiments can be performed from one 15 cm dish. Dependent on the quality of the antibody and abundance of the target protein, ChIP-seq experiments can usually be performed by pooling immunoprecipitated material from 10 to 20 ChIP experiments (2–5 ChIP's for some histone modifications).

14. Protein A (or Protein G for antibodies raised in goat) sepharose beads (5 µl beads per ChIP reaction) are washed 3 × in ChIP lysis buffer. Cut the pipette tip to avoid damaging the beads. After each wash, beads are pelleted by centrifugation (1 min at $700 \times g$) and resuspended in lysis buffer. Alternatively, magnetic beads can be used.
15. Protein A beads are incubated 2 h at 4 °C rotating in ChIP lysis buffer (volume of beads to lysis buffer is 1:5) containing 1 µg/µl BSA.
16. Thaw chromatin on ice and take out 1–5% for input control - save for decross-linking and purification.
17. Use 100 µl chromatin per ChIP representing chromatin from approximately 1×10^6 cells.
18. Add antibody (1–2 µg - do initial titration test for new antibodies), BSA (1 µg/µl), and ChIP lysis buffer to total of 400 µl.
19. Incubate 3 h at 4 °C rotating to generate the immunocomplex.
20. Add 25 µl prepared beads/ChIP.

21. Incubate overnight (ON) at 4 °C rotating.
22. Wash ChIP reactions with 1000 µl of each of the wash buffers 1–4 for 5 min rotating (2× in wash buffer 1, 1× in wash buffer 2, 1× in wash buffer 3, and 2× in wash buffer 4). Remember to dilute the ChIP wash buffer 4 (10× stock) ten times before use. Spin down at $700 \times g$ between each wash.
23. After the last washing step remove as much washing buffer as possible without disturbing the beads.
24. Incubate beads with 400 µl elution buffer (1% SDS, 0.1 M NaHCO$_3$) rotating at RT.
25. Spin down beads and transfer supernatant to a new 1.5 ml tube.
26. Add 16 µl 5 M NaCl to supernatant and vortex briefly.
27. Decross-link the chromatin at 65 °C for at least 4 h (or ON). Decross-link the input samples including 16 µl 5 M NaCl and elution buffer to total volume of 400 µl.
28. Add 1 volume (400 µl) of phenol chloroform isoamylalcohol.
29. Vortex at least 30 s and mix the tube by inversion every 10 s and spin down ($10,000 \times g$) 5 min at RT.
30. Take aqueous (Upper!!) phase, add 1 volume (400 µl) chloroform, vortex at least 30 s, mix by inversion and spin down ($10,000 \times g$) 5 min at RT.
31. Take the upper aqueous phase and add 20 µg glycogen, 40 µl 3 M NaAc, pH 4.5, and 1 ml ice cold 100% EtOH.
32. Precipitate the DNA with a 1 h to ON incubation at −20 °C.
33. Pellet the DNA by centrifugation at maximum speed for 25 min at 4 °C, remove supernatant, and wash pellet with 200 µl 70% EtOH.
34. Spin down ($10,000 \times g$) 5 min at 4 °C, remove supernatant, dry the sample at the bench with open lid, and resuspend pellet in 60 µl ultrapure water.

Note: At the end of the purification step it is important to keep the volume of the immunoprecipitated and input DNA samples low in order to fit enough material into library preparation. From the 60 µl of resuspended ChIP material, 10 µl can be diluted (dependent on the number of ChIP reactions performed) for qPCR evaluation of the ChIP experiment and 10 µl for concentration measurements (e.g., Pico Green or Qubit® (Life Technologies)). This leaves 40 µl of material for the library preparation.

2.2.4 qPCR validation of ChIP experiment

The last step of the ChIP module is validation of the ChIP experiment itself by qPCR. The immunoprecipitated DNA is examined for enrichment of specific regions corresponding to regions containing histone modifications or transcription factor-binding sites. While many histone modifications show a characteristic distribution around either promoters or enhancers of regulated genes (Mikkelsen et al., 2010; Steger et al., 2010) preexisting knowledge about the exact location of transcription factor binding is needed for validation using such target sites. We routinely evaluate the percentage recovery (i.e., qPCR signal of immunoprecipitated DNA over input DNA) for a few positive target sites and the fold over background for at least two negative regions before proceeding with library preparation. Fold over background should be minimum 15–20 to produce high quality ChIP-seq data.

2.3. Library preparation for Illumina sequencing

Library preparation of fragmented and immunoprecipitated DNA for use with the Illumina® sequencing platform, involves repair of 3'- and 5'-ends and addition of dA-tails followed by ligation to an adaptor and PCR amplification (Fig. 14.1). Although the overall workflow is straightforward, many variations of the protocol are used in different laboratories and provided by various companies (e.g., Illumina, New England Biolabs, Bioo Scientific). We have optimized our library preparation procedure to fit the needs for the low amount of DNA material precipitated by TF ChIP. Using this protocol, laboratories with low-to-medium number of samples will be able to perform library preparation of 8–16 samples using standard laboratory equipment in less than two working days. Automated library preparation platforms are provided by several companies (e.g., Diagenode, Beckman Coulter). In our protocol, preassembled kits are replaced by reagents purchased separately, which reduces the price of library preparation significantly. The procedure allows for robust library preparation from as little as 10 ng of DNA as starting material. However, more starting material (preferably 20–50 ng) will generally produce libraries of greater quality and reproducibility.

2.3.1 End repair

The first part of the library preparation module converts the DNA overhangs into phosphorylated blunt ends using T4 DNA polymerase and *Escherichia coli* DNA Pol I large fragment (Klenow polymerase). The 3'–5'-exonuclease activity of these enzymes removes 3'-overhangs and the polymerase activity fills in the 5'-overhangs.

1. Dilute Klenow DNA polymerase 1:5 in ultrapure water.
2. Prepare the following reaction mix in a 1.5 ml or PCR tube and incubate for 30 min at 20 °C.

End repair

ChIP-enriched DNA	40 µl
T4 DNA ligase buffer (NEB B0202)	5 µl
dNTP mix (NEB N0447S)	2 µl
T4 DNA polymerase (NEB M0203)	1 µl
Diluted Klenow DNA polymerase (NEB M0210)	1 µl
T4 PNK (NEB M0201)	1 µl
Total volume	50 µl

2.3.2 Size selection

Size selection is an important step as the Illumina sequencing technology is limited to a certain fragment size (150–800 bp) for proper cluster generation before sequencing. Furthermore, sequencing smaller fragments will make binding site identification more accurate as the sequencing reads will be centered more tightly around the site. Size selection can be performed by gel electrophoresis either manually (2% low-melt agarose gel) or by automated solutions (e.g., E-Gel® (Life Technologies, Inc.) or Pippin Preparation (Sage Science)). Alternatively, SPRI (solid phase reversible immobilization)-based AMPure XP® beads (Beckman Coulter, Inc.) can be used with great success for size selection of ChIP-seq samples with low DNA concentrations as they provide a higher recovery than agarose gels. Furthermore, AMPure XP beads makes it easier to increase the number of samples processed simultaneously. Different library preparation protocols perform size selection of the sample library at different steps in the protocol. We have successfully adopted the "double-SPRI beads" size selection method after the end repair step as described below.

3. Bring AMPure XP beads to RT and vortex to resuspend.
4. Add 40 µL of AMPure XP beads (0.8 × ratio) to each sample (50 µl end-repaired DNA), gently pipette up and down 10 times and incubate the samples at RT for 5 min.
5. Place the tubes on a magnetic stand at RT until the sample appears clear (5 min) and gently transfer 85 µl of the *supernatant* to a new tube making sure not to transfer any beads. Some liquid may remain in the wells. This selectively removes large DNA fragments above approx. 300 bp.
6. Add 17 µL of AMPure XP beads (0.2 × ratio) to the supernatant and gently pipette up and down 10 times. Incubate at RT for 5 min.

7. Place the tubes on the magnetic stand at RT until the sample appears clear (5 min), and gently remove and discard the clear sample making sure not to disturb the beads. Some liquid may remain in the wells. This selectively removes DNA below 200 bp.
8. While tubes are in the magnetic stand, gently add 180 µL of freshly prepared 80% ethanol to each tube and incubate at RT for 30 sec. Carefully, remove ethanol with a pipette. Repeat for a total of 2 ethanol washes and ensure all ethanol has been removed.
9. Dry beads at RT for 5 min while the tubes are on the magnetic stand with the lids open.
10. Remove tubes from magnetic stand and resuspend the dried beads with 42 µL ultra-pure water. Gently, pipette up and down 10 times mixing thoroughly and ensure beads are properly resuspended. Incubate resuspended beads at RT for 2 min.
11. Place the tubes on the magnetic stand at RT until the sample appears clear (5 min), and gently transfer 39 µL of clear sample to a new tube without transfer of beads.
12. Leave samples on ice or store at −20°C before continuing the adenylation of 3′-ends.

Note: For the "double-SPRI beads" size selection, it is crucial to make fresh 80% ethanol as low percentage ethanol will elute DNA rather than wash the beads. In addition, the nature of the AMPure XP beads allows them to more easily flow with the liquid than other beads, and it is therefore important not to disturb or transfer the beads during washing and elutions.

2.3.3 Adenylation of 3′-ends

This part of the library preparation module adds an "A" base to the 3′-end of the blunt-ended, phosphorylated DNA fragments using the polymerase activity of Klenow fragment. This prepares the DNA fragments for ligation to the adapters, which have a single "T" base overhang at their 3′-end.

13. Prepare the following reaction mix in a 1.5 ml or PCR tube and incubate for 30 min at 37 °C.

Adenylation of 3′-ends	
Size-selected DNA	39 µl
NEB buffer 2 (×10) (NEB B7002S)	5 µl
dATP 1 mM (NEB N0440S)	5 µl
Klenow exo (3′–5′-exo-) (NEB M0212)	1 µl
Total volume	50 µl

14. Clean up using AMPure XP beads protocol. Bring AMPure XP beads to RT and vortex to resuspend.
15. Add 90 µl of AMPure XP beads (1.8 × ratio) to each sample. Mix thoroughly by pipetting up and down at least 10 × and incubate sample at RT for 5 min.
16. Place the tubes on the magnetic stand at RT until the sample appears clear (5 min) and gently remove and discard the clear sample making sure not to disturb beads. Some liquid may remain in wells.
17. While tubes are in the magnetic stand, gently add 180 µl of freshly prepared 80% ethanol to each tube and incubate at RT for 30 s. Carefully, remove ethanol with a pipette. Repeat for a total of two ethanol washes and ensure all ethanol has been removed.
18. Dry beads at RT for 5 min while the tubes are on the magnetic stand with the lids open.
19. Resuspend dried beads in 12 µl ultrapure water. Gently, pipette up and down 10× mixing thoroughly and ensure beads are properly resuspended. Remove tubes from stand and incubate at RT for 2 min.
20. Place the tubes on the magnetic stand at RT until the sample appears clear (5 min), and gently transfer 10 µl of clear sample to a new tube without transfer of beads. Leave samples on ice before continuing the adaptor ligation.

Note: After the cleanup of the adenylated DNA, it is recommended to continue directly to the adaptor ligation step.

2.3.4 Adaptor ligation

The adaptor ligation step adds indexing adapters to the ends of the DNA fragments, preparing them to be hybridized to an Illumina flow cell. Indexing adaptors provide the opportunity to multiplex several samples into one lane of the flow cell. Multiplexing is particular useful as sequencing capacity has greatly increased (six- to eightfold) from the Illumina GAIIx to the latest HiSeq instrument. A persistent problem with the ligation step is selfannealing adapters. Therefore, after the adaptor ligation step it is important to remove all nonligated adaptor dimers as these will form clusters on the flow cell and reduce the number of sequencing reads from the sample. The AMPure XP beads cleanup step after the ligation is therefore crucial for eliminating such competing adaptors at the flow cell. Gel purification is not recommended here, as it may result in loss of material and sample complexity before the PCR amplification step. The Bioanalyzer quality control step (Section 2.3.6) will reveal if adaptor dimers are present and whether additional cleanup is needed. It is recommended to adjust the amount of adaptors to a 10:1 ratio of adaptor to DNA. Indexed adaptors are included in the Illumina TruSeq

ChIP sample preparation kit. However, for this protocol, adaptors were ordered separately thereby reducing the cost of the library preparation significantly. When ordering your own adaptors it is strongly recommended to acquire in-depth knowledge about how the adaptors work, their sequence and structure in order to obtain successful libraries and sequencing results.[3]

21. Prepare the following reaction mix in a 1.5 ml or PCR tube and incubate for 15 min at room temperature.

Starting material	10 ng	25 ng	50 ng
DNA sample	10 μl	10 μl	10 μl
Quick ligation buffer (2×) (NEB B2200)	12.5 μl	12.5 μl	12.5 μl
Adaptor oligo mix (1.5 μM)	0.5 μl	1.25 μl	2.5 μl
Quick T4 DNA ligase (NEB M2200)	2 μl	2 μl	2 μl
Ultra pure water	2.5 μl	1.75 μl	0.5 μl
Total volume	27.5 μl	27.5 μl	27.5 μl

22. Clean up using the AMPure XP beads protocol as in Section 2.3.3 but adjust the volume of beads and water during elution (27.5 μl beads (1.0 × ratio) and 23 μl water during elution).
23. Gently transfer 20 μl of clear sample to a new tube without transfer of beads. Leave samples on ice or store at −20 °C before continuing the amplification step.

2.3.5 Enrich adapter-ligated DNA fragments by PCR

PCR amplification is used to selectively enrich for DNA fragments with adaptors correctly added in both ends. Note that carry over of nonligated adaptors from the ligation step to the PCR reaction will decrease the amplification of adaptor-ligated sample and generate adaptor products compromising the generation of clusters. The number of PCR cycles needed to obtain enough material for sequencing will vary depending on the amount of starting material of the right size. PCR amplification should always be run with as few cycles as possible to avoid unwanted PCR bias of the library. In this procedure, half of the material from the adaptor ligation step is used for the PCR amplification. This allows for adjustments of the number of PCR cycles after the evaluation of the Bioanalyzer quality control step (see Section 2.3.6). Alternatively, library

---

[3] http://tucf-genomics.tufts.edu/documents/protocols/TUCF_Understanding_Illumina_TruSeq_Adapters.pdf.

enrichment can be monitored real-time by qPCR allowing precise control of the optimal number of PCR cycles (KAPA Biosystems—KK2701).

24. Prepare the following PCR reaction mix:

PCR reaction mix	
DNA sample	10 μl
Phusion HF master mix (2×)	10 μl
Multiplex PCR primer 1.0 (6.25 μM)	1 μl
Multiplex PCR primer 2.0 (6.25 μM)	1 μl
Total volume	22 μl

25. Amplify using the following PCR program.

PCR program	
98 °C	30 s
98 °C	10 s
65 °C 10–16 cycles	30 s
72 °C	30 s
72 °C	5 min

PCR cycles	Starting material (ng)
10–12	50
12–14	25
14–16	10

26. Clean up using AMPure XP beads protocol as in Section 2.3.3 but adjust the volume of beads and water during elution (22 μl beads (1×) and 22 μl water during elution).
27. Gently transfer 20 μl of clear sample to a new tube without transfer of beads. Leave samples on ice or store at −20 °C.

2.3.6 Sample library quality control

The PCR-amplified sample library must be run on a gel or a Bioanalyzer 2100 (Agilent Technologies) using the High Sensitivity DNA or DNA 1000 kit to verify the fragment size distribution, to estimate the approximate concentration, and to check for adaptor dimers and potential PCR artifacts (Fig. 14.2).

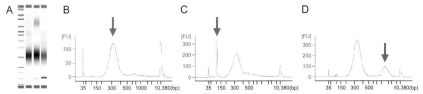

Figure 14.2 Bioanalyzer quality control. (A) Bioanalyzer data represented as gel image. (B) Bioanalyzer trace showing the DNA fragments present in the libraries. Red arrow indicates library peak ready for sequencing. (C) Sample containing adaptor dimers (red arrow) indicating the need for further cleanup. (D) Library with extra high molecular band. This will not affect sequencing results but indicates over amplification of the sample (i.e., need to run fewer PCR cycles). FU, fluorescence units. (See the color plate.)

If the Bioanalyzer trace shows two bands, one at the expected size and one higher molecular weight band, a portion of the adaptor-ligated DNA fragments have annealed to each other. This is due to the nature of the long adaptors and is more predominant after many PCR cycles. This type of double band will not affect the sequencing output as these products will be denatured before clustering of the flow cell.

To check the quality of the sample library, it is advised to perform ChIP-PCR using the same primers as for the quality control of the ChIP module (see Section 2.2.4). Fold over background and the ratio between the ChIP signals in the different samples should be comparable before and after library preparation.

2.3.7 Sample library quantification
The accurate quantification of the number of enriched DNA molecules in a library is critical for optimal sequencing results on any Illumina sequencing platform. Overestimation of library concentration results in lower cluster density, while underestimation of library concentration results in too many clusters on the flow cell, which can lead to poor cluster resolution and loss of sequencing data. Currently, the golden standard for measuring the precise number of adaptor-ligated DNA molecules in a library is by qPCR, for example, using the KAPA Library Quantification kit (KK4824).

2.3.8 Illumina sequencing
The Illumina sequencing technology, including technical aspects and biochemistry, has been described previously (Bentley et al., 2008). It has been shown that the number of positive target sites identified by ChIP-seq in general increases with the amount of sequencing reads (Myers et al., 2011).

In general, we aim for 10×10^6 unique reads for each ChIP-seq, which is in accordance with the recently published recommendations from the ENCODE project (Landt et al., 2012).

The recent advances in the Illumina sequencing platforms (HiSeq1000/1500/2000/2500) have increased the sequencing capacity and speed of sequencing significantly. This allows for increasing sequencing depth and/or multiplexing several samples per lane resulting in a significantly reduced cost per sample. The protocol presented here can be used for multiplexing 8–10 ChIP experiments in a single lane at the required sequencing depth.

2.4. Downstream data analyses

ChIP-seq data require processing in a pipeline of computational steps. Recently, the single steps of such pipelines and the various software used have been described in great detail (Furey, 2012; Landt et al., 2012; Schmidt et al., 2009). Here, we provide a short overview of the workflow used in our studies including a graphical representation (Fig. 14.1).

The Illumina sequencing output is base called nucleotide sequence reads. Preprocessing of these sequencing reads by filtering of low quality bases and trimming of reads can be performed with different software (e.g., the FASTX-toolkit). Subsequent mapping of the reads back to the reference genome can be done using different short-read aligners such as Bowtie, BWA, and MAQ (for comprehensive list of short-read aligners and references, see http://en.wikipedia.org/wiki/Sequence_alignment_software). Visual inspection of the ChIP-seq data in a genome browser (e.g., UCSC or Ensemble) is a good initial quality control of the data set allowing a quick evaluation of the read distribution and the enrichment relative to a control sample. The overall goal in the downstream data analysis is identification of ChIP signal enrichment, that is, peak detection (e.g., HOMER or MACS). Finally, most downstream analysis pipelines include a motif search step (e.g., MEME/DREME) of the peak detected regions in order to determine whether certain motifs are enriched at the sites where a particular factor binds. Furthermore, motif discovery can be used to elucidate the combination of factor co-occupancy which might integrate regulatory action at multiple sites in the genome (as demonstrated in Nielsen et al., 2008; Lefterova et al., 2008; Steger et al., 2010; Mikkelsen et al., 2010; Siersbæk et al., 2011). Finally, in order to fully exploit the information in ChIP-seq datasets, many additional analyses need to be applied.

3. FUTURE CHALLENGES

Data obtained from ChIP-seq experiments have shown that most TFs bind to thousands of sites, some of which are close to genes whereas most are far away from annotated genes. A major future challenge is to link these binding sites to function. Currently, proximity to regulated genes is used as an indicator that sites are functionally relevant, but very often multiple binding sites are found in the vicinity of a regulated gene and the relative importance of these is difficult to assess. The recent chromatin conformation-based assays such as 4C (Zhao et al., 2006), Hi-C (Lieberman-Aiden et al., 2009), and ChIA-PET (Li et al., 2010) that detect looping between distal elements and promoters (Li et al., 2012) will help predict which sites are functionally most important; however, proof of function awaits the development of high-throughput genome-editing analyses.

Another challenge in ChIP-seq analyses is the limited resolution that only allows detection of binding regions of ∼200 bp rather than actual binding sequence. Techniques such as ChIP-exo (Rhee & Pugh, 2011) combined with digital genomic footprinting (Hesselberth et al., 2009) may allow detection of exact binding sequences at a genome-wide level.

Finally, a third challenge lies in distinguishing direct from indirect binding. Further development of technologies, such as UV laser cross-linking, that only cross-links DNA and protein in close proximity will be a major breakthrough.

In conclusion, although ChIP-seq technologies have revolutionized our understanding of transcriptional regulation there are a number of future challenges that need to be solved before we fully understand transcriptional regulation at a genome-wide level.

ACKNOWLEDGMENTS

The authors thank members of the Mandrup group for valuable input to this chapter and for their work to continuously refine ChIP-seq protocols. This work was supported by the Novo Nordisk Foundation, the Lundbeck Foundation, and the Danish Independent Research Council | Natural Science.

REFERENCES

Adli, M., & Bernstein, B. E. (2011). Whole-genome chromatin profiling from limited numbers of cells using nano-ChIP-seq. *Nature Protocols, 6,* 1656–1668.

Barski, A., Cuddapah, S., Cui, K., Roh, T. Y., Schones, D. E., Wang, Z., et al. (2007). High-resolution profiling of histone methylations in the human genome. *Cell, 129,* 823–837.

Bentley, D. R., Balasubramanian, S., Swerdlow, H. P., Smith, G. P., Milton, J., Brown, C. G., et al. (2008). Accurate whole human genome sequencing using reversible terminator chemistry. *Nature, 456*, 53–59.

Cristancho, A. G., & Lazar, M. A. (2011). Forming functional fat: A growing understanding of adipocyte differentiation. *Nature Reviews Molecular Cell Biology, 12*, 722–734.

Farmer, S. R. (2006). Transcriptional control of adipocyte formation. *Cell Metabolism, 4*, 263–273.

Furey, T. S. (2012). ChIP-seq and beyond: New and improved methodologies to detect and characterize protein-DNA interactions. *Nature Reviews Genetics, 13*, 840–852.

Green, H., & Kehinde, O. (1974). Sublines of mouse 3T3 cells that accumulate lipid. *Cell, 1*, 113–116.

Helledie, T., Grontved, L., Jensen, S. S., Kiilerich, P., Rietveld, L., Albrektsen, T., et al. (2002). The gene encoding the acyl-CoA-binding protein is activated by peroxisome proliferator-activated receptor gamma through an intronic response element functionally conserved between humans and rodents. *Journal of Biological Chemistry, 277*, 26821–26830.

Hesselberth, J. R., Chen, X., Zhang, Z., Sabo, P. J., Sandstrom, R., Reynolds, A. P., et al. (2009). Global mapping of protein-DNA interactions in vivo by digital genomic footprinting. *Nature Methods, 6*, 283–289.

Landt, S. G., Marinov, G. K., Kundaje, A., Kheradpour, P., Pauli, F., Batzoglou, S., et al. (2012). ChIP-seq guidelines and practices of the ENCODE and modENCODE consortia. *Genome Research, 22*, 1813–1831.

Lefterova, M. I., & Lazar, M. A. (2009). New developments in adipogenesis. *Trends in Endocrinology & Metabolism, 20*, 107–114.

Lefterova, M. I., Steger, D. J., Zhuo, D., Qatanani, M., Mullican, S. E., Tuteja, G., et al. (2010). Cell-specific determinants of peroxisome proliferator-activated receptor γ function in adipocytes and macrophages. *Molecular and Cellular Biology, 30*, 2078–2089.

Lefterova, M. I., Zhang, Y., Steger, D. J., Schupp, M., Schug, J., Cristancho, A., et al. (2008). PPARγ and C/EBP factors orchestrate adipocyte biology via adjacent binding on a genome-wide scale. *Genes & Development, 22*, 2941–2952.

Li, G., Fullwood, M., Xu, H., Mulawadi, F. H., Velkov, S., Vega, V., et al. (2010). ChIA-PET tool for comprehensive chromatin interaction analysis with paired-end tag sequencing. *Genome Biology, 11*, R22.

Li, G., Ruan, X., Auerbach, R. K., Sandhu, K. S., Zheng, M., Wang, P., et al. (2012). Extensive promoter-centered chromatin interactions provide a topological basis for transcription regulation. *Cell, 148*, 84–98.

Lieberman-Aiden, E., van Berkum, N. L., Williams, L., Imakaev, M., Ragoczy, T., Telling, A., et al. (2009). Comprehensive mapping of long-range interactions reveals folding principles of the human genome. *Science, 326*, 289–293.

Mikkelsen, T. S., Xu, Z., Zhang, X., Wang, L., Gimble, J. M., Lander, E. S., et al. (2010). Comparative epigenomic analysis of murine and human adipogenesis. *Cell, 143*, 156 169.

Myers, R. M., Stamatoyannopoulos, J., Snyder, M., Dunham, I., Hardison, R. C., & Bernstein, B. E. (2011). A user's guide to the encyclopedia of DNA elements (ENCODE). *PLoS Biology, 9*, e1001046.

Nielsen, R., Pedersen, T. Å., Hagenbeek, D., Moulos, P., Siersbæk, R., Megens, E., et al. (2008). Genome-wide profiling of PPARg:RXR and RNA polymerase II occupancy reveals temporal activation of distinct metabolic pathways and changes in RXR dimer composition during adipogenesis. *Genes & Development, 22*, 2953–2967.

Nowak, D. E., Tian, B., & Brasier, A. R. (2005). Two-step cross-linking method for identification of NF-kB gene network by chromatin immunoprecipitation. *Biotechniques, 39*, 715–725.

Ren, B., Robert, F., Wyrick, J. J., Aparicio, O., Jennings, E. G., Simon, I., et al. (2000). Genome-wide location and function of DNA binding proteins. *Science, 290*, 2306.

Rhee, H. S., & Pugh, B. F. (2011). Comprehensive genome-wide protein-DNA interactions detected at single-nucleotide resolution. *Cell, 147*, 1408–1419.

Robertson, G., Hirst, M., Bainbridge, M., Bilenky, M., Zhao, Y., Zeng, T., et al. (2007). Genome-wide profiles of STAT1 DNA association using chromatin immunoprecipitation and massively parallel sequencing. *Nature Methods, 4*, 651–657.

Schmidt, D., Wilson, M. D., Spyrou, C., Brown, G. D., Hadfield, J., & Odom, D. T. (2009). ChIP-seq: Using high-throughput sequencing to discover protein–DNA interactions. *Methods, 48*, 240–248.

Schoppee Bortz, P. D., & Wamhoff, B. R. (2011). Chromatin immunoprecipitation (ChIP): Revisiting the efficacy of sample preparation, sonication, quantification of sheared DNA, and analysis via PCR. *PLoS ONE, 6*, e26015.

Shankaranarayanan, P., Mendoza-Parra, M. A., Walia, M., Wang, L., Li, N., Trindade, L. M., et al. (2011). Single-tube linear DNA amplification (LinDA) for robust ChIP-seq. *Nature Methods, 8*, 565–567.

Siersbæk, M. S., Loft, A., Aagaard, M. M., Nielsen, R., Schmidt, S. F., Petrovic, N., et al. (2012). Genome-wide profiling of peroxisome proliferator-activated receptor g in primary epididymal, inguinal, and brown adipocytes reveals depot-selective binding correlated with gene expression. *Molecular and Cellular Biology, 32*, 3452–3463.

Siersbæk, R., Nielsen, R., John, S., Sung, M. H., Baek, S., Loft, A., et al. (2011). Extensive chromatin remodelling and establishment of transcription factor 'hotspots' during early adipogenesis. *EMBO Journal, 30*, 1459–1472.

Siersbæk, R., Nielsen, R., & Mandrup, S. (2010). PPARg in adipocyte differentiation and metabolism—Novel insights from genome-wide studies. *FEBS Letters, 584*, 3242–3249.

Siersbæk, R., Nielsen, R., & Mandrup, S. (2012). Transcriptional networks and chromatin remodeling controlling adipogenesis. *Trends in Endocrinology & Metabolism, 23*, 56–64.

Solomon, M. J., & Varshavsky, A. (1985). Formaldehyde-mediated DNA-protein crosslinking: A probe for in vivo chromatin structures. *Proceedings of the National Academy of Sciences of the United States of America, 82*, 6470.

Steger, D. J., Grant, G. R., Schupp, M., Tomaru, T., Lefterova, M. I., Schug, J., et al. (2010). Propagation of adipogenic signals through an epigenomic transition state. *Genes & Development, 24*, 1035–1044.

Zhao, Z., Tavoosidana, G., Sjolinder, M., Gondor, A., Mariano, P., Wang, S., et al. (2006). Circular chromosome conformation capture (4C) uncovers extensive networks of epigenetically regulated intra- and interchromosomal interactions. *Nature Genetics, 38*, 1341–1347.

Zwart, W., Koornstra, R., Wesseling, J., Rutgers, E., Linn, S., & Carroll, J. (2013). A carrier-assisted ChIP-seq method for estrogen receptor-chromatin interactions from breast cancer core needle biopsy samples. *BMC Genomics, 14*, 232.

CHAPTER FIFTEEN

Analysis and Isolation of Adipocytes by Flow Cytometry

Susan M. Majka*, Heidi L. Miller[†], Karen M. Helm[‡], Alistaire S. Acosta[‡], Christine R. Childs[‡], Raymond Kong[§], Dwight J. Klemm[†,1]

*Department of Medicine, Vanderbilt University, Nashville, Tennessee, USA
[†]Department of Medicine, University of Colorado Anschutz Medical Campus, Aurora, Colorado, USA
[‡]Cancer Center Flow Cytometry Core, University of Colorado Anschutz Medical Campus, Aurora, Colorado, USA
[§]Amnis Corporation, Seattle, Washington, USA
[1]Corresponding author: e-mail address: dwight.klemm@ucdenver.edu

Contents

1. Introduction	282
2. Preparation of Adipocytes by Collagenase Digestion of Adipose Tissue	286
3. Staining of Single Cell Suspensions from Adipose Tissue for Flow Cytometry Analysis	287
3.1 LipidTOX Deep Red neutral lipid staining of lipid droplets	289
3.2 Fluorescent antibody staining to exclude stromal/vascular cells from adipocytes	290
3.3 DyeCycle Violet staining to identify events with nuclei and distinguish singlets from aggregates	290
4. Analysis and Sorting of Adipocytes by Flow Cytometry	291
4.1 MoFlo XDP settings	292
4.2 Isolation of adipocytes from stromal cells based on cell size: FSC versus SSC gating	292
4.3 Separation of single cells (singlets) from cell aggregates	293
4.4 Identification of events containing lipid droplets	294
4.5 Exclusion of stromal/vascular cells	294
4.6 Validation of adipocyte isolation strategy	294
5. Summary	294
Acknowledgment	295
References	295

Abstract

Analysis and isolation of adipocytes via flow cytometry is particularly useful to study their biology. However, the adoption of this technology has often been hampered by the presence of stromal/vascular cells in adipocyte fractions prepared from collagenase-digested adipose tissue. Here, we describe a multistep staining method and gating strategy that effectively excludes stromal contaminants. Initially, we set a gate optimized to the size and internal complexity of adipocytes. Exclusion of cell aggregates is then performed based on fluorescence of a nuclear stain followed by positive selection to collect only those cell events containing lipid droplets. Lastly, negative selection of cells expressing stromal or vascular lineage markers removes any remaining stromal contaminants. These procedures are applicable to simple analysis of adipocytes and their subcellular constituents by flow cytometry as well as isolation of adipocytes by flow sorting.

1. INTRODUCTION

Flow cytometry is a powerful technique that provides the ability to rapidly measure multiple cellular parameters across single cell suspensions and large cell populations with tremendous precision. During cytometry, analysis of light scatter can distinguish different cells based on their size, shape, and internal complexity. The presence and amount of specific intracellular and cell surface molecules can be measured with antibodies or ligands conjugated with fluorescent probes. Likewise, fluorescent indicators are available to measure the transport of ions across cellular membranes, as well as assess mitochondrial activity and other metabolic parameters. Specific cell subtypes can be isolated from mixed populations of single cell suspensions on specialized cytometers called "sorters." In spite of the tremendous analytical power and advantages afforded by flow cytometry, this technology has been underutilized in the study of adipocyte biology.

A general assumption hampering the broad scale adoption of flow cytometry to the study of adipocytes is that they are too large (50–200 μm) and fragile to effectively analyze and sort using modern benchtop flow cytometers. However, the internal diameter of the fluidics of modern flow cytometers ranges from 150 to 250 μm—large enough to accommodate all but the largest fat cells. Moreover, typical flow pressures range from only 5 to 10 psi and the laminar flow within the cytometer exerts minimal shear stress on cells. In addition, adipocytes are fairly deformable. Their shape is largely defined by large cytoplasmic lipid droplets, which contain liquid triglyceride at room or higher temperature, rather than cytoskeletal elements. In collaboration with Amnis Corporation (Seattle, WA) we have acquired brightfield images of intact unilocular and multilocular adipocytes (and cell debris) in the fluidics stream of an ImageStream X Imaging Flow Cytometer (Fig. 15.1).

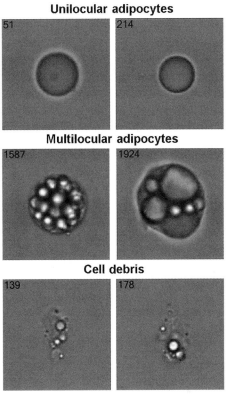

Figure 15.1 Brightfield images of unilocular and multilocular adipocytes, and cell debris acquired with an ImageStream X Imaging Flow Cytometer at Amnis Corporation. The images demonstrate the feasibility of using flow cytometry to study intact adipocytes.

Selecting adipocytes and collecting them for study via flow sorting is a bit more complex than analysis alone. Sorting exposes cells to higher flow pressures, smaller fluidic diameters, and therefore, increased shear stress. Within the sorter, cells are encapsulated within droplets of sheath fluid. The trajectory of the droplets is controlled by imparting them with an electric charge. The path of the charged droplets is deflected toward an oppositely charged plate so that individual droplets may be captured in separate tubes. However, the high flow pressures and relatively small "tip" diameters (70–100 μm) required to produce the droplets may result in shear-induced adipocyte lysis. In spite of these conditions we find that a small percentage of fat cells survive sorting as determined by microscopy of the sorted cells (Fig. 15.2). Moreover, lysis of adipocytes during sorting can actually facilitate recovery of

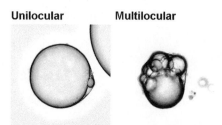

Figure 15.2 Brightfield images of an intact unilocular and an intact multilocular adipocyte after flow sorting. Free-floating adipocytes were sorted on a MoFlo XDP instrument equipped with a 70-μm tip. Adipocytes were collected in a 12 × 75-mm polypropylene tube containing 150 μl of flow buffer. Images were acquired with a conventional light microscope.

intact nuclei (Majka et al., 2010), nucleic acids, and other cellular constituents for further analysis (Majka et al., 2010; Schaedlich, Knelangen, Santos, Fischer, & Santos, 2010).

Applying these concepts, we and other laboratories have begun to exploit flow cytometry to study aspects of adipocyte biology with considerable success. For example, Shi and Kandror (2008) and Bruzzone et al. (2012) developed flow cytometry techniques to measure the degree of translocation of the insulin-stimulated glucose transporter, Glut4, to the membrane of 3T3-L1 adipocytes. Zhu et al. (2012) measured mitochondrial membrane potential and production of reactive oxygen species in Lyrm1-depleted 3T3-L1 adipocytes, while Festy et al. (2005) used flow cytometry to assess surface protein expression between human primary adipocytes and stromal cells. Other laboratories have employed flow cytometry to measure lipid accumulation in adipocytes (Lee, Chen, Wiesner, & Huang, 2004), assess production of Toll-like receptor 2 and tumor necrosis factor α in primary adipocytes (Murakami, Bujo, Unoki, & Saito, 2007), and track the differentiation of rat primary preadipocytes in response to anandamide (Karaliota, Siafaka-Kapadai, Gontinou, Psarra, & Mavri-Vavayanni, 2009). We exploited flow cytometry and fluorescence deconvolution microscopy to demonstrate the production of adipocytes from bone marrow-derived progenitor cells (Crossno, Majka, Grazia, Gill, & Klemm, 2006; Majka et al., 2010, 2012). Additionally, we found that flow sorting of primary mouse adipocytes was a highly efficient means to isolate intact nuclei for cytogenetic analysis (Majka et al., 2010) and RNA for global gene expression analysis. Schaedlich et al. (2010) used similar methods to recover RNA from embryonic stem cell-derived adipocytes for RT-PCR analysis.

As obesity and metabolic diseases linked to adipose tissue continue to place health and financial burdens on society, scientists have begun to address complicated questions about adipocyte production and lineage, fat cell turnover, apoptosis and senescence. Flow cytometry is an important addition to the arsenal of sophisticated techniques that are now employed in these studies. The ability of flow cytometry to screen multiple adipocyte parameters over large cell populations is an advantage over microscopic methods. Moreover, even the best commercial laser confocal systems are limited to a nominal resolution of approximately 200 nm, making it difficult to discern the orientation and relationship between structures and markers in tissue sections and whole mounts. Flow cytometry overcomes this limitation by facilitating analysis of individual cells or events. Imaging flow cytometers, including the ImageStream X (Amnis, Seattle, WA) combine the large population/multiple parameter advantages of traditional flow cytometry, with the ability to acquire brightfield images of adipocytes in the flow stream. Finally, modern flow cytometers may exceed the ability of laser confocal microscopes to detect diffuse fluorescent signals (e.g., cytosolic EGFP), since fluorescence is integrated over the entire cell or event instead of individual pixels.

A final remaining challenge to using flow cytometry in the study of adipocytes is the presence of contaminating stromal/vascular cells. These additional cell types pose a complication in adipocyte fractions prepared by collagenase digestion. Repeated wash steps fail to remove all stromal cell contamination. We previously published a multistep flow cytometry procedure (Majka et al., 2012) that effectively excludes stromal/vascular cells from adipocyte preparations. The strategy involved gating or acquisition of cells whose size and internal complexity were greater than stromal/vascular cells, exclusion of cell aggregates, and exclusion of cells or events expressing stromal lineage markers. The procedure was validated by demonstrating distinct marker gene expression between adipocytes and stromal cells sorted by this strategy. More recently, we have made two important improvements to this strategy. First, we identify and collect single cells and exclude aggregates based on fluorescence of the nuclear stain, Vybrant DyeCycle Violet (Molecular Probes/Life Technologies, Eugene, OR). Second, we collect only events containing lipid droplets as determined by staining with HSC LipidTOX Neutral Lipid stain (also Molecular Probes).

In the following sections we provide detailed methods for preparation of free-floating adipocytes, staining of adipocytes with DyeCycle Violet, LipidTOX Deep Red, and fluorescent antibodies to stromal lineage markers. We further delineate our multistep gating regimen to capture and analyze adipocytes free of stromal contaminants.

2. PREPARATION OF ADIPOCYTES BY COLLAGENASE DIGESTION OF ADIPOSE TISSUE

A single cell suspension of free-floating adipocytes is required for flow cytometry. Free-floating fat cells are isolated from adipose tissue by digesting the tissue in a buffered collagenase solution. Following digestion, undigested tissue fragments are removed by filtration, and the adipocytes are separated from stromal/vascular cells by centrifugation and/or flotation. There are numerous variations on this procedure in the literature; typically reflecting differences in digestion buffer composition and centrifugation/flotation conditions. In general, most methods produce adipocyte suspensions of comparable quality. The procedure outlined below is routinely used in our laboratory however, it may be modified as appropriate based on specific assay parameters for successful flow cytometry of adipocytes.

1. Adipose tissue is weighed, rinsed with phosphate-buffered saline, and minced with scissors or razor blades to produce fragments of approximately 1 mm.
2. The tissue fragments are transferred to a 50-ml screw cap conical polypropylene centrifuge tube and digestion buffer (Krebs–Ringers–HEPES + 2.5 mM glucose + 2% fetal bovine serum + 200 μM adenosine + 1 mg/ml collagenase (Sigma, C2139), pH 7.4) is added at roughly 1 ml/0.25 g tissue. Krebs–Ringers–HEPES contains 120 mM NaCl, 4.7 mM KCl, 2.2 mM CaCl$_2$, 10 mM HEPES, 1.2 mM KH$_2$PO$_4$, and 1.2 mM MgSO$_4$.
3. The suspension is incubated for 1 h at 37 °C at 100 rpm on an orbital shaker.
4. Following tissue digestion the suspension is passed through a 150-μm mesh Celltrics filter (Partec GmbH) followed by an equal volume of wash buffer (Hanks balanced salt solution [Cellgro 21-022-CV], 2% fetal bovine serum (FBS), 200 μM adenosine, pH 7.4).
5. The filtered suspension is centrifuged at 150 × g for 8 min. During centrifugation, the adipocytes will form a layer at the top of the liquid.
6. Transfer adipocytes to a clean 15-ml conical tube and add 3–4 volumes of wash buffer. Gently resuspend the adipocytes by rocking the sealed tube.

7. Centrifuge the suspension at 150 × g for 8 min and transfer the adipocyte layer to a clean 12 × 75-mm polypropylene tube.

Comments and troubleshooting
1. Following collagenase digestion, filtration of the cell suspension through the Celltrics filter can be facilitated by placing a double layer of gauze above the filter mesh to retain large tissue fragments that may impede filtration.
2. Some protocols suggest centrifugation of adipose cell suspensions at 300–500 × g. We find this results in considerable adipocyte lysis evident by the presence of large, clear triacylglycerol droplets in the adipocyte layer following centrifugation. Centrifugation at 150 × g prevents mechanical damage to the fat cells.
3. We generally transfer the floating adipocyte layer between tubes using a 1-ml PipetMan pipettor. We snip 1–2 mm off the end of the plastic pipette tip to enlarge the bore of the tip. This facilitates transfer of the viscous adipocyte layer.
4. Polypropylene plasticware is preferred over polystyrene and other materials.
5. Rocking rather than vortexing is preferred for resuspending adipocytes in wash buffer as it minimizes cell damage.

3. STAINING OF SINGLE CELL SUSPENSIONS FROM ADIPOSE TISSUE FOR FLOW CYTOMETRY ANALYSIS

To isolate and analyze adipocytes free from contamination by stromal/vascular cells we have devised a flow cytometry strategy based on the (1) the significant size difference between large adipocytes and considerably smaller stromal cells, (2) exclusion of cell aggregates, (3) presence of large lipid droplets in adipocytes, and (4) exclusion of cells bearing stromal lineage markers. Initial "separation" of adipocytes and stromal cells by their size is based on differences in forward scatter (FSC) and side scatter (SSC) distributions between the two populations (described in Section 4). Single cells containing nuclei are distinguished from cell aggregates by DyeCycle Violet fluorescence, and events containing lipid droplets are identified by LipidTOX Deep Red fluorescence. Finally, any remaining stromal contaminants are excluded based on their staining with phycoerythrin (PE)-conjugated antibodies to stromal cell lineage markers. Figure 15.3A shows an example of this gating scheme. The following sections describe the

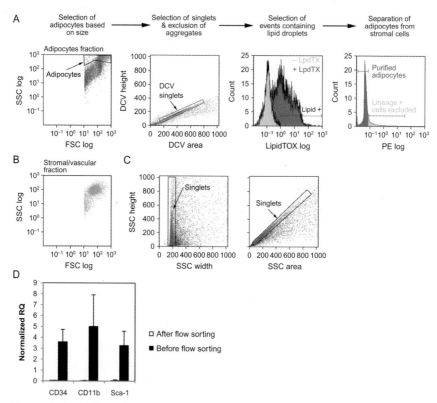

Figure 15.3 Multistep flow cytometry/sorting strategy. (A) Gating strategy for adipocyte isolation is diagramed from left to right. In the first step, adipocytes are identified by their large size and refractile properties in a plot of FSC versus SSC. A gate is placed around the population of cells that are larger and more refractile than those present in the stromal/vascular fractions (B). In the next step, DyeCycle Violet (DCV) fluorescence of the gated adipocytes is evaluated in a plot of peak height versus peak area (note the linear scale). The singlets that form a diagonal distribution are gated, while cell aggregates are ignored. Singlets can also be identified optically by comparing SSC (or FSC) peak height to peak width or peak area as shown in (C). In the third step, LipidTOX fluorescence of the singlets is evaluated on a histogram. We compare the fluorescence signal distribution of LipidTOX-stained cells to a small portion of unstained cells to positively identify events containing a lipid droplet. Finally, any remaining stromal contaminants are excluded based on their labeling with PE-conjugated antibodies to stromal markers. The purified adipocytes can undergo further analysis and/or sorting. (D) QRT-PCR was used to verify the staining and gating strategy. The results show the presence of several stromal cells markers in adipocyte fractions prior to sorting. These markers were virtually undetectable in flow-purified adipocytes. (See the color plate.)

procedures we routinely use to stain free-floating adipocytes with these reagents, which generally follow the recommendations of the reagent suppliers.

3.1. LipidTOX Deep Red neutral lipid staining of lipid droplets

Lipid droplets are often stained with reagents like Oil Red O or Nile Red. Oil Red-stained lipid droplets are easily identified by their intense red coloration under visible light or strong red fluorescence (Reusch & Klemm, 2002). However, staining with Oil Red O requires fixation of cells and staining in the presence of organic solvents (e.g., 60% isopropanol) and is, therefore, limited to the analysis of dead cells. Nile Red, while soluble in DMSO and other organic solvents, can be diluted into aqueous buffers for the staining of live or dead adipocytes. However, its emission maximum at 528 nm may overlap with green (FITC, GFP) and yellow–orange (PE) fluorophores in flow cytometry. For our flow cytometry applications, we have found LipidTOX stains useful as they can be applied in aqueous solutions to either live or fixed/dead cells. LipidTOX dyes are available with green (emission maximum 505 nm) or red (609 nm) fluorescence for applications where other green- or red-emitting fluorophores are not an issue, or in far-red emission (655 nm) to prevent overlap with green, yellow, or orange fluorophores. The following procedure details the steps we use to stain free-floating adipocytes with LipidTOX Deep Red.

1. LipidTox Deep Red is added to the adipocyte suspension at a 1:125 dilution and mixed gently by flicking the tube with a finger or by gentle inversion.
2. Incubate cells at 37 °C for 25 min with gentle inversion after 10–15 min.
3. Centrifuge cell suspension at $150-200 \times g$ for 2 min. Remove the subnatant from the layer of floating adipocytes.
4. Wash the cells by resuspending the adipocytes in an equal volume of wash buffer and centrifugation at $150-200 \times g$ for 2 min. Remove subnatant.
5. Resuspend cells to the previous volume with flow buffer (HBSS + 5% FBS + 200 μM adenosine).

Comment
    At this stage we remove the subnatant rather than transferring the adipocyte layer (supernatant) between tubes to limit handling and lysis of the fragile fat cells.

3.2. Fluorescent antibody staining to exclude stromal/vascular cells from adipocytes

A crucial step in isolating or analyzing adipocytes by flow cytometry is exclusion of stromal/vascular cells from the adipocyte population. Our strategy to identify and exclude stromal cells is based on staining the free-floating adipocyte fraction with fluorescent antibodies to stromal lineage markers. For most experiments with mouse adipocyte preparations we use PE-conjugated antibodies to myeloid cells (CD11b, BD Pharmingen 557397), T lymphocytes (CD3, BioLegend 100206), B lymphocytes (CD45R, BioLegend 103208), red blood cells (TER-119, BioLegend 116208), granulocytes (Gr-1, BioLegend 108408), megakaryocytes and platelets (CD61, BioLegend 104308), endothelial cells (CD34, BD Pharmingen 551387), and neurons (NCAM, R&D Systems FAB5674P). In practice, it is possible to substitute a PE-conjugated antibody to the pan-leukocyte marker CD45 (BD Pharmingen 553081) for the cocktail of individual lineage-specific antibodies and exclude over 99% of stromal contaminants.

Antibody staining is done concurrent with LipidTOX Deep Red. Antibodies are used at a concentration of 0.25 µg per 10^6 cells. Since accurate counts are often difficult to obtain with adipocytes, suspension volume is used as a proxy for cell number (i.e., if cells can be suspended easily in 0.5 ml it is assumed there are 10×10^6 or fewer cells. If a larger volume is needed the antibody amount is increased). Antibodies are added after LipidTOX Deep Red and the procedure above is followed. If not staining with LipidTOX Deep Red the antibody incubation is done on ice rather than at 37 °C, as adipocytes are somewhat more susceptible to lysis at warm temperatures needed for LipidTOX staining.

3.3. DyeCycle Violet staining to identify events with nuclei and distinguish singlets from aggregates

The study of adipocytes has often been thwarted by the inability to distinguish or separate adipocytes from stromal/vascular cells present in fat tissue. With its ability to analyze individual "events," flow cytometry and sorting affords investigators a powerful tool to distinguish individual adipocytes from stromal contaminants based on optical and/or fluorescent properties. Single cells can be distinguished from cell aggregates by optical methods during flow cytometry. However, we have found that staining the adipocyte suspension with DNA stains allows us to distinguish single cells from aggregates, and unambiguously identify events that contain single nuclei. The following procedure describes our steps for staining free-floating adipocytes with the DNA stain, DyeCycle Violet.

1. After incubating adipocytes with LipidTOX and fluorescent antibodies and washing, the cells are resuspended to the previous volume and DyeCycle Violet is added to a dilution of 1:333 to the suspension and mixed gently.
2. Incubate at 37 °C for 30 min with one gentle mix after 15 min.
3. Cells are kept warm and transported for flow cytometric analysis with no further washing steps.

Comments

1. Hoechst 33342 can also be used to identify single cells as it stains nuclei in live or dead cells, and has an emission maximum close to that of DyeCycle Violet. When used with unfixed cells, DAPI and propidium iodide stain nuclei in dead cells or free-floating nuclei making them useful for distinguishing live versus dead cells. DyeCycle Violet is recommended for staining of nuclei in live cells by the manufacturer, but we find that it provides intense nuclear fluorescence in both live and dead cells, and can be used with unfixed or paraformaldehyde-fixed cells. We also find that DyeCycle Violet is generally more "forgiving" than DAPI, Hoechst, or propidium iodide, affording more flexibility with regard to concentration and incubation times.
2. The DyeCycle manufacturer recommends that cells be maintained at 37 °C following staining. We find this also prevents the adipocytes from congealing into a semisolid "clump" that is difficult for cytometer fluidics to aspirate. We transport cells between our laboratory and the flow cytometry core in an insulated container filled with glass beads preheated to 37 °C.

4. ANALYSIS AND SORTING OF ADIPOCYTES BY FLOW CYTOMETRY

The following flow cytometry steps and gating regimen represent our day-to-day strategy performed on a MoFlo XDP Flow Cytometer/Sorter (Beckman Coulter, Inc., Brea, CA). However, we have used these same parameters on a CyAn ADP Analyzer and a Gallios Flow Cytometer (both

Beckman Coulter), and they should be applicable to most modern flow cytometer systems.

4.1. MoFlo XDP settings

Adipocytes are sorted using a Moflo XDP cell sorter with Summit 5.3 software (Beckman Coulter). A 100-μm nozzle tip is used with a sheath pressure of 30 psi and a drop drive frequency of 46,700 Hz and amplitude of 15 V. The sheath fluid consists of Isoflow (Beckman Coulter, Fullerton, CA). The sample and collection tubes are maintained at 15 °C using an attached Haake recirculating water bath. To keep cells in suspension the Moflo is equipped with a SmartSampler sample station with the sample agitation set to maintain an agitation cycle of 4 s on and 5 s off. Sort mode is set to Purify 1 to prevent the capture of cells close to unwanted events (e.g., clusters, debris, events lacking a nucleus or lipid droplet, events positive for lineage markers) in the droplet queue. Appropriate signal compensation is set using single color control samples.

Samples are analyzed using the following excitation lasers and emission filters: PE—552-nm laser and 580/23 bandpass filter; LipidTOX Deep Red—641-nm laser and 670/28 filter; DyeCycle Violet—UV laser and 447/60 bandpass filter. We use UV as an alternate excitation wavelength for DyeCycle Violet in lieu of the standard 405 nm violet excitation, because the MoFlo XDP is equipped with a Co-Lase (Propel Labs, Ft. Collins, CO) colinear red–violet laser system and spillover of the DyeCycle Violet signal into the LipidTOX Deep Red detector precludes the use of the 405-nm laser for excitation of the DyeCycle Violet.

4.2. Isolation of adipocytes from stromal cells based on cell size: FSC versus SSC gating

Adipocytes are substantially larger (50–200 μm) than stromal cell populations (typically <20 μm), and the presence of highly refractile lipid droplets in their cytoplasm endows them with tremendous light scattering character. These features make it possible to separate adipocytes from the majority of stromal cells and cell debris during initial FSC versus SSC gating. Light scattered in line with the incident laser beam, the FSC, is influenced by cell size, whereas light scattered to the side, the SSC parameter, by organelles and lipid droplets is influenced by internal complexity. Analysis of flow cytometry data generally begins by plotting these two parameters on a scatter diagram. Because of the large size and refractile nature of adipocytes, we plot the log value for each parameter rather than their linear values.

Figure 15.3B shows a typical FSC/SSC plot for stromal/vascular cells. When compared to FSC/SSC distribution for free-floating adipocytes in Fig. 15.3A, it is evident that there is some overlap in the distribution of the events in each population. However, there are clearly cells with very high FSC and SSC values present in the adipocyte fraction that are not observed in the stromal/vascular fraction. We begin our isolation of adipocytes from stromal cells by gating solely on these large, refractile cells as indicated in the figure. There are some small adipocytes that do not fall within this gate, and if it necessary to analyze the entire fat cell population, one may gate on the entire cell population for further analysis.

4.3. Separation of single cells (singlets) from cell aggregates

Aggregates of adipocytes and stromal cells are common in fat cell fractions following collagenase digestion. To characterize or sort single cells, cell aggregates must be excluded from the analysis or sort. During flow cytometry this may be accomplished optically, by comparing the fluorescence pulse peak height (either FSC or SSC) to the pulse peak width or area. The ratio between peak height and peak width or peak area is proportional for single cells, which form a vertical distribution on a scatter plot of peak height versus peak width (Fig. 15.3C) or a diagonal distribution when peak height is plotted against peak area. Cell aggregates exhibit increased peak width and area compared to peak height and, therefore, do not fall on the distinct vertical (width) or horizontal (area) distributions of singlet events. Setting a gate around the singlets effectively excludes cell aggregates from further analysis and sorting.

Because of the unique optical properties of adipocytes (single and multiple refractile cytoplasmic lipid droplets), we have found that a similar gating strategy based on DyeCycle Violet fluorescence parameters is a more reliable method of identifying single cells. Figure 15.3A, second panel from the left, shows a plot of DyeCycle Violet fluorescence height versus fluorescence area for free-floating adipocytes. The diagonal distribution of single cells is easily distinguished from aggregates, and is gated for further analysis. The use of DyeCycle Violet in this step also ensures that all gated events contain a nucleus and are not free-floating lipid droplets or other debris.

4.4. Identification of events containing lipid droplets

Adipocytes contain one or more large triglyceride droplets, which distinguish them from relatively lipid-deficient stromal/vascular cells. This characteristic provides another means of separating fat cells from stromal contaminants. Lipid droplets (both within intact adipocytes and free-floating) stain intensely with LipidTOX dyes and are easily distinguished from lipid-free events. For gating purposes we compare the fluorescence profile (histogram) of a small portion of unstained adipocytes to the fluorescence of LipidTOX-stained cells (Fig. 15.3A, second panel from the right). The lipid-positive cells are gated for further analysis or sorting.

4.5. Exclusion of stromal/vascular cells

The final gating step in our strategy removes any remaining stromal contaminants based on their staining with PE-conjugated antibodies to stromal cell lineage markers. This step is necessary to remove any stromal cells that may adhere to free lipid droplets and thus not be excluded based on DyeCycle and LipidTOX staining in the previous gating steps. As with LipidTOX staining and analysis, we compare the PE fluorescence profile of a small aliquot of unstained cells to that of adipocytes stained with a cocktail of PE-conjugated antibodies to lineage markers on a histogram of PE fluorescence (Fig. 15.3A, far right panel). The small percentage of PE-positive events is excluded from the gated PE-negative adipocytes.

4.6. Validation of adipocyte isolation strategy

QRT-PCR for various stromal-specific factors was performed on RNA purified from adipocytes before and after flow sorting with our multistep gating strategy. Figure 15.3D shows that stromal-specific RNAs are absent in flow-purified adipocytes, but are readily detectable in the preflow adipocyte fraction.

5. SUMMARY

This chapter presents a method for the separation of free-floating adipocytes from stromal/vascular cell contaminants during flow cytometry or flow sorting. The technique is eminently suited to the rapid analysis of multiple parameters in a large number of adipocytes, and can be adapted to suit a range of assay requirements. We routinely use this method to follow the production of adipocytes from progenitors expressing green fluorescent protein (detected with a 488-nm laser and 529/28 bandpass filter), and to obtain purified adipocyte nuclei for cytogenetic analysis. Obviously, antibodies to lineage markers conjugated to a wide variety of fluorophores are available commercially, and LipidTOX stains are available in green, red, and deep red versions. Therefore, a variety of color combinations compatible with many experimental requirements and instrument laser configurations are available. Given its many advantages, we encourage investigators to adopt flow cytometry to their arsenal of techniques as they unravel the intricacies of adipocyte biology.

ACKNOWLEDGMENT
This work was supported by a National Institutes of Health grant R01-DK078966 (to D. J. K.).

REFERENCES
Bruzzone, S., Ameri, P., Briatore, L., Mannino, E., Basile, G., Andraghetti, G., et al. (2012). The plant hormone abscisic acid increases in human plasma after hyperglycemia and stimulates glucose consumption by adipocytes and myoblasts. *FASEB Journal, 26*, 1251–1260.

Crossno, J. T., Jr., Majka, S. M., Grazia, T., Gill, R. G., & Klemm, D. J. (2006). Rosiglitazone promotes differentiation of bone marrow-derived circulating progenitor cells to multilocular adipocytes in adipose tissue. *Journal of Clinical Investigation, 116*, 3220–3228.

Festy, F., Hoareau, L., Bes-Houtmann, S., Pequin, A.-M., Gontheir, M.-P., Munstun, A., et al. (2005). Surface protein expression between human adipose tissue-derived stromal cells and mature adipocytes. *Histochemistry and Cell Biology, 124*, 113–121.

Karaliota, S., Siafaka-Kapadai, A., Gontinou, C., Psarra, K., & Mavri-Vavayanni, M. (2009). Anandamide increases the differentiation of rat adipocytes and causes PPARgamma and CB1 receptor upregulation. *Obesity (Silver Spring), 17*, 1830–1838.

Lee, Y.-H., Chen, S.-Y., Wiesner, R. J., & Huang, Y.-F. (2004). Simple flow cytometric method used to assess lipid accumulation in fat cells. *Journal of Lipid Research, 45*, 1162–1167.

Majka, S. M., Fox, K. E., Psilas, J. C., Helm, K. M., Childs, C. R., Acosta, A. S., et al. (2010). De novo generation of white adipocytes from the myeloid lineage via mesenchymal intermediates is age, adipose depot, and gender specific. *Proceedings of the National Academy of Sciences of the United States of America, 107*, 14781–14786.

Majka, S. M., Miller, H. L., Sullivan, T., Erickson, P. F., Kong, R., Weiser-Evans, M., et al. (2012). Adipose lineage specification of bone marrow-derived myeloid cells. *Adipocyte, 1*(4), 215–229.

Murakami, K., Bujo, H., Unoki, H., & Saito, Y. (2007). High fat intake induces a population of adipocytes to co-express TLR2 and TNFalpha in mice with insulin resistance. *Biochemical and Biophysical Research Communications, 354*, 727–734.

Reusch, J. E. B., & Klemm, D. J. (2002). Inhibition of CREB activity decreases protein kinase B/Akt expression in 3T3-L1 adipocytes and induces apoptosis. *Journal of Biological Chemistry, 277*, 1426–1432.

Schaedlich, K., Knelangen, J. M., Santos, A. N., Fischer, B., & Santos, A. (2010). A simple method to sort ESC-derived adipocytes. *Cytometry. Part A, 77A*, 990–995.

Shi, J., & Kandror, K. V. (2008). Study of glucose uptake in adipose cells. In K. Yang (Ed.), *Adipose tissue protocols*: *Vol. 456.* (pp. 307–315). Totowa, NJ: Humana Press.

Zhu, G.-Z., Zhang, M., Kou, C.-Z., Ni, Y.-H., Ji, C.-B., Cao, X.-G., et al. (2012). Effects of Lyrm1 knockdown on mitochondrial function in 3T3-L1 murine adipocytes. *Journal of Bioenergetics and Biomembranes, 44,* 225–232.

CHAPTER SIXTEEN

Flow Cytometry Analyses of Adipose Tissue Macrophages

Kae Won Cho[*], David L. Morris[†], Carey N. Lumeng[*,1]

[*]Department of Pediatrics and Communicable Diseases, University of Michigan Medical School, Ann Arbor, Michigan, USA
[†]Department of Medicine, Indiana University School of Medicine, Indianapolis, Indiana, USA
[1]Corresponding author: e-mail address: clumeng@umich.edu

Contents

1. Introduction 298
2. Materials 299
    2.1 Isolation and preparation of SVCs from mice 299
    2.2 Staining SVCs for cell surface markers to identify ATMs 300
    2.3 Flow cytometry analysis 301
    2.4 FACS procedure to purify ATM populations 301
    2.5 Using magnetic beads and positive selection to enrich for $CD11b^+$ SVCs 302
    2.6 Data analysis 302
3. Methods 302
    3.1 Isolation and preparation of SVCs from mice 302
    3.2 Staining SVCs for cell surface markers to identify ATMs 304
    3.3 Flow cytometry analysis 306
    3.4 FACS procedure to purify ATM populations 307
    3.5 Using magnetic beads and positive selection to enrich for $CD11b^+$ SVCs 309
    3.6 Data analysis 310
4. Discussion 310
Acknowledgments 313
References 313

Abstract

Within adipose tissue, multiple leukocyte interactions contribute to metabolic homeostasis in health as well as to the pathogenesis of insulin resistance with obesity. Adipose tissue macrophages (ATMs) are the predominant leukocyte population in fat and contribute to obesity-induced inflammation. Characterization of ATMs and other leukocytes in the stromal vascular fraction from fat has benefited from the use of flow cytometry and flow-assisted cell sorting techniques. These methods permit the immunophenotyping, quantification, and purification of these unique cell populations from multiple adipose tissue depots in rodents and humans. Proper isolation, quantification, and characterization of ATM phenotypes are critical for understanding their role in adipose tissue function and obesity-induced metabolic diseases. Here, we present the flow cytometry protocols for phenotyping ATMs in lean and obese mice employed by our laboratory.

1. INTRODUCTION

Obesity induces a low-grade inflammatory state that contributes to insulin resistance, diabetes, and metabolic syndrome (Glass & Olefsky, 2012; Gregor & Hotamisligil, 2011; Lumeng & Saltiel, 2011; Xu et al., 2003). Obesity-induced inflammation is characterized by chronic elevations in circulating inflammatory cytokines, adipokines, and monocytes (Gregor & Hotamisligil, 2011). At the tissue level, inflammatory pathways are induced in visceral adipose tissue due in part to dynamic quantitative and phenotypic changes in adipose tissue leukocytes, which include macrophages, neutrophils, mast cells, T cells, and eosinophils (Liu et al., 2009; Nishimura et al., 2009; Strissel et al., 2010; Talukdar et al., 2012; Wu et al., 2011). Among these, adipose tissue macrophages (ATMs) are the predominant leukocyte population in fat (Nishimura et al., 2009; Wentworth et al., 2010). In both mouse models and human subjects, obesity leads to increased ATM accumulation in visceral adipose depots (Harman-Boehm et al., 2007; Weisberg et al., 2003; Xu et al., 2003). In mouse models, ATM content can increase from ~10% to 15% of nonadipocyte cells in fat in lean mice to ~50% of cells in obese mice (Weisberg et al., 2003; Xu et al., 2003). ATM content positively correlates with the metabolic derangements associated with obesity in rodent and humans (Kanda et al., 2006; Wentworth et al., 2010; Xu et al., 2003).

Obesity is also associated with qualitative changes in the phenotype and function of ATMs. In lean mice, resident ATMs are distributed between adipocytes in healthy adipose tissue and express anti-inflammatory markers typical of "alternatively activated" or M2-polarized macrophages (e.g., arginase 1, CD301/Mgl1, and CD206) (Lumeng, Bodzin, & Saltiel, 2007). Dietary obesity triggers the accumulation of ATMs into "crown-like structures" around dead adipocytes (Cinti et al., 2005; Strissel et al., 2007). These infiltrating ATMs express the dendritic cell marker CD11c and genes typical of "classically activated" or M1-polarized macrophages (Lumeng, Bodzin, & Saltiel, 2007). Recruited $CD11c^+$ ATMs secrete proinflammatory cytokines such as TNF-α and IL-6 and generate reactive oxygen species via inducible nitric oxide synthase (NOS2) (Lumeng, DelProposto, Westcott, & Saltiel, 2008;

Lumeng, Deyoung, Bodzin, & Saltiel, 2007). Collectively, these and other observations have led to the paradigm that ATMs undergo a "phenotypic switch" from an anti-inflammatory M2 state to a proinflammatory M1 state (Lumeng, Bodzin, & Saltiel, 2007). While this is an oversimplification of a complex regulatory system, evidence from knockout mice support the general model M1/M2 balance in macrophages can play a pivotal role in the development of adipose tissue inflammation in obesity (Chawla, Nguyen, & Goh, 2011; Lumeng & Saltiel, 2011).

The limited number and complex heterogeneity of stromal vascular cells (SVCs) isolated from fat depots poses a challenge on the types of analyses that can be applied to directly study ATMs. *In vitro* assays using bone marrow-derived macrophages or macrophage cell lines are limited in that they may not recapitulate the ATM microenvironment. The use of flow cytometry has emerged as the preferred method to interrogate ATM content and heterogeneity in mouse models. When done properly flow cytometry allows investigators to simultaneously examine both general cell properties (e.g., relative size and granularity) and expression of extracellular and intracellular proteins on individual cells isolated from fat. In concert with purification schemes such as immunomagnetic cell enrichment and flow-assisted cell sorting (FACS), flow cytometry becomes an invaluable tool for studying ATMs and other leukocytes in adipose tissue. Technical approaches vary from laboratory to laboratory making it somewhat of a challenge to interpret data across studies. Much of this may stem from the use of collagenase digestion protocols originally developed for adipocyte isolation that may not adequately capture all leukocytes for downstream analysis (Rodbell, 1964). In this chapter, we provide the protocol used by our group optimized for detection and purification of ATM subsets by flow cytometry. We provide several practical considerations for optimizing cell yield, for selecting proper reagents and flow cytometry controls, and for gating SVCs to characterize distinct ATM subsets.

2. MATERIALS

2.1. Isolation and preparation of SVCs from mice

1. C57BL/6J mice (The Jackson Laboratory, Bar harbor, ME; stock #00064)
2. Sterile or ethanol-cleaned surgical instruments
3. 70% Ethanol
4. 10 cc Luer-Lok syringes with needles (25G × 1″)

5. 1× Phosphate-buffered saline (PBS)
6. Digestion buffer: Hanks' balanced salt solution with Ca^{2+} and Mg^{2+} supplemented with 0.5% bovine serum albumin (BSA)
7. 10× Collagenase solution (10 mg/ml in digestion buffer). Type II collagenase (Sigma-Aldrich; Catalog #C6885). Solution should be prepared freshly for optimal results and can be filter-sterilized (0.22 μm)
8. 100 μm Nylon cell strainers (BD Falcon; Catalog #352360)
9. 1× RBC lysis buffer: 155 mM NH_4Cl, 10 mM $KHCO_3$, 0.1 M EDTA; sterile filtered through 0.22-μm filter and store in aliquots at 4 °C
10. 0.5 M EDTA
11. FACS buffer: PBS with 1 mM EDTA, 25 mM HEPES, and 1% heat-inactivated fetal bovine serum (FBS)
12. Trypan blue solution, 0.4% (Invitrogen; Catalog #15250)
13. Refrigerated clinical centrifuge with swing-bucket rotors and adapters for 15- and 50-ml conical tubes
14. Incubator set to 37 °C
15. Test tube rocker/mixer
16. Hemocytometer for counting cells

2.2. Staining SVCs for cell surface markers to identify ATMs

1. 5-ml Polystyrene round-bottom test tubes for flow cytometry (12 × 75 mm)
2. Fluorochrome-labeled anti-mouse monoclonal antibodies against indicated cell surface molecules (Table 16.1)

Table 16.1 Basic antibody cocktail for ATM staining

Antibody	Clone	Supplier	Titrated concentration (μg/10^6 cells)
F4/80 PE	BM8	eBioscience	0.2
CD11b APC eFluor 780	M1/70	eBioscience	0.16
CD11c PE-Cy7	N418	eBioscience	0.2
CD45.2 PerCP-Cy5.5	104	eBioscience	0.2
CD301 Alexa Fluor 647	ER-MP23	AbD Serotec	0.2

3. Viability dye: DAPI or Live/Dead Fixable Dead Cell Kit (Life Technologies; Catalog #L34955)
4. 16% Paraformaldehyde (PFA) (EM grade)

2.3. Flow cytometry analysis

1. Flow cytometer (e.g., FACSCanto II; BD Biosciences) equipped with three lasers (405-nm violet laser, 488-nm blue laser, and 640-nm red laser) and detectors for the indicated fluorochromes
2. SVCs stained with each individual fluorochrome-labeled antibody (single stained (SS) controls) and fluorescence minus one (FMO) controls (Table 16.2)

2.4. FACS procedure to purify ATM populations

1. Cell sorter (e.g., FACSAria II; BD Biosciences) equipped with three lasers (405-nm violet laser, 488-nm blue laser, and 640-nm red laser) and detectors for the indicated fluorochromes

Table 16.2 Single stained (SS) and fluorescence minus one (FMO) controls for ATM staining

	Pacific Blue	PE	PE-Cy7	PerCP5.5	APC	APC-Cy7
Unstained						
SS-Pacific	Live/dead					
SS-PE		F4/80				
SS-PECy7			CD11c			
SS-PerCP5.5				CD45		
SS-APC					CD301	
SS-APCCy7						CD11b
FMO Pacific	Isotype	F4/80	CD11c	CD45	CD301	CD11b
FMO PE	Live/dead	Isotype	CD11c	CD45	CD301	CD11b
FMO PE-Cy7	Live/dead	F4/80	Isotype	CD45	CD301	CD11b
FMO PerCP5	Live/dead	F4/80	CD11c	Isotype	CD301	CD11b
FMO-APC	Live/dead	F4/80	CD11c	CD45	Isotype	CD11b
FMO-APC-Cy7	Live/dead	F4/80	CD11c	CD45	CD301	Isotype

2. FACS buffer: PBS with 1 mM EDTA, 25 mM HEPES, and 1% heat-inactivated FBS

2.5. Using magnetic beads and positive selection to enrich for CD11b$^+$ SVCs

1. 5-ml Polystyrene round-bottom test tubes for flow cytometry (12 × 75 mm)
2. MACS buffer: PBS (without Ca^{2+} and Mg^{2+}) supplemented with 0.5% BSA
3. CD11b microbeads (Miltenyi Biotec; Catalog #130-049-601)
4. MACS MS cell separation columns with corresponding MACS separator magnet (Miltenyi Biotec, Bergisch Gladbach, Germany)

2.6. Data analysis

1. Flow cytometry analysis software (e.g., FlowJo, TreeStar Inc., Ashland, OR)

3. METHODS

3.1. Isolation and preparation of SVCs from mice

In this section, we describe the dissection of mouse adipose tissue and subsequent recovery of SVCs by collagenase digestion. We recommend using at least one entire perigonadal fat pad for flow cytometry analysis and FACS because regional differences in the distribution of leukocytes in visceral fat from mouse models have been noted (Cho et al., 2007). Experimental design should also account for the need for additional SVCs for compensation controls, which will be explained in detail in Section 3.3. We have found that minced tissue size, shaking intensity, and incubation time in collagenase buffer are critical parameters that should be optimized to maximize the yield of viable SVCs from visceral fat. Increasing surface area by finely mincing the tissue and frequent manual shaking during the digestion tends to increase SVC yield. Importantly, the duration of collagenase exposure should be empirically determined for each lot of collagenase. Increasing incubation times in collagenase buffer, especially longer than 1 h, significantly decrease SVC yield and increase cell death and should be avoided.

3.1.1 Isolation of adipose tissue

1. Euthanize mouse according to approved procedures and disinfect the skin with 70% ethanol.

2. Open the thoracic cavity to expose the heart. Perform a cardiac perfusion to remove blood from tissues. Using a 10 cc syringe and 25G needle, slowly perfuse the left ventricle with 10 ml of PBS. At the same time, puncture the right atria to allow blood and perfusate to escape the circulation.
3. Isolate the perigonadal adipose depot using sterile technique. Remove any visible gonadal tissue. When isolating mesenteric or inguinal adipose depots, care should be taken to remove lymph nodes.
4. Weigh the isolated fat pads and note the value. This weight will be used to normalize flow cytometry data.
5. Wash the adipose tissue with PBS to remove any contaminants, such as fur, and place tissue in a plastic weigh boat on ice.

3.1.2 Isolation of SVCs by collagenase digestion

1. Mince adipose tissue into small pieces with scissors (approx. 3–5 mm in size) in a weigh boat on ice.
2. Transfer minced tissues ($\leq\sim 1$ g) into 10-ml round-bottom tube containing 7 ml of ice-cold digestion buffer and keep on ice. For fat pads >1 g, mince the tissue and transfer into a 50-ml conical tube containing 10 ml of ice-cold digestion buffer. When tissue is added to buffer, mince further, and return the tube to ice until all samples have been harvested.
3. Add 1 ml (or 1.5 ml for >1 g) of $10\times$ collagenase buffer. Adjust the final concentration of collagenase to 1 mg/ml by adding additional digestion buffer.
4. Incubate at 37 °C for 20–45 min with vigorous shaking using a test tube rocker. Higher cell yields can be achieved when the tubes are vigorously shaken by hand every 10 min. After 30 min, 10 µl of the digestion mixture should be examined microscopically. At this point, adipocytes should appear as large single cells; leukocytes and other stromal cells will be much smaller. If leukocytes are still attached to adipocytes, the digestion should continue; if not, proceed to the next step.
5. After digestion, add EDTA to a final concentration of 10 mM and incubate at 37 °C for an additional 5–10 min. This step is necessary to facilitate full dissociation of SVCs; EDTA exposure should be limited to avoid adversely effecting cell viability.
6. Prewet a 100-µm nylon filter with PBS and place onto a 50-ml conical tube. Using a transfer pipette, transfer the bottom layer of cell slurry onto the filter, followed by the adipocyte containing upper layer. This

prevents larger adipocytes from clogging the filter before SVCs can pass freely through. Wash the filter twice gently by adding 10 ml of FACS buffer.

7. Centrifuge cell slurry at $500 \times g$ for 10 min at 4 °C to separate adipocytes and SVCs. After centrifugation, adipocytes form a white layer on top while SVCs form a red/white pellet on bottom of the tube.
8. Gently aspirate and discard adipocytes and supernatant. At this point, the SVC-containing pellet will be easy to disturb, so great care should be taken. If necessary, adipocytes (floating layer) can be collected using a transfer pipette prior to aspiration.
9. Manually disrupt pellet by flicking the tube and resuspend in 0.5 ml RBC lysis buffer. Incubate 5 min at room temperature with occasional gently shaking.
10. Neutralize RBC lysis by adding 5 ml of FACS buffer.
11. Centrifuge at $500 \times g$ for 10 min at 4 °C. Resuspend cell pellets in 3–5 ml FACS buffer and incubate on ice.
12. Mix 10 μl of each sample 1:1 with trypan blue solution. Count viable cells carefully using hemocytometer. The resulting cell number and total volume give an estimate of the number of SVCs per fat pad. Typical yields are $1-3 \times 10^6$ cells/g fat from lean mice, $2-5 \times 10^6$ cells/g fat from obese mice.

3.2. Staining SVCs for cell surface markers to identify ATMs

For multiparameter flow cytometry, the selection of antibodies and fluorochrome conjugates is a critical step (Baumgarth & Roederer, 2000; Maecker, Frey, Nomura, & Trotter, 2004). It has been our experience that ATMs from obese mouse fat have a significant amount of autofluorescence when excited by the blue laser (488 nm); thus, we try to avoid using FITC- or Alexa488-conjugated antibodies in our staining protocols for ATMs. The first step in selecting fluorochromes is to consider the flow cytometry or FACS instrument on which the samples will be analyzed. We recommend using instruments with at least a three-laser configuration (e.g., FACSCanto II equipped with 405 nm violet laser, 488 nm blue laser, and 640 nm red laser) as this will provide the greatest flexibility and allow for five to seven colors to be detected; however, if such an instrument is not available, two-laser configurations (e.g., 488/640 nm or 561/640 nm) are the next best option. When selecting fluorochrome conjugates, it is generally recommended to use those with the highest staining index (e.g., PE,

PE-Cy5, APC) for rare cellular events, but care must be taken to minimize spectral overlap. Detailed discussions of these issues are presented elsewhere (Baumgarth & Roederer, 2000; Maecker et al., 2004). Once the antibodies and fluorochrome conjugates have been chosen, preliminary experiments should be performed to titer the antibody concentrations to maximize separation of positive and negative cell populations. Based on the antibody selection and titration data, SVCs can be stained with antibody cocktails (containing antibodies at predetermined concentrations in staining buffer) as described in the basic protocol below. For controls, additional SVCs are stained with each antibody individually (SS controls) and the antibody cocktail minus one antibody (FMO controls). Table 16.2 shows a summary of the control tubes needed in this antibody panel.

1. Transfer 1×10^6 SVCs from Section 3.1 into a 5-ml polystyrene round-bottom tube and resuspend in 100 µl of FACS buffer to obtain a final concentration 1×10^7 SVCs/ml. *Note*: When preparing cells for sorting, the total number of cells can be increased, but each reagent should be linearly increased as well.
2. Add 0.5–1 µg of Fc-block (anti-CD16/32) and incubate on ice for 10 min.
3. Repeat steps 1–2 for each control tube listed in Table 16.2. SVCs can be pooled from multiple mice for the SS and FMO controls. If cells are limiting, as few as 1×10^5 SVCs can be used for control tubes.
4. For each sample, prepare an antibody staining cocktail for ATMs as indicated in Table 16.1. Bring the cocktail to 100 µl/sample by adding FACS buffer.
5. Add 100 µl of antibody cocktail to each sample tube.
6. Flick tube to mix.
7. For SS and FMO controls, add the antibody or antibodies as indicated in Table 16.2.
8. Incubate all tubes for 30 min at 4 °C protected from light.
9. Wash cells with 2 ml FACS buffer and centrifuge 10 min at $500 \times g$ at 4 °C.
10. Carefully, aspirate supernatants.
11. Wash cells again with 2 ml FACS buffer and centrifuge 10 min at $500 \times g$ at 4 °C.
12. Carefully, aspirate supernatants.
13. Optional (for flow cytometry analysis only): stain the samples and appropriate controls with viability dye (fixable live/dead stain). Dilute dye (1:500–1:1000) in 200 µl PBS/sample. Incubate the cells in dye

solution at room temperature for 30 min. Repeat steps 9–12 to wash cells. Proceed to step 14.

14. For flow cytometry analysis, fix cells by adding 200 µl of 0.1% PFA. Cells can be stored in PFA at 4 °C in the dark until data acquisition. Prolonged exposure in PFA may increase autofluorescence; therefore, it is advisable to remove PFA after 1 h and replace with 200–300 µl of FACS buffer if cells will not be analyzed within 24 h.
15. For sorting of viable cells by FACS, suspend cells in FACS buffer to obtain a final concentration suitable for sorting (generally $1–10 \times 10^7$/ml). Add DAPI (0.2 µg/ml) to each sample and appropriate controls to allow for live/dead cell discrimination. Protect tubes from light and place on ice for transport to the FACS instrument.

3.3. Flow cytometry analysis
3.3.1 Compensation procedures
As noted above, care should be taken to minimize emission spectra overlap when selecting fluorochrome-conjugated antibodies for ATMs. However, it is difficult to completely eliminate spectral overlap in multicolor staining. Therefore, it is necessary to perform a compensation procedure during each experiment before acquiring data from the samples of interest. Each instrument has a different compensation procedure, with some software packages generating a compensation matrix automatically from SS controls. Therefore, we recommend consulting the manual and following the manufacturer's recommendations for your instrument. When compensation is correctly applied, the median fluorescence intensities (MFIs) of the positive and negate populations of the individual SS controls are aligned in all neighboring channels. It is important to use SS controls prepared from SVCs (Section 3.2, Step 3) because these cells will have the same properties as the samples of interest. Here, we provide a brief guide for getting started.

1. Using an unstained control sample, adjust side scatter and forward scatter so that the cell populations of interest are on scale.
2. Using the unstained control sample, adjust the photomultiplier tube (PMT) gain for each fluorochrome detector so that the peak MFI of the unstained cells on a histogram is within $10^1–10^2$ on a log scale.
3. Acquire all SS compensation controls. If necessary, adjust the PMT gains for each detector so that the positive (stained) population can be discriminated from the negative (unstained) cells. We acquire and save 10,000–30,000 events for each SS control.

4. Calculate compensation values across all included detector according to instructions for your instrument. Importantly, apply compensation values to all SS controls, FMO controls, and samples of interest.

3.3.2 Data acquisition

1. Using the same instrument settings, acquire all FMO controls individually. FMO controls contain all the antibodies in the staining cocktail, but one antibody is replaced with an isotype control. For example, FMO-APC contains the APC-isotype antibody rather than APC-CD301 (Table 16.2). This is necessary for discriminating positive from negative populations during subsequent gating of the samples of interest. Acquire and save 10,000–30,000 events.
2. Acquire and save 10,000–50,000 events from the samples of interest. To generate statistically sound data for frequency determination, a sufficient number of events need to be obtained; this may require recording more than 50,000 events per sample.

3.3.3 Gating strategy to identify and characterize ATMs

Figure 16.1 depicts our general strategy for identifying M1 ($CD11c^+$) and M2 ($CD301^+$) ATMs for quantification and purification by FACS. Cell aggregates, dead cells ($DAPI^+$), and cellular debris are first excluded (Fig. 16.1A). This step greatly reduces, but does not eliminate, autofluorescence in SVC preparations. Next, adipose tissue leukocytes ($CD45^+$) are selected. ATMs from both lean and obese mice coexpress F4/80 and CD11b (Fig. 16.1A). However, $F4/80^{mid}$ cells (gate 2) contain eosinophils (Siglec-F^+) and neutrophils ($GR1^+$; Fig. 16.1B); therefore, care should be taken to gate ATMs as $F4/80^{high}CD11b^{high}$ (gate 1) to minimize contamination with these cell types. Finally, viable $CD45^+F4/80^{high}CD11b^{high}$ ATMs are analyzed for surface expression of CD11c (M1 marker) and CD301 (M2 marker), which identify discrete M1 and M2 ATM subsets in both lean and obese mice (Figs. 16.1A and 16.3B).

3.4. FACS procedure to purify ATM populations

1. Prior to FACS, decide the number of desired target cells to be collected. Cells should be sorted directly into the appropriate media (e.g., cell culture media, lysis buffer) for the downstream application. Prepare enough collection tubes for the cells of interest.
2. Prepare SVCs from adipose tissue as described in Section 3.1.

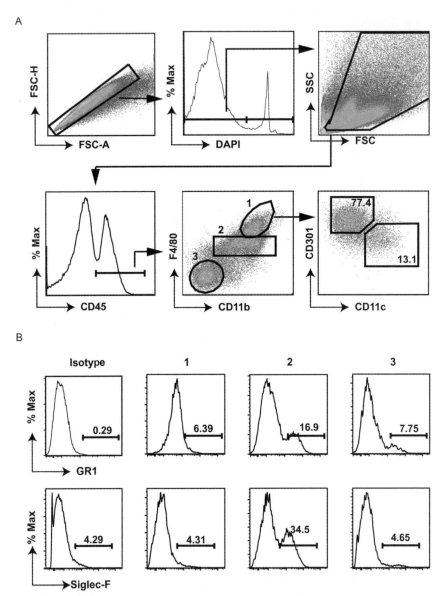

Figure 16.1 Gating strategy to identify and characterize ATMs. (A) Gating strategy for the identification of M1 (CD11c$^+$) and M2 (CD301$^+$) ATMs according to the flow cytometry protocol described herein. (B) Histograms of Gr-1 and Siglec-F expression on CD11b^{high}F4/80^{high} (gate 1), CD11b$^{mid/high}$F4/80^{dim} (gate 2), and CD11b$^-$F4/80$^-$ (gate 3) SVCs demonstrates that the CD11b$^{mid/high}$F4/80^{dim} population (gate 2) can be contaminated with neutrophils (Gr-1) and eosinophils (Siglec-F). (See the color plate.)

3. Stain SVCs with antibody cocktail as described in Section 3.2. Cells should be transported to the FACS facility on ice and protected from light.
4. Set up the FACS instrument according to the manufacturer's directions.
5. Run SS and FMO controls to setup compensation matrix and sorting gates.
6. Optimize the FACS instrument for sorting according to the manufacturer's protocol and adjust speed of sample acquisition to optimize purity of the sample.
7. Vortex the source cells and run a test sort.
8. Continue sorting cells until the desired cell number is achieved or the source cells are exhausted.

3.5. Using magnetic beads and positive selection to enrich for CD11b$^+$ SVCs

In this section, we describe the enrichment of CD11b$^+$ leukocytes from SVCs using magnetic beads and positive selection (Fig. 16.2). The majority of the cells in this pool will be ATMs but contamination with neutrophils and eosinophils will be likely. Downstream *in vitro* studies can further enrich for ATMs by adhesion to plastic.

1. In a 5-ml polystyrene round-bottom tube, suspend 10^7 SVCs in 90 μl of MACS buffer.
2. Add 10 μl of CD11b microbeads.

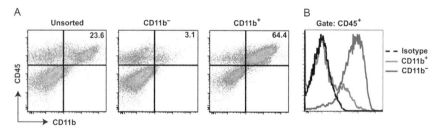

Figure 16.2 Enrichment of CD11b$^+$ SVCs using magnetic beads. SVCs were isolated as described in Section 3.1 and CD11b$^+$ cells were positively selected using Miltenyi microbeads as described in Section 3.5. Unsorted and sorted cells were analyzed by flow cytometry. (A) Scatterplots showing CD45 versus CD11b expression on unsorted (left), CD11b-depleted (CD11b$^-$; middle), and CD11b-enriched (CD11b$^+$; right) SVCs. The percentage of total cells for the CD45$^+$CD11b$^+$ population is indicated. (B) Histogram showing enrichment efficiency of CD11b$^+$ SVCs following positive selection. (See the color plate.)

3. Mix well by flicking tube or very gently vortexing. Incubate for 15 min at 4 °C.
4. Wash cells by adding 1 ml of MACS buffer and centrifuge at 500 × g for 10 min at 4 °C. Discard supernatant.
5. Resuspend cell pellet in 500 μl of MACS buffer.
6. Place MS separation column in magnetic field of MACS separator. Equilibrate column by adding 500 μl of MACS buffer.
7. Apply cell suspension to center of column. Collect unlabeled (CD11b$^-$) cells that pass through the column in a 5-ml tube.
8. Wash column three times with 500 μl of MACS buffer, allowing the column to empty with each wash. Do not allow the column to dry. Collected flow through (CD11b$^-$ fraction) can be assayed by flow cytometry if necessary.
9. Remove MS column from the magnetic field. Place in a 5-ml tube.
10. Add 1 ml of MACS buffer to the column. Using the provided plunger, expel the CD11b$^+$ fraction into the tube.
11. Centrifuge cell fractions at 500 × g for 5 min at 4 °C. Discard supernatant, and resuspend cell pellets in 300 μl of FACS buffer and stain for cell surface antigens or further analysis.

3.6. Data analysis

1. Analyze flow cytometry data with FlowJo software.
2. Determination the frequency of ATMs and ATM subsets as a percentage of total viable cells (Fig. 16.3C and D).
3. Based on the % of SVCs, ATM content can also be reported as absolute cell number (%ATMs × total SVC isolated), or normalized to adipose tissue mass (%ATMs × total SVCs/fat pad weight (g) (Fig. 16.3C and D).
4. Perform statistical analysis.

4. DISCUSSION

The approach presented here has evolved over the years in our laboratory and has undergone a variety of refinements to best assess ATM in a variety of mouse models (Lumeng et al., 2011; Morris et al., 2013; Morris, Oatmen, Wang, DelProposto, & Lumeng, 2012; Singer et al., 2013; Westcott et al., 2009). Such techniques provide the most reliable quantitation of ATM content and assess the balance between proinflammatory M1 (CD11c$^+$) ATMs and anti-inflammatory M2 (CD301$^+$) ATMs in lean and obese mice. This FACS strategy has also been

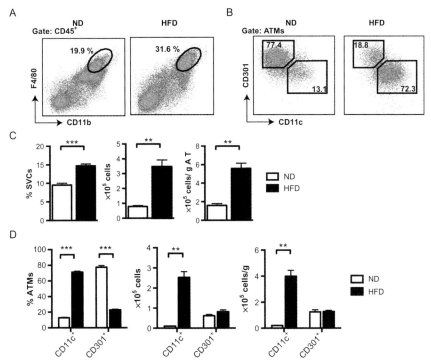

Figure 16.3 High-fat diet increases ATM content and alters the distribution of M1 and M2 ATMs in visceral fat. C57BL/6J mice were fed a normal diet (ND) or high-fat diet (HFD; 60 kcal from fat) for 12 weeks before stromal vascular cells (SVCs) were isolated from epididymal adipose tissue and examined by flow cytometry as described in these protocols. (A) Representative scatterplots of SVCs from ND (left) and HFD (right) mice stained with CD11b and F4/80. Cells are gated on viable $CD45^+$ SVCs as shown in Fig. 16.1A. (B) Representative scatterplots showing gates for M1 ($CD11c^+CD301^-$) and M2 ($CD11c^-CD301^+$) ATMs. (C) ATM content in epididymal adipose tissue shown as percentage of total SVCs, absolute number per fat pad, and normalized to visceral adipose tissue mass. (D) Distribution of ATM subsets in visceral fat from lean and obese mice shown as percentage of total ATMs, absolute number per fat pad, and normalized to adipose tissue mass. $**P<0.01$, $***<0.001$ versus ND. (See the color plate.)

successfully used to purify ATMs for gene expression analysis and functional assays (Morris et al., 2013; Singer et al., 2013). While similar principles may apply for analysis of human ATMs, Hagman et al. (2012) provide an excellent strategy specifically tailored to human samples.

In our experience, the most critical steps in the protocol are efficient digestion of adipose tissue by collagenase and proper staining and compensation of SVCs. We have found that gentle extraction techniques optimized for study of adipocyte function do not efficiently remove ATMs from the floating adipocytes during the differential centrifugation step. The upper layer can be contaminated with lipid-laden ATMs, which are buoyant, with ATMs that remain attached to adipocytes without vigorous digestion. Attempts to combine analysis of SVCs with adipocyte function have been challenging, as digestion strategies need to be optimized for each application.

Our strategy to identify ATMs is based on initial observations that macrophages in adipose tissues coexpress F4/80 and CD11b (Weisberg et al., 2003). Since then, it has been established that F4/80^{mid} cells in fat from mice can also contain eosinophils (Siglec-F$^+$) and neutrophils (GR-1$^+$), although they are minor populations compared to ATMs (Fig. 16.1B) (Talukdar et al., 2012; Wu et al., 2011). In our staining protocols, ATMs are best identified as F4/80^{high}CD11b^{high} (Fig. 16.1A; gate 1).

How to appropriately normalize ATM data from flow cytometry continues to be a challenge in fat depots that massively expand in size due to adipocyte hypertrophy. In our example data (Fig. 16.3C), SVCs from lean mice contains 9% ATMs and high-fat diet (HFD) exposure for 12 weeks increased ATM content to 15% (1.6-fold increase). However, the absolute and relative (cells/g fat) number of ATMs increased approximately threefold (Fig. 16.3C). In agreement with previous reports (Li et al., 2010; Lumeng et al., 2008), HFD-induced obesity also altered the distribution of M1 and M2 ATM subtypes in visceral fat (Fig. 16.3B and D). ATM content and the frequencies of M1 versus M2 ATMs obtained here are comparable to other studies (Weisberg et al., 2003; Xu et al., 2003), indicating that our flow cytometry method provides a simple and reproducible method for assessing the content and heterogeneity of ATMs in mice models.

One obstacle that limits flow analysis is the significant cellular autofluorescence of adipose SVCs and ATMs. Especially in obese mice, ATMs are embedded in the autofluorescent SSC^{hi}FSC^{hi} populations of cells that are prominent in the SVC. Analyses should not exclude these cells as this will significantly underreport ATM content. Gating to exclude dead cells, debris, and cell aggregates can improve results, but a significant amount of autofluorescence from ATMs remains likely due to high lipid content. In our experience, these limitations almost completely exclude the routine use of FITC- and Alexa488-conjugated monoclonal antibodies for macrophages in staining cocktails. This limitation can be partially overcome by using flow cytometers with yellow–green (561 nm) lasers versus the conventional blue (488 nm) laser configuration (Morris, D.L., unpublished observation).

For those starting such a procedure for the first time, consultation with flow cytometry experts is highly recommended prior to performing experiments. Over time this protocol has been shown to give highly reproducible results and we feel is a good starting point for those looking to examine ATMs in their mouse models. Additional refinements of the protocol should continue by all groups in this field as it continues in our laboratory.

ACKNOWLEDGMENTS
K. W. C. and C. N. L. are supported by NIH grants DK090262 and DK092873 and a Career Development Award from the American Diabetes Association. D. L. M. was supported by NIH NRSA award DK091976.

REFERENCES
Baumgarth, N., & Roederer, M. (2000). A practical approach to multicolor flow cytometry for immunophenotyping. *Journal of Immunological Methods*, *243*(1–2), 77–97.

Chawla, A., Nguyen, K. D., & Goh, Y. P. (2011). Macrophage-mediated inflammation in metabolic disease. *Nature Reviews. Immunology*, *11*(11), 738–749. http://dx.doi.org/10.1038/nri3071.

Cho, C. H., Koh, Y. J., Han, J., Sung, H. K., Jong Lee, H., Morisada, T., et al. (2007). Angiogenic role of LYVE-1-positive macrophages in adipose tissue. *Circulation Research*, *100*(4), e47–e57.

Cinti, S., Mitchell, G., Barbatelli, G., Murano, I., Ceresi, E., Faloia, E., et al. (2005). Adipocyte death defines macrophage localization and function in adipose tissue of obese mice and humans. *Journal of Lipid Research*, *46*(11), 2347–2355.

Glass, C. K., & Olefsky, J. M. (2012). Inflammation and lipid signaling in the etiology of insulin resistance. *Cell Metabolism*, *15*(5), 635–645.

Gregor, M. F., & Hotamisligil, G. S. (2011). Inflammatory mechanisms in obesity. *Annual Review of Immunology*, *29*, 415–445.

Hagman, D. K., Kuzma, J. N., Larson, I., Foster-Schubert, K. E., Kuan, L. Y., Cignarella, A., et al. (2012). Characterizing and quantifying leukocyte populations in human adipose tissue: Impact of enzymatic tissue processing. *Journal of Immunological Methods*, *386*(1–2), 50–59.

Harman-Boehm, I., Bluher, M., Redel, H., Sion-Vardy, N., Ovadia, S., Avinoach, E., et al. (2007). Macrophage infiltration into omental versus subcutaneous fat across different populations: Effect of regional adiposity and the comorbidities of obesity. *The Journal of Clinical Endocrinology and Metabolism*, *92*(6), 2240–2247.

Kanda, H., Tateya, S., Tamori, Y., Kotani, K., Hiasa, K., Kitazawa, R., et al. (2006). MCP-1 contributes to macrophage infiltration into adipose tissue, insulin resistance, and hepatic steatosis in obesity. *The Journal of Clinical Investigation*, *116*(6), 1494–1505.

Li, P., Lu, M., Nguyen, M. T., Bae, E. J., Chapman, J., Feng, D., et al. (2010). Functional heterogeneity of CD11c-positive adipose tissue macrophages in diet-induced obese mice. *The Journal of Biological Chemistry*, *285*(20), 15333–15345.

Liu, J., Divoux, A., Sun, J., Zhang, J., Clement, K., Glickman, J. N., et al. (2009). Genetic deficiency and pharmacological stabilization of mast cells reduce diet-induced obesity and diabetes in mice. *Nature Medicine*, *15*(8), 940–945.

Lumeng, C. N., Bodzin, J. L., & Saltiel, A. R. (2007). Obesity induces a phenotypic switch in adipose tissue macrophage polarization. *The Journal of Clinical Investigation*, *117*(1), 175–184.

Lumeng, C. N., DelProposto, J. B., Westcott, D. J., & Saltiel, A. R. (2008). Phenotypic switching of adipose tissue macrophages with obesity is generated by spatiotemporal differences in macrophage subtypes. *Diabetes*, *57*(12), 3239–3246.

Lumeng, C. N., Deyoung, S. M., Bodzin, J. L., & Saltiel, A. R. (2007). Increased inflammatory properties of adipose tissue macrophages recruited during diet-induced obesity. *Diabetes, 56*(1), 16–23.

Lumeng, C. N., Liu, J., Geletka, L., Delaney, C., Delproposto, J., Desai, A., et al. (2011). Aging is associated with an increase in T cells and inflammatory macrophages in visceral adipose tissue. *Journal of Immunology, 187*(12), 6208–6216.

Lumeng, C. N., & Saltiel, A. R. (2011). Inflammatory links between obesity and metabolic disease. *The Journal of Clinical Investigation, 121*(6), 2111–2117.

Maecker, H. T., Frey, T., Nomura, L. E., & Trotter, J. (2004). Selecting fluorochrome conjugates for maximum sensitivity. *Cytometry. Part A: The Journal of the International Society for Analytical Cytology, 62*(2), 169–173.

Morris, D. L., Cho, K. W., Delproposto, J. L., Oatmen, K. E., Geletka, L. M., Martinez-Santibanez, G., et al. (2013). Adipose tissue macrophages function as antigen presenting cells and regulate adipose tissue CD4+ T cells in mice. *Diabetes, 62*(8), 2762–2772.

Morris, D. L., Oatmen, K. E., Wang, T., DelProposto, J. L., & Lumeng, C. N. (2012). CX3CR1 deficiency does not influence trafficking of adipose tissue macrophages in mice with diet-induced obesity. *Obesity, 20*(6), 1189–1199.

Nishimura, S., Manabe, I., Nagasaki, M., Eto, K., Yamashita, H., Ohsugi, M., et al. (2009). CD8+ effector T cells contribute to macrophage recruitment and adipose tissue inflammation in obesity. *Nature Medicine, 15*(8), 914–920.

Rodbell, M. (1964). Metabolism of isolated fat cells. I. Effects of hormones on glucose metabolism and lipolysis. *The Journal of Biological Chemistry, 239*, 375–380.

Singer, K., Morris, D. L., Oatmen, K. E., Wang, T., DelProposto, J., Mergian, T., et al. (2013). Neuropeptide Y is produced by adipose tissue macrophages and regulates obesity-induced inflammation. *PLoS One, 8*(3), e57929.

Strissel, K. J., DeFuria, J., Shaul, M. E., Bennett, G., Greenberg, A. S., & Obin, M. S. (2010). T-cell recruitment and Th1 polarization in adipose tissue during diet-induced obesity in C57BL/6 mice. *Obesity, 18*(10), 1918–1925.

Strissel, K. J., Stancheva, Z., Miyoshi, H., Perfield, J. W., 2nd, DeFuria, J., Jick, Z., et al. (2007). Adipocyte death, adipose tissue remodeling, and obesity complications. *Diabetes, 56*(12), 2910–2918.

Talukdar, S., Oh da, Y., Bandyopadhyay, G., Li, D., Xu, J., McNelis, J., et al. (2012). Neutrophils mediate insulin resistance in mice fed a high-fat diet through secreted elastase. *Nature Medicine, 18*(9), 1407–1412.

Weisberg, S. P., McCann, D., Desai, M., Rosenbaum, M., Leibel, R. L., & Ferrante, A. W., Jr. (2003). Obesity is associated with macrophage accumulation in adipose tissue. *The Journal of Clinical Investigation, 112*(12), 1796–1808.

Wentworth, J. M., Naselli, G., Brown, W. A., Doyle, L., Phipson, B., Smyth, G. K., et al. (2010). Pro-inflammatory CD11c+CD206+ adipose tissue macrophages are associated with insulin resistance in human obesity. *Diabetes, 59*(7), 1648–1656.

Westcott, D. J., Delproposto, J. B., Geletka, L. M., Wang, T., Singer, K., Saltiel, A. R., et al. (2009). MGL1 promotes adipose tissue inflammation and insulin resistance by regulating 7/4hi monocytes in obesity. *The Journal of Experimental Medicine, 206*(13), 3143–3156.

Wu, D., Molofsky, A. B., Liang, H. E., Ricardo-Gonzalez, R. R., Jouihan, H. A., Bando, J. K., et al. (2011). Eosinophils sustain adipose alternatively activated macrophages associated with glucose homeostasis. *Science, 332*(6026), 243–247.

Xu, H., Barnes, G. T., Yang, Q., Tan, G., Yang, D., Chou, C. J., et al. (2003). Chronic inflammation in fat plays a crucial role in the development of obesity-related insulin resistance. *The Journal of Clinical Investigation, 112*(12), 1821–1830.

AUTHOR INDEX

Note: Page numbers followed by "*f*" indicate figures and "*t*" indicate tables.

A

Aagaard, M. M., 261–262, 263–265
Aaron, J., 135
Abe, Y., 245
Abel, E. D., 4–5
Abu-Elheiga, L., 3–4
Acosta, A. S., 283–284, 285
Adams, D. J., 129
Adkison, M. G., 200–201
Adler, E. S., 200–201
Adli, M., 263–265
Agarwal, A. K., 162–163
Aggbao, P. C., 147–148
Ahfeldt, T., 96–97, 113–118
Ahlfeldt, F. E., 126
Ailhaud, G., 94
Akerblad, P., 3–4
Akpulat, S., 245
Albarado, D. C., 76
Albrektsen, T., 263
Aleksandrova, K., 258
Alibegovic, A. C., 236
Almer, S., 149
Alonso, M. N., 20–21
Alt, E. U., 9
Alvarez, R. E., 150–153
Amador, A. G., 207–208
Ameri, P., 284
Amodaj, N., 83–86
An, D., 178–179, 196
Andaloussi, S. E., 3–4
Anderson, C., 200–201
Andersson, H., 162
Andersson, J., 228
Andraghetti, G., 284
Andres, A., 9
Angelin, B., 22–23
Antonio, G. E., 127
Antunes-Rodrigues, J., 200–201
Aoi, N., 32

Aoki, V. T., 200–201
Aparicio, O., 262, 277
Aplin, A. C., 76–77
Apotheker, S., 77, 78–79
Aprikian, O., 94
Arce, C. H., 236
Ardilouze, J. L., 228–230, 235, 236, 237
Arganda-Carreras, I., 86
Arner, E., 5–6, 32, 114–118
Arner, P., 5–6, 94–95, 228–229, 230, 237
Arribas, M., 9
Asano, S., 95–96
Ashjian, P. P., 18
Asola, M., 145–146, 238–239
Astrup, A., 228–230
Aubry, E. M., 162–165, 169–171, 173
Auerbach, R. K., 277
Auwerx, J., 96–97, 113–114
Avinoach, E., 298
Ayers, C. R., 244–245, 258

B

Baba, S., 153–154
Babaev, V. R., 4–5
Bae, E. J., 312
Bae, Y., 76
Baek, S., 262, 276
Baessler, A., 244–245, 258
Bailey, S. T., 5
Bainbridge, M., 262
Baker, M., 76–77
Balasubramanian, S., 275–276
Baldini, G., 256–257
Bamshad, M., 200–201, 207
Bando, J. K., 298, 312
Bandyopadhyay, G., 298, 312
Banerjee, R. R., 5
Barak, Y., 4–5
Barbatelli, G., 298–299
Barber, P. R., 76–77

Barlow, C., 4–5
Barnard, T., 207–208
Barnes, G. T. G., 20–21, 298, 312
Barnett, A. H., 245
Baro, D. J., 212, 218, 220
Barski, A., 262
Bartke, A., 207–208
Bartness, T. J., 94, 200–201, 202–203, 205, 207–208, 209, 211, 212, 218, 220
Bartz, R., 162–163
Basile, G., 284
Batchelor, B. B., 32, 59–63
Bathija, A., 127–128
Batzoglou, S., 262, 275–276
Baumgarth, N., 37–38, 304–306
Bazin, R., 94–95
Bazuine, M., 3–4
Beard, C., 10
Beatty, P. J., 148
Becker, D. M., 48
Behrens, W. A., 213
Beilfuss, A., 76
Bellinger, L. L., 209–210
Ben Salem, D., 125–126
Benarroch, E. E., 207–208
Bengtsson, T., 146
Bennett, G., 298
Benoist, C., 18
Bentley, D. R., 275–276
Berg, A. H., 244–245, 246
Berg, C., 3–4
Berge, R. K., 96
Berger, J. P., 244–245
Bergmann, O., 5–6, 32, 114–118
Bergstrom, R. W., 96–97, 113–114
Bernard, S., 5–6, 32, 114–118
Bernstein, B. E., 263–265, 275–276
Berry, R., 32–33, 37, 48–49, 59–63, 128–129
Bes-Houtmann, S., 284
Beznak, A. B. L., 200–201
Bhanot, S., 3–4
Bhat, S., 5
Bielby, C., 162–163, 164–165
Bilenky, M., 262
Billestrup, N., 3
Billington, C. J., 202, 203

Binns, D., 162–163, 169, 173
Birsoy, K., 32–33, 37, 42, 48, 59–63, 128–129
Bishop, J. M., 6–8
Bjorndal, B., 96
Bjornheden, T., 95–96
Blaak, E. E., 228–230
Blanchette-Mackie, E. J., 19–20
Blasberg, R. G., 145–146
Blocksom, B. H. Jr., 126
Blondin, D. P., 144t
Bloom, F. E., 212
Bluher, M., 9, 298
Bobbert, T., 245
Bodzin, J. L., 20–21, 298–299
Boeing, H., 258
Bolinder, J., 228–229
Boll, D. T., 152–153, 154–156
Boord, J. B., 4–5
Booten, S. L., 3–4
Borga, M., 149
Bornert, J. M., 5–6
Borowiak, M., 184–185
Borowiecki, M., 32
Bosnakovski, D., 48
Boss, O., 4–5
Bostrom, P., 9
Bouchard, C., 48
Boucher, J., 6–9
Boushel, R., 228–229
Boutati, E., 228
Bouvy, N. D., 142, 146
Bowers, R. R., 201, 202, 203, 205, 207–208, 209, 211, 218
Boyko, E. J., 96–97, 113–114
Bradshaw, A. D., 95–96
Brakenhielm, E., 22–23
Branca, R. T., 147–148
Brasier, A. R., 263
Brassard, P., 228–230, 237
Brau, A. C., 148
Bredella, M. A., 125–126, 128
Brennand, K., 114–118
Briatore, L., 284
Brito, M. N., 212, 218, 220
Brito, N. A., 212, 218, 220
Britton, T., 32
Brix, G., 238–239

Brochu, M., 96–97, 113–114
Brodie, B. B., 217–218
Brown, C. G., 275–276
Brown, D. M., 202, 203
Brown, E. J., 5
Brown, G. D., 276
Brown, W. A., 298
Bruder, H., 152–153, 154–156
Bruning, J., 9
Bruzzone, S., 284
Bucci, M., 228–229
Buchholz, B. A., 5–6, 32, 114–118
Bueno-de-Mesquita, H. B., 258
Bujo, H., 284
Bulow, J., 228–230, 236, 239
Burant, C. F., 94–95
Burdi, A. R., 94
Burkart, A., 77
Burri, L., 96
Bydder, M., 147–148
Byron, K. J., 236

C

Calder, P. C., 96
Calise, D., 96–97
Callaghan, M. F., 148
Canbay, A., 76
Cancello, R., 18–19, 53
Cañete, M., 28
Cannon, B., 22–23, 146, 200–201
Cantu, R. C., 200–201
Cao, Q., 94–95
Cao, R., 22–23
Cao, X.-G., 284
Caplin, J., 96–97, 113–118
Carle, F., 218
Carleton, R. A., 48
Carlsson, C., 3
Caron, L., 184–185
Carpenter, A. E., 59–63
Carpentier, A. C., 228–230, 237
Carragher, D. M., 21–22
Carrascosa, J. M., 9
Carroll, J., 263–265
Carter, N., 9
Casteilla, L., 68, 96–97
Castonguay, T. W., 209–210
Cawthorn, W. P., 94–95

Cawthorne, M. A., 95–96
Ceresi, E., 18–19, 53, 298–299
Cesari, M., 96–97, 113–114
Ceyhan, O., 48, 59–63
Chaffin, A. E., 9
Chambon, P., 5–6
Chan, J. L., 150–153
Chan, L., 162
Chan, W. S., 163
Chang, B. H., 162–163
Chapman, J., 312
Chawla, A., 4–5, 298–299
Cheeseman, J., 228–229
Chen, C. W., 68
Chen, H. C., 95–96
Chen, K. W., 96–97, 113–114
Chen, S. Y., 95–96, 284
Chen, W,.
Chen, W. W., 59–63, 162–163, 202–203
Chen, X., 277
Chia, J. M., 147–148
Chiang, S. H., 3–4, 94–95
Childs, C. R., 283–284
Chirala, S. S., 3–4
Chiu, P. C., 3–4, 5, 6–8
Chiueh, C. C., 213
Cho, C. H., 18–19, 22–23, 76, 302
Cho, K. W., 21–22, 310–311
Choi, C. S., 3–4
Choi, J. H., 9
Choi, S., 76
Chou, C. J. C., 20–21, 298, 312
Chouinard, M., 77
Christensen, N. J., 228–229
Christiaens, V., 76
Chung, S. A., 153–154
Cignarella, A., 310–311
Cinti, S., 18–19, 20, 32, 53, 96, 200–201, 202, 203, 205, 207–208, 209, 218, 298–299
Clampit, J., 3–4
Clark, M. L., 228–229, 231
Clarke, C., 59–63
Clarke, I. J., 200–201
Clement, K., 298
Cloutier, A. M., 128
Coenen, H. H., 145

Cohade, C., 146
Cohen, P., 2, 9, 48
Colditz, G. A., 48
Colvill, L. M., 200–201
Combs, T. P., 244–245, 246
Conroe, H. M., 48
Cooper, J. R., 212
Correll, J. W., 203
Corselli, M., 68
Corvera, S., 77
Costa, E., 217–218
Coursindel, T., 3–4
Cousin, B., 96–97
Covington, J. D., 5–6
Cowan, C. A., 114–118
Crampes, F., 236
Crandall, D. L., 18, 22–23, 76, 238
Crisan, M., 68
Crissman, H. A., 37–38
Cristancho, A. G., 262, 276
Cristóbal, J., 28
Crossno, J. T. Jr., 284
Cruce, J. A., 59–63
Cuddapah, S., 262
Cui, J., 201, 209–210
Cui, K., 262
Cui, X., 163
Currie, J. M., 228, 235, 236
Custer, R. P., 126
Cypess, A. M., 142

D

Dahlström, N., 149
Daley, G. Q., 126
Dalla Nora, E., 162
D'Amico, G., 76–77
Danielian, P. S., 5–6
Darimont, C., 94
Darzynkiewicz, Z., 37–38
Davidson, M. G., 20–21
Davis, S., 127–128
Dawson, K. L., 135
de Ferranti, S., 94–95
de Souza, C. J., 59–63
De Ugarte, D. A. D., 18
DeFuria, J., 298–299
DeGrado, T. R., 145
Dela, F., 236

Delaney, C., 22, 298–299, 310–311
Delepine, M., 162
Delproposto, J., 22, 298–299, 310–311
DelProposto, J. B., 18–19, 20, 298–299, 310–311, 312
Delproposto, J. L., 21–22, 310–311
Demas, G. E., 201–202, 207–208
Demontiero, O., 125–126
Deng, C. X., 4–5
Deng, J., 163
Deng, Y., 5, 6–8
Dennis, A. L., 228–229
Dennis, R. J., 162–165, 169–171, 173
Dentz, E., 5–6
Deo, R. C., 96–97, 113–118
Depocas, F., 203, 207–208, 213
Des Rosiers, M. H., 145–146
Desai, A., 22, 298–299, 310–311
Desai, M., 298, 312
Desruisseaux, M. S., 94–95
Devlin, M. J., 128
Deyoung, S. M., 20–21, 298–299
Dietrich, A., 96–97, 113–114
Dietz, W. H., 48
DiGirolamo, M., 94, 95–96
Dilioglou, S., 10
Dimitriadis, G., 228
Dionne, I. J., 96–97, 113–114
Divoux, A., 298
Dixon, W. T., 147–148
Dlabar, A., 217–218
Dobbins, R. L., 207–208
Doebber, T., 244–245
Doh, K. O., 76
Dolinsky, V. W., 94–95
Dong, M., 59–63
Döppen, W., 32
Douris, N., 3–4
Doyle, L., 298
Drossaerts, J. M., 142, 146
Du, X., 162–163, 164–165, 244–245, 246
Dubois, N. C., 6–8
Duijnhoven, F. J., 258
Dunham, I., 275–276
Dupin, C. L., 9
Duque, G., 125–126
Dushay, J., 3–4, 5, 6–8
Dusold, D., 97–98

E

Eberhardt, N. L., 3–4
Ebert, A., 145
Ebinuma, H., 257
Eckhardt, M., 59–63
Edelstein, A., 83–86
Eden, S., 95–96
Edouard, C., 135
Edwardson, J. M., 162–165, 169–171, 173
Egan, J. J., 19–20
Eguchi, J., 3–4, 5, 6–8
Ehlen, J. C., 207–208, 209
Elghetany, M. T., 3
Elia, M., 228
Ellerfeldt, K., 94–95
Eltabbakh, G. H., 96–97, 113–114
Emery, J. L., 126
Enevoldsen, L. H., 236
Engles, J. M., 153–154
Engström, E., 153–154
Engvall, J., 153–154
Enoksson, S., 228–229
Erez, O., 258
Erickson, P. F., 284, 285
Eriksson, J. W., 59–63
Estall, J., 48
Eto, H., 32
Eto, K., 18, 20–21, 298
Evans, J. D., 96, 127–128

F

Fahey, F., 126
Faloia, E., 298–299
Faouzi, M., 200–201
Farese, R. V. Jr., 95–96
Farmer, S. R., 261–262
Farnebäck, G., 149
Farnier, C., 94–95
Fasshauer, M., 96–97, 113–114
Faust, I. M., 94–95
Fawcett, D. W., 203
Fazeli, P. K., 125–126, 128, 129
Fedrigo, M., 147–148
Fei, W., 162–163, 164–165
Feil, R., 8
Feil, S., 8
Feng, D. D., 19–20, 312
Ferguson, C., 162–163, 164–165
Ferrante, A. W. Jr., 298, 312
Ferrara, P., 18–19, 53
Ferrell, R. E., 6
Ferron, M., 6–8
Festuccia, W. T. L., 202, 203, 205, 211, 218
Festy, F., 284
Fielding, B. A., 228–229, 230, 235, 236, 237
Finegold, D. N., 6
Firoozan, S., 239
Fischer, B., 283–284
Fisher, F. M., 3–4, 245
Fitch, M., 5–6
Fitz-Gerald, C. S., 184–185
Fitzpatrick, J., 148
Flohr, T. G., 152–153, 154–156
Floyd, Z. E., 135
Fogel, E., 76–77
Fogliano, M., 256–257
Follett, G. F., 126
Ford, B. J., 23–24
Forsgren, M. F., 149
Forster, C. D., 228–229
Foster, D. O., 203, 207–208
Foster, M. T., 200–201, 202–203, 207–208, 209, 218
Foster-Schubert, K. E., 310–311
Fox, C. H., 97–98
Fox, E. C., 3–4
Fox, K. E., 283–284
Frank, P. G., 19–20
Frayn, K. N., 76, 228–229, 230, 231, 235, 236, 237
Fretz, J. A., 129
Frey, T., 304–306
Freyschuss, U., 228
Fried, S. K., 2, 3–4, 94–95
Friedman, J. M., 32–33, 37, 42, 48, 59–63, 128–129
Friedrich, G., 10
Friman, O., 59–63
Frise, E., 86
Frisén, J., 32
Frontini, A., 48, 205, 207–208, 209
Fu, Y., 4–5
Fujimoto, T., 19
Fujimoto, W. Y., 96–97, 113–114

Fujioka, T., 180
Fujisawa, T., 163
Fujita, H., 18, 95–96
Fujita, K., 257
Fukuhara, A., 76
Fullwood, M., 277
Furey, T. S., 276
Furth, S. L., 244–245, 258

G

Gaddis, C. A., 5, 6–9
Gaeta, B., 162–163
Gagen, K., 59–63
Gagnon-Auger, M., 228–230, 237
Gait, M. J., 3–4
Galante, P., 3
Galitzky, J., 228–229, 237
Gallardo, N., 9
Gallian, E., 95–96
Gao, B., 22–23
Gao, X. Y., 202–203
Garfield, A. S., 163
Garn, S. M., 94
Garty, N. B., 19–20
Garvey, W. T., 4–5, 245
Gavrilova, O., 129
Gealekman, O., 48, 77, 78–79
Gedde-Dahl, T. Jr., 162
Geisler, J. G., 3–4
Geletka, L. M., 3–4, 21–22, 298–299, 310–311
Gerin, I., 94–95
Gesuita, R., 218
Giang, A. H., 9
Gibney, M. A., 236
Giddings, J. A., 231
Giles, M. E., 200–201
Gill, R. G., 284
Gilsanz, V., 147–148, 153–154
Gimble, J. M., 135, 262, 268–269, 276
Giordano, A., 48, 200–201, 202, 203, 207–208, 209, 218
Girardier, L., 201
Glass, C. K., 298
Glatting, G., 238–239
Glickman, J. N., 298
Godfrey, C., 3–4

Goh, Y. P., 4–5, 298–299
Goldfien, A., 200–201
Goldfine, A. B., 18, 142
Goldstein, M. S., 200–201
Gondor, A., 277
Gontheir, M.-P., 284
Gontinou, C., 284
Goodman, H. M., 200–201
Goodman, J. M., 162–163, 169, 173
Goossens, G. H., 228–229, 236
Goralski, K. B., 94–95
Goran, M. I., 147–148
Gore, M. O., 244–245, 258
Gorgoglione, M., 77, 78–79
Gotsch, F., 258
Granneman, J. G., 3, 18, 48, 144t, 146, 147, 178–179
Grant, G. R., 262, 268–269, 276
Grasruck, M., 152–153
Grattan, D. R., 200–201
Graves, D. C., 95–96
Graves, R. A., 4–5
Gray, S. L., 162
Grazia, T., 284
Green, H., 263
Greenberg, A. S., 19–20, 298
Greenstein, A., 4–5
Gregersen, H., 238
Gregor, M. F., 298
Gries, F. A., 32
Griffin, M. J., 3–4
Griffith, J. F., 127
Grimaldi, P., 94
Grimsey, N., 162
Grishin, N. V., 162–163
Grönroos, T., 145–146
Grontved, L., 263
Gruber, K., 152–153, 154–156
Gu, Z., 3–4
Guérin, B., 144t
Guerra, C., 9
Guiu, B., 125–126
Gum, R. J., 3–4
Gupta, R. K., 3–4, 48
Gurav, K., 77, 78–79
Gurley, C. M., 20
Gusev, A., 77
Guseva, N., 77, 78–79

H

Haaparanta, M., 146–147
Hadfield, J., 276
Hadro, E., 4–5
Haesemeyer, R., 95–96
Hagenbeek, D., 262, 276
Hagihara, K. I., 201
Hagman, D. K., 310–311
Hagstrom-Toft, E., 228–229
Hajnal, J. V., 148
Halberg, N., 245
Haman, F., 144t
Hamilton, G., 147–148
Hamilton, J. M., 202, 203
Hamilton, M. P., 244–245, 258
Hammer, R. E., 48
Hammer, T., 236
Han, J., 18–19, 22–23, 76, 94, 302
Han, Y., 202–203
Hanif, W., 245
Hara, K., 257
Hardison, R. C., 275–276
Harman-Boehm, I., 298
Harrington, A., 6
Harris, R. B., 200–201, 207–208, 209
Hartig, S. M., 162–163
Hartigan, C., 77, 78–79
Hartson, L., 21–22
Hasch, Z., 200–201
Hasegawa, K., 180
Hasegawa, M., 179, 184, 189f, 191f, 192f, 194f
Hatlebakk, J. G., 238
Hauschka, P. V., 126
Hausman, D. B., 94
Hausman, G. J. G., 18, 22–23, 76, 94, 238
Havel, R. J., 200–201
Hawkins, M. M., 19–20, 244–245
He, W., 4–5
Heglind, M., 142, 144t, 145–146
Heidendal, G. A., 229–230
Heinonen, I., 228–229
Heisler, L. K., 163
Heldmaier, G., 203
Helledie, T., 263
Hellerstein, M., 94
Helm, K. M., 283–284
Hemon, P., 201

Henagan, T. M., 5–6
Henderson, R. M., 172
Henrich, M. M., 145
Herb, S. F., 127–128
Herberg, L., 32
Hermann, S., 162–163
Herzog, H., 145
Hesselberth, J. R., 277
Hevener, A., 4–5
Hiasa, K., 298
Higashiyama, H., 95–96
Hilton, C. L., 162–163, 169, 173
Himms-Hagen, J., 200–201, 209–210
Hioki, K., 193–194, 196
Hiramatsu, H., 193–194, 196
Hirano, Y., 22–23
Hirsch, J. J., 32, 59–63, 94–96
Hirsch, L. J., 236
Hirst, M., 262
Hitchcox, K. M., 178–179, 196
Hiuge, A., 257
Hjemdahl, P., 228
Ho, L. M., 152–153, 154–156
Hoareau, L., 284
Hochedlinger, K., 10
Hodson, L., 76, 228–229
Hoff, D. A., 238
Hoffman, E. J., 144, 145
Hofmann, D., 6–8
Hojbjerre, L., 236
Hollis, J. H., 200–201
Holzer, P., 209–210
Honda, H., 153–154
Hoover, K., 83–86
Horii, S., 18, 20–21
Horowitz, M. C., 128, 129
Hosogai, N., 76
Hosoya, Y. Y., 18, 22–23, 28, 95–96
Hotamisligil, G. S., 298
Hu, H. H., 147–148, 149, 153–154
Huang, J. I. J., 18
Huang, S.-C., 144, 145
Huang, Y. F., 95–96, 284
Huggins, C., 126
Hultin, M., 228
Humphreys, S. M., 76, 228–229, 235, 236
Hurwitz, L. M., 152–153
Hymel, D. T., 76

I

Iida, H., 146, 163
Ikawa, T., 21–22
Ilic, V., 228–229, 230, 235, 236, 237
Imai, T., 5–6
Imakaev, M., 277
Inamoto, T. T., 22–23
Inoue, H., 95–96
Inoue, K., 32
Inoue, N., 3–4
Ishibashi, J., 48
Ishii, T., 22–23
Ishikawa, M., 3–4
Ishimori, N., 18, 20–21
Ishitake, T., 244–245, 258
Ito, D., 163
Ito, M., 193–194, 196
Iwai, K., 201
Iwakoshi, N. N., 94–95

J

Jacene, H. A., 153–154
Jackson, H. A., 153–154
Jackson, R. M., 200–201
Jaenisch, R., 10
Jakubowicz, B., 95–96
Jancso, N., 209–210
Jancso-Gabor, A., 209–210
Janisse, J., 178–179
Jansco, G., 209
Jansco-Gabor, A., 209
Jansen, E., 258
Jansson, P. A., 228–230
Jarver, P., 3–4
Jasmin, J.-F., 19–20
Jayaweera, A. R., 239
Jee, W. S., 126
Jelnes, R., 229–230
Jenab, M., 258
Jennings, E. G., 262, 277
Jensen, M. D., 3–4, 94–95
Jensen, S. S., 263
Ji, C.-B., 284
Ji, S., 257
Jiang, M., 5–6
Jiang, Q. X., 162–163, 169, 173
Jick, Z., 298–299

Jin, J., 22–23, 94
Johansson, A., 149
John, S., 262, 276
Johnson, F. B., 97–98
Johnson, J. E. Jr., 201
Johnson, P. R., 59–63, 94–95
Johnson, T. R., 152–153
Jones, T. R., 59–63
Jong Lee, H., 18–19, 22–23, 76, 302
Joo, F., 209
Joost, H. G., 207–208
Jouihan, H. A., 298, 312
Juhlin-Dannfelt, A., 228
Jun Koh, Y., 18–19, 22–23
Jung, R. T., 202

K

Kaartinen, J. M., 94–95
Kadowaki, T., 257
Kahn, B. B., 9
Kajimura, S., 2, 3–4, 48
Kalender, W. A., 152–153
Kaloulis, K., 6–8
Kanaya, A. M., 96–97, 113–114
Kanda, H., 298
Kandror, K. V., 284
Kanematsu, M., 95–96
Kang, I. H., 59–63
Kang, S., 3–4, 94–95
Kanno, I., 146
Kaplan, A., 96–97, 113–118
Kapterian, T. S., 162–163, 164–165
Karaliota, S., 284
Karastergiou, K., 94–95
Karlsson, M., 228
Karpe, F., 76, 228–229, 230, 235, 236, 237
Karsenty, G., 6–8
Kassem, M., 135
Kato, H., 32
Kaul, S., 239
Kawada, T., 201
Kawahata, M., 193–194, 196
Kawamoto, H., 21–22
Kaynig, V., 86
Kechagias, S., 149
Keen-Rhinehart, E., 200–201
Kehinde, O., 263
Kelso, E., 201, 207–208, 209

Kemerink, G. J., 142, 146, 229–230
Kemppainen, J., 228–229
Kennedy, C., 145–146
Kennedy, R. T., 94–95
Kepczynska, M. A., 95–96
Kerman, I. A., 200–201
Kern, M., 96–97, 113–114
Kerppola, T. K., 167, 168
Kershaw, E. E., 3–4, 5, 6–8
Khallouf, E., 162
Khandekar, M. J., 48
Khanna, A., 148
Kheradpour, P., 262, 275–276
Kihara, S., 245, 257
Kiilerich, P., 263
Kim, J. K., 4–5
Kim, K., 22–23, 94
Kim, Y. B., 4–5
Kimak, M. A., 6
Kimura, Y. Y., 22–23
Kinoshita, M., 95–96
Kiraly, E., 209
Kitazawa, R., 298
Kitsis, R. N. R., 19–20
Kiviranta, R., 3–4
Klein, R. L., 245
Kleiner, S., 3–4, 48
Klemm, D. J., 284, 289–290
Klingenspor, M., 203
Kloting, N., 96–97, 113–114
Klotz, E., 152–153
Knelangen, J. M., 283–284
Knutsson, H., 149
Knuuti, J., 146–147, 228–229
Kobayashi, H., 245
Kobayashi, K., 193–194, 196
Kodukula, S., 3–4
Koh, Y. J., 76, 302
Kohlwein, S. D., 162–163
Koivisto, V. A., 146–147
Kolb, D., 162–163
Komura, N., 257
Kong, R., 284, 285
Koning, R. I., 162–163
Koornstra, R., 263–265
Kopin, I. J., 213
Kordari, P., 3–4
Kostaridou, L., 152–153

Kotani, K., 298
Kou, C.-Z., 284
Kovacs, P., 96–97, 113–114
Kozak, L. P., 9
Krahmer, N., 162–163, 164–165
Kral, J. G., 18, 22–23, 76, 238
Kramer, B., 124–125
Krause, D., 125–126
Krauss, B., 152–153, 154–156
Krief, S., 94–95
Kritchevsky, S. B., 96–97, 113–114
Kroeger, P., 3–4
Krylov, D., 129
Kuan, L. Y., 310–311
Kuerschner, L., 162–163
Kuhajda, F. P., 3–4
Kuhl, D. E., 144, 145
Kumada, M., 245, 257
Kumar, A., 178–179
Kumar, T. R., 94–95
Kundaje, A., 262, 275–276
Kurpad, A., 228
Kusanovic, J. P., 258
Kusminski, C. M., 48, 244
Kusser, K., 21–22
Kuzma, J. N., 310–311
Kwon, H. H., 19–20
Kyba, M., 48
Kylefjord, H., 4–5

L

Labbé, S. M., 144*t*
Lacelle, S., 207–208
Lafontan, M., 228–229, 230, 237
Lagercrantz, H., 207–208
Laine, H., 145–146
Lam, K. S., 246
Lambadiari, V., 228
Lamprecht, M. R., 59–63
Lander, E. S., 262, 268–269, 276
Landt, S. G., 262, 275–276
Lane, M. D., 3–4
LaNoue, K. F., 94–95
Laque, A., 200–201
Lara-Castro, C., 245
Larmer, J., 172
Larsen, O. A., 229–230
Larson, I., 310–311

Larsson, A., 228–229
Lassen, N. A., 229–230
Lau, F. H., 96–97, 113–118
Laurent, D., 59–63
Laville, M., 96–97, 113–114
Lawrence, E. C., 6
Lazar, M. A., 5, 6–9
Lazzarini, S. J., 200–201
Le, J., 4–5
Le Lay, S., 94–95
Le Liepvre, X., 94–95
Lechel, U., 238–239
Lechertier, T., 76–77
Lee, A. H., 94–95
Lee, F. K., 127
Lee, J. E., 18, 22–23, 94
Lee, K. Y., 6–9
Lee, M. J., 2, 94–95
Lee, S., 162–163, 169, 173
Lee, Y. H., 3, 48, 95–96, 284
Lefebvre, A. M., 96–97, 113–114
Lefrere, I., 94–95
Lefterova, M. I., 261–262, 268–269, 276
Lehman, S., 142
Leibel, R. L., 298, 312
Leinhard, O. D., 149
Lekstrom-Himes, J., 4–5
Leonard, W. R., 178–179
Leonetti, D. L., 96–97, 113–114
Leong, H. X., 20–21
Leung, P. C., 127
Lever, J. D., 202
Levin, B. E., 201, 213, 214
Levin, M., 95–96
Levine, A. S., 202, 203
Levine, B., 200–201
Levine, J. A., 3–4
LFerrante, A. W. Jr., 298, 312
Li, D., 298, 312
Li, G., 277
Li, H., 162–163, 164–165
Li, L., 162–163
Li, M. V., 162
Li, N., 263–265
Li, P., 162–163, 312
Li, W. P., 125–126, 162–163
Liang, H. E., 298, 312

Liao, D., 4–5
Liaw, L., 6
Lidell, M. E., 142, 144t, 145–147
Lieberman-Aiden, E., 277
Lijnen, H. R., 76
Lillie, R. D., 124–125
Lim, J. S., 22–23, 94
Lindberg, K., 3
Lindbjerg, J., 229
Linde, B., 228
Lindmark, S., 59–63
Linka, A., 239
Linn, S., 263–265
Lisanti, M. P., 19–20
Lithell, H., 94–95
Liu, J., 22, 298–299, 310–311
Liu, L., 22–23
Lo, J. C., 48
Lo, M. M., 244–245, 258
Lodish, H. F., 256–257
Loft, A., 261–262, 263–265, 276
Loftus, T. M., 3–4
Londos, C., 19–20
Longair, M., 86
Longo, K. A., 94–95
Lönnroth, P., 145–146, 228–230
Lopes-Virella, M. F., 4–5
Lotinun, S., 128
Louis, G. W., 94–95, 200–201
Low, P. A., 207–208
Lu, M., 312
Lucas, P. C., 94–95
Lumeng, C. N., 3–4, 18–19, 20–21, 22, 298–299, 310–311, 312
Lunati, E., 147–148
Lundberg, P., 149
Lundgren, M., 59–63
Lundin, C., 162
Luo, N., 4–5, 245
Lustgarten, J., 94–95

M

Macdonald, I. A., 228–229, 230, 231, 236, 237
MacDougald, O. A., 3–4, 94–95, 128, 129
Macovski, A., 150–153
Madsen, J., 228–230
Maecker, H. T., 37–38, 304–306

Maeda, K., 4–5
Magre, J., 162
Mai, K., 245
Majka, S. M., 284, 285
Major, E., 32
Makowski, L., 4–5
Manabe, I. I., 18, 20–21, 22–23, 28, 95–96, 298
Mandrup, S., 3–4, 261–262, 263–265
Mangner, T. J., 144t, 146, 147, 178–179
Mannino, E., 284
Maratou, E., 228
March, K. L., 162
Marchand, S., 5–6
Marchesi, G. F., 218
Marcus-Samuels, B., 129
Mariano, P., 277
Marin, D., 152–153, 154–156
Marinov, G. K., 262, 275–276
Marjamäki, P., 145–146, 238–239
Maroni, B. J., 95–96
Marques, M. A., 228–229
Marquet, E., 201
Martin, E., 228–230, 237
Martin, L. F., 94–95
Martin, R. J., 94
Martinez, C., 9
Martinez, R. J., 184–185
Martinez-Santibanez, G., 21–22, 310–311
Marzola, P., 147–148
Mason, M. M., 129
Mathis, D., 18
Matsumoto, D., 32
Matsumoto, Y., 244–245, 258
Matsushita, M., 142
Matsuzaka, T., 3–4
Mattacks, C. A., 96
Mavri-Vavayanni, M., 284
Mayer, A., 76
Mayerhofer, A., 207–208
Mazaki-Tovi, S., 258
McCann, D., 298, 312
McCollough, C. H., 152–153, 154–156
McGarry, J. D., 207–208
McGuire, D. K., 244–245, 258
McKenzie, C. A., 148
McKinley, M. J., 200–201
McMahon, A. P., 5–6

McNelis, J., 298, 312
McNulty, P. H., 145–146
McQuaid, S. E., 228–229
McTernan, P. G., 245
Mefford, I. N., 211
Megens, E., 262, 276
Mellovitz, B., 4–5
Melnyk, A., 209–210
Mendez, J. M., 202–203
Mendoza-Parra, M. A., 263–265
Meng, L., 163
Mepani, R. J., 3–4, 48
Mergian, T., 298–299, 310–311
Messaddeq, N., 5–6
Metzger, D., 5–6
Meunier, P., 135
Meydani, M., 22–23
Meywirth, A., 203
Michael, M. D., 9
Michael, S. K., 5–6
Michailidou, Z., 76
Michelsen, B., 3
Middelbeek, R. J. W., 178–179, 196
Migliorini, R. H., 200–201, 202, 203, 205, 211, 218
Mikkelsen, T. S., 262, 268–269, 276
Miller, E., 76
Miller, H. L., 284, 285
Miller, K. K., 125–126, 128
Miller, M. A., 97–98
Milton, J., 275–276
Misra, M., 125–126, 128
Mitchell, G., 298–299
Mitrou, P., 228
Mitsnefes, M. M., 244–245, 258
Miura, S., 146
Miyao, M., 22–23
Miyata, Y., 76
Miyazaki, O., 257
Miyoshi, H., 298–299
Mizuno, H. H., 18
Moitra, J., 129
Molofsky, A. B., 298, 312
Monia, B. P., 3–4
Moore, S. G., 135
Moquin, A., 21–22
Morbeck, D. E., 94–95
Morgan, M. A., 4–5

Morhard, D., 152–153
Mori, H., 94–95
Mori, M. A., 6–9
Morisada, T., 18–19, 22–23, 76, 302
Morley, J. E., 202, 203
Moro, K., 21–22
Morris, D. L., 21–22, 310–311
Morris, R. J., 228, 235, 236
Morroni, M., 48, 218
Mory, G., 201
Moser, M., 94
Motamed, K., 95–96
Mottillo, E. P., 3, 48
Moulos, P., 262, 276
Mourtzikos, K. A., 146
Mowers, J., 3–4
Mowrer, G., 96–97, 113–118
Moyron-Quiroz, J. E., 21–22
Mozaffarian, D., 94–95
Muccino, D., 5–6
Mukherjee, S., 202
Mulawadi, F. H., 277
Mulder, A. H., 239
Muller, S. P., 238–239
Mullican, S. E., 5, 6–9, 262
Munck, O., 229
Munstun, A., 284
Munzing, W., 238–239
Murakami, K., 284
Murakami, M., 146
Murano, I., 298–299
Murphy, K. T., 202–203
Murray, S. F., 3–4
Murthy, S. N., 9
Muzik, O., 144t, 146, 147, 178–179
Myers, M. G. Jr., 94–95
Myers, R. M., 275–276

N

Nadeau, B., 207–208
Nagajyothi., 94–95
Nagasaki, M. M., 18, 20–21, 22–23, 28, 95–96, 298
Nagy, A., 209
Nagy, T. R., 147–148
Nakahara, M., 179, 184, 189f, 191f, 192f, 194f
Nakatsuji, N., 180

Naoumova, R. P., 148
Nardi, V., 126
Naselli, G., 298
Navarro, P., 9
Naveiras, O., 126
Nayak, K. S., 147–148, 153–154
Nechad, M., 201
Nedergaard, J., 146, 200–201
Neff, H., 217–218
Negrel, R., 94
Nelson, D. K., 207–208
Nelson, R. C., 152–153, 154–156
Nelson, T., 129
Neville, M. J., 228–229, 236
Newell-Morris, L., 96–97, 113–114
Newman, A. B., 96–97, 113–114
Ng, C. K., 145–146
Ng, C. W., 32
Nguyen, K. D., 4–5, 298–299
Nguyen, M. T., 312
Nguyen, N. L., 202–203
Nguyen, P., 200–201
Ni, Y.-H., 284
Nicklas, B. J., 96–97, 113–114
Nicolato, E., 147–148
Nicoloro, S. M., 77
Nicosia, R. F., 76–77
Nielsen, N. B., 236
Nielsen, R., 261–262, 263–265, 276
Nielsen, S. L., 235
Niemi, T., 142, 144t, 145–146
Niijima, A., 202–203
Nio-Kobayashi, J., 142
Nishimura, S., 18, 20–21, 22–23, 28, 95–96, 298
Nishio, M., 179, 184, 189f, 191f, 192f, 194f
Nomura, L. E., 304–306
Noponen, T., 144t, 146–147
Nordenstrom, J., 237
Nordstrom, R., 162
Norman, D., 202
Nowak, D. E., 263
Nozawa, A., 18, 20–21
Nuttall, M. E., 135
Nuutila, P., 144t, 146–147, 228–229
Nygaard, E. B., 178–179, 196

O

Oatmen, K. E., 21–22, 310–311
Oberleithner, H., 172
Obin, M. S., 298
O'Brien, F., 201
O'Brien, T., 3–4
Oden, B., 95–96
Odom, D. T., 276
O'Dowd, J. F., 95–96
Oehlbeck, L. W. F., 124–125
Ogawa, Y., 94–95
Oh da, Y., 298, 312
Oh, N., 22–23, 94
Oh, W., 3–4
Ohisalo, J. J., 94–95
Ohmura, K., 18, 20–21
Ohmura, Y., 18, 20–21
Ohsaki, Y., 19
Ohsugi, M., 18, 20–21, 298
Oikonen, V., 144t, 145–147
Okamatsu-Ogura, Y., 142
Okamoto, Y., 95–96
Oldfield, B. J., 200–201
Olefsky, J. M., 4–5, 298
Olive, M., 129
Olson, P., 4–5
O'Rahilly, S., 162–165, 169–171, 173
Orava, J., 142, 144t, 145–147
O'Regan, D. P., 148
Ormerod, M. G., 37–38
Osafune, K., 184–185
Osborn, O., 4–5
Oshima, K., 76
Osman, O. S., 95–96
Otsu, M., 18
Ouchi, N., 245
Ouellet, V., 144t
Ovadia, S., 298
Ozcan, U., 94–95
Ozdelen, E., 94–95
Ozeki, M. M., 22–23

P

Pajvani, U. B., 244–245, 246
Pakbiers, M. T., 229–230
Palade, G. E., 129
Panagiotakos, D. B., 228
Panus, D. A., 128
Parini, P., 22–23
Park, J. G., 76
Park, T. S., 68
Parkkola, R., 145–146, 238–239
Parlee, S. D., 94–95
Pasarica, M., 76
Paschoalini, M. A., 200–201
Patel, J., 3
Patlak, C. S., 145–146
Pauli, F., 262, 275–276
Paulson, E. K., 152–153, 154–156
Pavlova, Z., 147–148
Payne, V. A., 162
Pedersen, T. A., 262, 276
Peed, L. C., 5, 6–9
Peltoniemi, P., 145–146, 238–239
Penicaud, L., 96–97
Penn, D. M., 201, 207–208, 209
Penninx, B. W., 96–97, 113–114
Pequin, A.-M., 284
Perdikari, A., 9
Perez-Matute, P., 236
Perfield, J. W., 298–299
Perkins, T. G., 147–148
Peroni, O. D., 4–5, 9
Persson, A., 153–154
Petersilka, M., 152–153, 154–156
Petkova, A. P., 3, 48
Petrovic, N., 261–262, 263–265
Pettersson, J., 149
Pettigrew, K. D., 145–146
Phelps, M. E., 144, 145
Phillips, T. G., 3–4
Phipson, B., 298
Phoenix, S., 144t
Pietzsch, T., 86
Pinz, I., 128
Pi-Sunyer, F. X., 48
Plath, K., 10
Platt, K. A., 4–5
Poissonnet, C. M., 94
Polak, J., 236
Pond, C. M., 96
Popper, C. W., 213
Poretsky, L., 94–95
Pouteau, E., 94
Poznanski, W. J., 32

Pramyothin, P., 94–95
Prestwich, T. C., 94–95
Psarra, K., 284
Psilas, J. C., 284, 285
Pugh, B. F., 277

Q
Qatanani, M., 262
Quaade, F., 229–230
Quentin, S. H., 207–208
Quisth, V., 228–229

R
Ragoczy, T., 277
Raitakari, M., 146
Rajala, M. W., 244–245, 246
Rajan, V., 238
Ramanathan, N., 162–165, 169–171, 173
Ramey, E. R., 200–201
Ramos-Vara, J., 97–98
Rangel-Moreno, J., 21–22
Ranvier, L. A., 126
Rasouli, N., 20
Ratcliffe, P. J., 76
Rausch, M. E., 76
Raybould, H. E., 209–210
Reddy, S. N., 209–210
Reddy, V. T., 162–163
Redel, H., 298
Redman, L. M., 76
Reeder, S. B., 149
Regis-Arnaud, A., 125–126
Reid, M. A., 94–95
Reimer, G., 32
Reivich, M., 145–146
Ren, B., 262
Renstrom, F., 59–63
Reue, K., 162–163
Reusch, J. E. B., 289–290
Reynisdottir, S., 94–95
Reynolds, A. P., 277
Rhee, H. S., 277
Ricardo-Gonzalez, R. R., 298, 312
Richard, D., 200–201
Richardson, R. L. R., 22–23
Ricolfi, F., 125–126
Ricquier, D., 201
Riegger, G., 244–245, 258

Riemenschneider, R. W., 127–128
Rietveld, L., 263
Rigamonti, A., 114–118
Riou, J. P., 96–97, 113–114
Roan, L. E., 76
Robert, F., 262
Robertson, G., 262
Robinson, J. P., 37–38
Robinson, S. D., 76–77
Robscheit-Robbins, F. S., 124–125
Rochford, J. J., 162–165, 169–171, 173
Rochlitz, H., 245
Rodbell, M., 299
Rodeheffer, M. S., 32–33, 37, 42, 48–49, 128–129
Rodman, D., 142
Roederer, M., 37–38, 304–306
Rogers, P. M., 5–6
Roh, T. Y., 262
Roller, P. P., 97–98
Roman, A. A., 94–95
Romanelli, A. J., 3–4
Romero, R., 258
Romu, T., 149
Rondinone, C. M., 3–4
Rooks, C. R., 201, 207–208, 209
Rosen, C. J., 128, 129
Rosen, E. D., 1–2
Rosenbaum, M., 298, 312
Rosenwald, M., 9
Ross, S. R., 4–5
Roth, R. H., 212
Rowitch, D. H., 5–6
Rowland, N. E., 202, 203
Ruan, X., 277
Ruge, T., 59–63
Rulicke, T., 9
Ruotsalainen, U., 146–147
Russell, S. J., 6–9
Rutgers, E., 263–265
Rydell, J., 149
Rydén, M., 32
Ryu, V., 202–203

S
Sabo, P. J., 277
Saeki, K., 179, 184, 189f, 191f, 192f, 194f

Sage, E. H., 95–96
Saha, P., 162–163
Saito, M., 142
Saito, Y., 284
Salisbury, S., 244–245, 258
Saltiel, A. R., 18–19, 20–21, 298–299, 310–311, 312
Samei, E., 152–153, 154–156
Samra, J. S., 228–229, 231, 235, 236
Samuel, V. T., 3–4
Sandhu, K. S., 277
Sandstrom, R., 277
Santone, G., 218
Santos, A. N., 283–284
Santosa, S., 94–95
Saris, W. H., 229–230
Sato, Y., 184–185
Saudek, V., 162–165, 169–171, 173
Sbarbati, A., 147–148
Schaedlich, K., 283–284
Scheller, E. L., 94–95, 128, 129
Scherer, P. E., 5, 6–8, 48, 94–95, 244–245, 256–257, 258
Schick, R. R., 76
Schindelin, J., 86
Schlossbauer, S., 244–245, 258
Schmelzer, J. D., 207–208
Schmidt, D., 276
Schmidt, S. F., 261–262, 263–265
Schneider, S. W., 172
Schon, M. R., 96–97, 113–114
Schones, D. E., 262
Schoppee Bortz, P. D., 265–267
Schrader, J., 76
Schraw, T., 245
Schroeder, M., 4–5
Schug, J., 262, 268–269, 276
Schulthess, T., 244–245, 246
Schunkert, H., 244–245, 258
Schupp, M., 262, 268–269, 276
Schwartz, G. J., 200–201, 202–203
Scrable, H., 94–95
Seale, P., 2, 48
Sedlmair, M., 152–153
Segawa, K., 76
Selin, C., 144, 145
Selway, J. L., 95–96
Sembongi, H., 162–165, 169–171, 173

Senard, J. M., 228–229
Sengenes, C., 228–229
Senst, C., 9
Seo, K. K., 22–23, 28
Sereda, O. R., 76
Serrano, R., 9
Seydoux, J., 201
Shackman, J. G., 94–95
Shaikenov, T., 3–4
Shankaranarayanan, P., 263–265
Shaughnessy, S., 3–4
Shaul, M. E., 298
Shi, H., 200–201, 202, 203, 205, 211, 218
Shi, J., 284
Shi, Z., 202–203
Shillingford, J. P., 126
Shimakawa, A., 148
Shinohara, Y., 19
Shipley, P. G., 124–125
Shoelson, S., 18
Shore, L. S., 127–128
Shui, G., 162–163, 164–165
Shyu, H. L., 4–5
Siafaka-Kapadai, A., 284
Sidman, R. L., 203
Siersbæk, M. S., 261–262, 263–265
Siersbæk, R., 261–262, 263–265, 276
Sigurdson, S. L., 209–210
Sim, M. F. M., 162–165, 169–171, 173
Simon, I., 262, 277
Simonsen, L., 228–229, 236, 239
Simpson, E. J., 228–229
Sims, E. A., 96–97, 113–114
Sinal, C. J., 94–95
Singer, K., 310–311
Sinha, M. K., 257
Sion-Vardy, N., 298
Sipilä, H., 146
Sirlin, C. B., 149
Sites, C. K., 96–97, 113–114
Sjolinder, M., 277
Sjostrom, L. G., 95–96, 228
Skorve, J., 96
Skyba, D. M., 239
Slakey, D. P., 9
Sloan, J. H., 257
Small, D., 6

Smedby, Ö., 153–154
Smith, D. L. Jr., 147–148
Smith, E. R., 3–4
Smith, G. P., 275–276
Smith, J. M., 126
Smith, M. L., 147–148
Smith, U., 228–229
Smits, P., 239
Smooker, H. H., 217–218
Smulders, N. M., 142, 146
Smyth, G. K., 298
Snyder, M., 275–276
Sobel, E., 162
Sokoloff, L., 144, 145–146
Solomon, M. J., 262
Son, J. E., 76
Song, C. K., 200–203, 205, 207–208, 209, 211, 212, 218, 220
Songer, T., 257
Sonne, M. P., 236
Sonoda, M., 257
Soriano, P., 6–8, 10
Sotornik, R., 228
Soufer, R., 145–146
Spalding, K. L., 5–6, 32, 114–118
Sparks, L. M., 9
Spencer, M., 20
Spiegelman, B. M., 1–2, 4–5, 9
Spyrou, C., 276
Staalesen, V., 96
Stamatoyannopoulos, J., 275–276
Stancheva, Z., 298–299
Stanford, K. I., 178–179, 196
Stark, K., 244–245, 258
Steenbergen, W., 238
Steger, D. J., 262, 268–269, 276
Steger, R. W., 207–208
Steppan, C. M., 5
Stern, J. S., 94–95
Sternberg, S. R., 109
Stierstorfer, K., 152–153, 154–156
Stocker, C. J., 95–96
Stockert, J. C., 28
Stöcklin, G. L., 145
Stohr, S., 203
Stonestrom, J. P., 150–153
Storch, J., 3–4
Strack, C., 244–245, 258

Straubhaar, J., 77
Strindberg, L., 145–146, 238–239
Strissel, K. J., 298–299
Stuurman, N., 83–86
Such, G., 209
Suemori, H., 180
Suess, C., 152–153
Suga, H., 32
Suganami, T., 94–95
Suh, J. M., 48
Sullivan, A. C., 201
Sullivan, T., 284, 285
Summers, L. K., 228–229, 230, 235, 236, 237
Sun, J., 298
Sun, K., 5, 6–8, 48
Sundaram, A., 5, 6–9
Sung, H. K., 18–19, 22–23, 76, 302
Sung, M. H., 262, 276
Süss, C., 152–153, 154–156
Suzue, K., 193–194, 196
Suzuki, M., 19
Suzuki, N., 163
Suzuki, S., 179, 184, 189f, 191f, 192f, 194f
Svensson, M., 59–63
Swerdlow, H. P., 275–276
Symons, D., 202
Szczepaniak, L. S., 207–208
Szolcsanyi, J., 209–210
Szymanski, K. M., 162–163
Szymczak, A. L., 10

T

Tabata, Y. Y., 22–23
Tache, Y., 209–210
Tack, C. J., 239
Takahashi, A., 146
Takahashi, K., 146
Takakuwa, R., 5–6
Takeuchi, T., 21–22
Tal, I., 142
Talukdar, S., 298, 312
Talukder, M. M. U., 162–165, 169–171, 173
Tamori, Y., 298
Tan, G. D., 236, 298, 312
Tan, G. G., 20–21
Tanabe, M., 21–22

Tanaka, S., 76
Tang, W., 48
Tang, Y. Y., 19–20
Tanowitz, H. B., 94–95
Tateya, S., 298
Tavassoli, M., 127, 128–129
Tavazoie, S., 48, 59–63
Tavoosidana, G., 277
Tavora, B., 76–77
Tchernof, A., 96–97, 113–114
Tchkonia, T., 94–95
Tchoukalova, Y. D., 5–6
Teixeira, V. L., 200–201
Telling, A., 277
Teräs, M., 146–147
Thalamas, C., 228–229
Thembani, E., 125–126
Thomas, B. J., 125–126, 128
Thomas, N. A., 128
Thompson, M., 77
Thureson-Klein, A., 207–208
Tian, B., 263
Tisdale, M. J., 94–95
Tobin, L., 228, 239
Tokorcheck, J., 3–4
Tokuhara, S., 18, 20–21
Tolvanen, T., 145–146
Tomaru, T., 5, 6–9, 262, 268–269, 276
Toncheva, G., 152–153
Tontonoz, P., 4–5
Tornehave, D., 3
Torriani, M., 125–126, 128
Tortoriello, D. V., 76
Tovar, J. P., 147–148
Townsend, K. L., 178–179, 196
Toyomasu, K., 244–245, 258
Tran, K. V., 48
Trevillyan, J. M., 3–4
Trindade, L. M., 263–265
Triscari, J., 201
Trotter, J., 37–38, 304–306
Trubowitz, S., 127–128
Trujillo, M. E., 94–95, 245
Trumpp, A., 6–8
Tsui, H., 20–21
Tsuneyoshi, N., 180
Turban, S., 76
Turer, A. T., 244

Turner, S., 94
Tuteja, G., 262
Tuthill, A., 162

U

Uchimura, N., 244–245, 258
Ueki, K., 9
Unal, R., 20
Unoki, H., 284
Urs, S., 6
Ussar, S., 6–9
Uysal, K. T., 4–5

V

Vaag, A., 236
Vaisbuch, E., 258
Vale, R., 83–86
Valtcheva, N., 8
Valverde, A. M., 9
van Baak, M. A., 228–230
van Berkum, N. L., 277
Van Deursen, J., 94–95
van Dijk, A. P., 239
van Duijnhoven, F. J., 258
van Gaal, L., 96–97, 113–114
van Leeuwen, T. G., 238
Van Maldergem, L., 162
van Marken Lichtenbelt, W. D., 142, 146
Vanhommerig, J. W., 142, 146
Vanin, E. F., 10
Vardhana, P., 76
Varghese, B., 238
Varshavsky, A., 262
Vaughan, C. H., 200–201
Vega, N., 96–97, 113–114
Vega, V., 277
Velkov, S., 277
Verdeguer, F., 3–4
Vernochet, C., 6–9
Vidal, H., 236
Vignali, K. M., 10
Vignon, G., 135
Vikman, H. L., 94–95
Viljanen, T., 144t, 146–147
Villa, M., 147–148
Villanueva, A., 28
Villar, M., 9
Vilmann, P., 229–230

Virtanen, K. A., 142, 144t, 145–146, 238–239
Virtue, S., 162
von Heijne, G., 162
Von Zglinicki, T., 94–95
Vu, A. T., 148

W

Wade, G. N., 200–202, 203
Wagner, K., 162
Wahl, R. L., 146, 153–154
Wahrenberg, H., 94–95
Walia, M., 263–265
Walker, P. M., 125–126
Wallace, P., 245
Walsh, K., 245
Wamhoff, B. R., 265–267
Wang, J., 257
Wang, L., 262, 263–265, 268–269, 276
Wang, P., 277
Wang, Q. A., 5, 6–8
Wang, S., 277
Wang, T., 48, 59–63, 310–311
Wang, X., 3–4, 5, 6–8
Wang, Y., 10, 163, 246
Wang, Z. V., 5, 6–8, 245, 262
Warady, B. A., 244–245, 258
Ward, K. K., 207–208
Warren, W. S., 147–148, 200–201
Watanabe, K., 3–4, 142
Watson, A., 200–201
Watts, L., 3–4
Webster, J. D., 97–98
Wegewitz, U., 245
Wei, K., 239
Weickert, M. O., 245
Weisberg, S. P., 76, 298, 312
Weiser-Evans, M., 284, 285
Wek, S. A., 19–20
Wentworth, J. M., 298
Wenzel, P. L., 126
Wesseling, J., 263–265
Westcott, D. J., 18–19, 20, 298–299, 310–311, 312
Westergren, M., 142, 144t, 145–146
Westermark, P. O., 5–6, 32, 114–118
Westphal, S., 76
Whipple, G. H., 124–125

White, N. M., 3–4
Whiting, J., 97–98
Wiesner, R. J., 95–96, 284
Wigström, L., 153–154
Wiklund, U., 228
Williams, G., 142
Williams, L., 277
Williams, S., 256–257
Wilson, L. K., 95–96
Wilson, M. D., 276
Windpassinger, C., 162
Wolfrum, C., 9
Wolinski, H., 162–163
Woo, J., 127
Wood, M. J., 3–4
Wool, I. B., 200–201
Workman, C. J., 10
Wree, A., 76
Wright, C. M., 5
Wright, W. S., 94–95
Wronski, T. J., 126
Wu, D., 298, 312
Wu, J., 2, 9
Wu, P., 20–21
Wu, Y., 2
Wutz, A., 10
Wylezinska-Arridge, M., 148
Wynshaw-Boris, A., 4–5
Wyrick, J. J., 262

X

Xi, Y., 129
Xiao, Q., 257
Xinyu, W., 244–245, 258
Xiong, X. Q., 202–203
Xu, A., 246
Xu, G., 163
Xu, H. H., 20–21, 277, 298, 312
Xu, J., 298, 312
Xu, Z., 262, 268–269, 276

Y

Yago, H., 257
Yahagi, N., 3–4
Yale, P., 228–230, 237
Yamada, T., 21–22
Yamamoto, T., 3–4
Yamaoki, Y., 22–23

Yamashita, H. H., 18, 20–21, 22–23, 28, 95–96, 298
Yamauchi, T., 257
Yang, D. D., 20–21, 298, 312
Yang, H., 209–210
Yang, Q. Q., 20–21, 298, 312
Yao-Borengasser, A., 20
Yap, S., 68
Yasuchika, K., 180
Yau, M. H., 246
Ye, J., 5–6
Ye, L., 9
Yechoor, V. K., 162
Yeung, D. K., 127
Yilmaz, E., 94–95
Yin, L., 147–148
Yoneda, M., 209–210
Yoneshiro, T., 142, 179, 184, 189f, 191f, 192f, 194f
Yoshizumi, T. T., 152–153
Youngstrom, T. G., 200–201, 203, 218
Yu, H., 148
Yu, S., 3–4, 5, 6–8
Yu, X. X., 3–4

Z

Zachrisson, H., 153–154
Zaror-Behrens, G., 201, 203, 207–208, 209–210
Zatz, L. M., 150–153
Zeng, T., 262
Zeve, D., 48
Zhang, F., 202–203
Zhang, J., 298
Zhang, M., 284
Zhang, W., 207–208
Zhang, X., 94–95, 262, 268–269, 276
Zhang, Y., 162–163, 164–165, 200–201, 276
Zhang, Z., 277
Zhao, Y., 262
Zhao, Z., 277
Zheng, M., 277
Zhou, Y. B., 202–203
Zhu, B., 20
Zhu, G.-Z., 284
Zhu, M. M., 18
Zhuo, D., 262
Ziegler, S. I., 238–239
Zingaretti, M. C., 18–19, 48, 53
Zong, H. H., 19–20
Zorzi, P., 76–77
Zou, X., 76
Zucker, L. M., 59–63
Zuk, P. A. P., 18
Zuker, M., 203
Zvonic, S., 135
Zwart, W., 263–265

SUBJECT INDEX

Note: Page numbers followed by "*f*" indicate figures and "*t*" indicate tables.

A

Adipocytes
  ChIP-seq (*see* Chromatin immunoprecipitation (ChIP))
  developmental stages, 3
  embryonic stem cell, 10–11
  flow cytometry (*see* Flow cytometry)
  hypertrophy, 94–95
  internal ribosomal entry site, 10
  knocking out genes
    adipose-specific recombination, 5–6
    brown and beige fat, 9
    fat pad, 3–4
    recombination efficiency, 8–9
    tissue specificity, 6–8
  methods, 11–13
  precursor cells
    and cell populations, 33
    cell surface marker, 32–33
    childhood and adolescence, 32
    DMEM media, 41, 42
    flow cytometry and FACS, 36–41
    intact adipocytes, 36, 44*f*
    lipid ghosts, 44–45, 44*f*
    Oil Red O lipid staining, 42–43
    stromal vascular fraction, 32
    SVF separation, 33–36
  ROSA26 locus, 10
  WAT, histomorphometry of (*see* White adipose tissue (WAT))
Adiponectin
  data presentation, 254–256
  FPLC gel fractionation
    customization and optimization, 250–251
    equilibration and regeneration step, 248–250, 249*t*
    flow rate and volume, 248, 249*t*
    improper elution collection setting, 248, 249*f*
    lower sample loading volumes, 248, 249*f*
    separation system, 248
    solutions, 248
    technical procedure, 248–250
    three primary complexes, 248, 249*f*
  HMW ELISAs, 257
  quantification, 254
  sample collection and preparation, 246
  SDS-PAGE fraction analysis, 256–257
  systemic metabolic health, 244
  western blot analysis (*see* Western blot analysis)
Adipose tissue
  angiogenesis assay (*see* Angiogenesis)
  brown adipose tissue (*see* Brown adipose tissue (BAT))
  cellular heterogeneity, 2–3
  paraffin-sectioned adipose tissue (*see* Paraffin-sectioned adipose tissue)
  role of, 1–2
  sectioned adipose tissue
    frozen sectioning, 68–72
    paraffin-sectioned adipose tissue, 53–68
  white adipose tissue (*see* White adipose tissue (WAT))
  whole mounted adipose tissue (*see* Whole mounted adipose tissue)
Adipose tissue blood flow (ATBF)
  ATBF values, 236
  baseline and fasting ATBF, 236–237
  complex metabolic and hormonal roles, 228
  consumable materials
    inset II infusion set, 232, 234*f*
    quick-set infusion set, 232, 233*f*
    syringes, 232
    ^{133}Xenon, 232
  contrast-enhanced ultrasound, 239
  Fick's equation, 228–229

Adipose tissue blood flow (ATBF) (*Continued*)
  laser Doppler flowmetry, 238
  microdialysis technique, 237
  microinfusion, 230
  positron emission tomography, 238–239
  procedure
    agent preparation, 232–233
    method and drawbacks, 236
    microinfusion protocol, 234–235
    subjects preparation, 233–234
    ^{133}Xe preparation, 232
  regulatory factors, 228–229, 229t
  stimulated (postglucose) ATBF, 237
  technical equipments
    microinfusion pumps, 232
    scintillation detectors, 231, 231f
  ^{133}Xenon washout techniques, 229–230
Adipose tissue macrophages (ATM), 20–21
  data analysis
    methods, 310, 311f
    software, 302
  FACS
    cell sorter equipment, 301
    FACS buffer, 302
    methods, 307–309
  flow cytometry analysis
    compensation procedures, 306–307
    data acquisition, 307
    gating strategy, 307, 308f
    single stained and fluorescence minus one controls, 301, 301t
  SVC (*see* Stromal vascular cells (SVC))
AFM. *See* Atomic force microscopy (AFM)
Alpha-methyl-*para*-tyrosine (AMPT) method, 212–214
Angiogenesis
  capillary growth area
    plugin, 83–86
    Zeiss Axio Observer Z1 microscope, 83–86
  capillary sprouts, 82–83
  embedding procedure, 80
  immunofluorescence analysis, 88–89
  isolated cell analysis, 89–90
  materials
    medium, instruments, and culture dishes, 77–78
    sample collection, 78–79
  method limitations, 90
  pro-angiogenic factors, 76
  sample preparation
    human adipose tissue, 79
    mouse adipose tissue, 79
  tissue vascularization, 76–77
ATBF. *See* Adipose tissue blood flow (ATBF)
Atomic force microscopy (AFM)
  advantages, 169–171
  contact mode and tapping-mode, 169–171
  definition, 169–171
  procedure, 171–173

B

BAT. *See* Brown adipose tissue (BAT)
BD FACSDiva™ software, 41
Bimolecular fluorescence complementation (BiFC)
  FRET, 166
  procedure, 167–169
  3T3-L1 preadipocytes, 166, 167f
  YFP, 166
Bioruptor® Twin, 265–267
Bone marrow adipose tissue. *See* Marrow adipose tissue (MAT)
Brown adipose tissue (BAT), 48
  chemical denervation
    intra-fat injections, 207–208
    6-OHDA, 207–208
    requirements, 208–209
  curved serrated forceps, 204
  data expression, 220
  dual energy computed tomography
    attenuation measurement, 150–152
    characterization, 153–154
    CT scanner protocol, 154–156
    definition, 150–152
    material decomposition, 152–153
    post processing, 156
  energy expenditure, 142, 143
  and ^{18}F-FDG uptake activity, 142
  hESC/hiPSC (*see* Human induced pluripotent stem cells/human embryonic stem cells (hESC/hiPSC))

local sensory denervation, 209–211
magnetic resonance imaging
  fat–water ratio, 147–148
  image acquisition, 148–149
  image calibration, 149
  intensity inhomogeneity correction, 149
  lipid fraction, 147–148
  multiecho chemical shift imaging, 148
  phase-sensitive reconstruction, 149
  visualization, 149–150
origin, 200–201
physiological response, 201
positron emission tomography
  hBAT perfusion, 147
  metabolic imaging, 144–146
  modeling principles, 144–146
  semiquantitative methods, 146–147
  uncoupling protein 1, 143–147
razor/depilatory cream, 204
sensory denervation, CGRP ELIA
  lyophilizer, 220
  requirements, 218–219
  tissue homogenization, 219
surgical denervation
  CGRP, 202–203
  EWAT, 205, 206f
  IBAT, 205, 205f
  identification of, 203
  IWAT, 205, 206f
  norepinephrine (NE) content, 202–203
  RWAT, 205
  sympathectomy, 202
  sympathetic activity assessment, 202–203
sympathetic denervation, NETO
  AMPT method, 212–214
  HPLC-EC, 217–218
  sample preparation, 214–217
tissue forceps, 204
uncoupling protein 1, 143
vicryl suture, 204

C

Calcitonin gene-related peptide (CGRP), 202–203
Chromatin immunoprecipitation (ChIP), 267–268
  (C/EBP)β and C/EBPδ, 262–263
  beads preparation, 267–268
  cross-linked chromatin, 263–265
  DNA purification, 267–268
  downstream data analyses, 266f, 276
  fragmentation, 265–267, 266f
  future aspects, 277
  illumina sequencing library preparation, 275–276
    adaptor ligation, 272–273
    end repair, 269–270
    3′-ends adenylation, 271–272
    PCR amplification, 273–274
    quality control, 274–275, 275f
    quantification, 275
    size selection, 270–271
  qPCR validation, 267–269
  quality and reproducible reaction, 262–263
Confocal microscopy
  adipocyte morphology, 19–20
  advantages, 23–24
  ATM and CLSs, 20–21
  blocking step, 26
  capillaries and blood vessels, 22–23
  cellular and noncellular components, 18
  3D reconstructions, 28
  fat-associated lymphoid clusters, 21–22
  materials, 24–25
  milky spots, 21–22
  nuclei and lipids staining, 27
  perfusion, collection, and fixation methods, 25–26
  primary antibody incubation, 26–27
  secondary antibody incubation, 27
Constitutive marrow adipose tissue (cMAT), 127
Contrast-enhanced ultrasound (CEU), 239
Crown-like structures (CLSs), 20–21
Cushing's syndrome, 94–95

D

3′3-Diaminobenzidine (DAB), 64–66
Dual energy computed tomography (DECT), 147
  attenuation measurement, 150–152
  characterization, 153–154
  CT scanner protocol, 154–156

Dual energy computed tomography
(DECT) (*Continued*)
definition, 150–152
material decomposition, 152–153
post processing, 156
DyeCycle Violet (DCV) staining, 290–291
Dynamic Data Exchange (DDE), 105

E

Enzyme-linked immunosorbent assays
(ELISA)
adiponectin complexes, 257
insulin concentrations, 193–194
Epididymal white adipose tissue (eWAT),
113–114, 115*f*, 116*f*, 203,
205, 206*f*

F

Fast protein liquid chromatography (FPLC),
245
equilibration and regeneration step,
248–250, 249*t*
flow rate and volume, 248, 249*t*
improper elution collection setting, 248,
249*f*
lower sample loading volumes, 248, 249*f*
separation system, 248
solutions, 248
technical procedure, 248–250
three primary complexes, 248, 249*f*
Fat-associated lymphoid clusters (FALCs),
21–22, 22*f*, 26, 26*t*
Fatty acids (FAs)
saturated, 127–128
unsaturated, 127
Flow-assisted cell sorting (FACS),
301–302
Flow cytometry
antibody staining, 36–37
ATM
compensation procedures, 306–307
data acquisition, 307
gating strategy, 307, 308*f*
SS and FMO controls, 301, 301*t*
cells collection and aggregates, 285
cell sorters, 39
cells survive, 283–284, 284*f*
cell suspensions, staining of, 288*f*

DyeCycle Violet staining, 290–291
fluorescent antibody staining, 290
LipidTOX neutral lipid stain, 289
collagenase digestion, 285
compensation errors, 37–38
data acquisition, 39
flow cytometry (*see* Flow cytometry)
FMO control, 38–39
FSC *vs*. SSC gating, 288*f*, 292–293
isolation strategy, 288*f*, 294
lipid droplets, 288*f*, 294
minimal markers, 39
MoFlo XDP settings, 288*f*, 292
obesity and metabolic diseases, 285
preparation, 286–287
single cells separation, 288*f*, 293
software analysis, 40–41
sorters, 282
stromal/vascular cells, 285, 288*f*, 294
unilocular and multilocular adipocytes,
282, 283*f*
FlowJO software, 40–41
Fluorescent-activated cell sorting (FACS)
and flow cytometry, 37–39
WAT depots, 41–42
Fluorescent-minus-one (FMO) control,
38–39
Forward scatter (FSC), 288*f*, 292–293
FPLC. *See* Fast protein liquid
chromatography (FPLC)
Frozen-sectioned adipose tissue
IHC, fluorochrome-conjugated
antibodies, 70–72
slide preparation, 68–72

H

High molecular weight (HMW) complexes,
244–245
Horseradish peroxidase (HRP), 63–68
Human induced pluripotent stem cells/
human embryonic stem cells (hESC/
hiPSC)
adherent culture, 190
basal medium, preparation of, 188
calorigenic assays, 192–193
culture medium, 186
dissociation liquid, 186
frozen stocks, 186

material
    BA differentiation media and culture vessels, 181
    calorigenic analyses, 183
    freezing, 181
    liquid dissociation, 180
    mouse embryonic fibroblasts (see Mouse embryonic fibroblasts (MEFs))
    Oil red O staining, 182
    oral fat tolerance tests, 183
    oral glucose tolerance tests, 183
    oxygen consumption analyses, 182–183
  oral fat tolerance tests, 194–195
  oral glucose tolerance tests, 193–194
  oxygen consumption analyses, 191–192
  sphere formation, 189–190
  supplements, 188, 189
Human lipodystrophy protein seipin. See Lipodystrophy
Hypogonadism, 94–95

I

Illumina® sequencing, 269
Immunofluorescence analysis
  adipose tissue angiogenesis, 88–89
  frozen sectioning, 68
Immunohistochemistry (IHC)
  fluorochrome-conjugated antibodies, 70–72
  horseradish peroxidase substrates, 70–72
  whole mounted adipose tissue, 48–49
Immunoprecipitation
  ChIP (see Chromatin immunoprecipitation (ChIP))
  and western blotting
    endogenous murine seipin, 163
    HEK293 cells, 163–164, 164f
    procedure, 165–166
    SDS-PAGE and analysis, 164–165, 164f
Inguinal white adipose tissue (iWAT), 113–114, 115f
Internal ribosomal entry site (IRES), 10
Interscapular BAT (IBAT), 203, 205, 205f

L

Laser Doppler flowmetry (LDF), 238
LipidTOX neutral lipid stain, 289
Lipodystrophy
  atomic force microscopy
    advantages, 169–171
    contact mode and tapping-mode, 169–171
    definition, 169–171
    procedure, 171–173
  bimolecular fluorescence complementation
    FRET, 166
    procedure, 167–169
    3T3-L1 preadipocytes, 166, 167f
    YFP, 166
  BSCL2 gene, 162
  immunoprecipitation and western blotting
    endogenous murine seipin, 163
    HEK293 cells, 163–164, 164f
    procedure, 165–166
    SDS-PAGE and analysis, 164–165, 164f
  N-terminus, 162
  PA phosphatase lipin 1, 162–163

M

Magnetic resonance imaging (MRI), 238
  fat–water ratio, 147–148
  image acquisition, 148–149
  image calibration, 149
  intensity inhomogeneity correction, 149
  lipid fraction, 147–148
  multiecho chemical shift imaging, 148
  phase-sensitive reconstruction, 149
  visualization, 149–150
Marrow adipose tissue (MAT), 137
  accumulation and distribution, 126–127
  cMAT, 127
  historical finding, 126
  materials
    decalcification solution, 130
    dissecting tools, 130
    glassware and plastic ware, 130
    mice, 129–130

Marrow adipose tissue (MAT) (*Continued*)
   osmium tetroxide staining solution, 130–131
   in medullary canal, 125f
   metabolic function, 127–128
   methods
      bone decalcification, 132–133
      bone fixation, 132
      microcomputed tomography (*see* Microcomputed tomography (Micro-CT))
      mouse dissection, 131–132
      mouse long bones, isolation of, 131–132
      osmium staining, decalcified bones, 133–134
   origin of, 128–129
   PFAS, 127
   quantitation of, 125–126
   rMAT, 127
   unsaturated fatty acids, 127
MEFs. *See* Mouse embryonic fibroblasts (MEFs)
Microcomputed tomography (Micro-CT)
   C57BL/6, 135f, 136f
   C3H/HeJ, 135f
   Ebf1 null tibia, 134f
   mouse femur, 136f
   segmentation, 134–135
   volumes of interest, 134–135, 136f
Microdialysis technique, 237
Mitomycin C (MMC), 195
Mouse embryonic fibroblasts (MEFs), 179–180
   cell thaw process, 187
   culture medium, 185
   gelatin-coated dishes, 187
   passage of, 187
   preparation, 185–186, 187

O

Oil Red O lipid staining
   adipocyte precursor cells, 42–43
   hESC/hiPSC, 182, 191
Osmium tetroxide
   cellular lipid content, 95–96
   MAT (*see* Marrow adipose tissue (MAT))

P

Paraffin-sectioned adipose tissue
   adipocyte size analysis, cell profiler, 59–63
   H&E staining, 58–59
   IHC, horseradish peroxidase substrates
      3′3-Diaminobenzidine, 64–66
      tyramide signal amplification, 66–68
   slide preparation
      deparaffinization and rehydration, 57
      paraffin embedding, 56
      sectioning, 56
      tissue processing, 55
Performic acid-Schiff (PFAS), 127
PET. *See* Positron emission tomography (PET)
Positron emission tomography (PET)
   ATBF, 238–239
   brown adipose tissue
      hBAT perfusion, 147
      metabolic imaging, 144–146
      modeling principles, 144–146
      semiquantitative methods, 146–147
      uncoupling protein 1, 143–147
Preadipocytes, 3–4, 32–33

Q

Qubit®, 268

R

Regulated marrow adipose tissue (rMAT), 127
Retroperitoneal WAT (RWAT), 203, 205

S

Sectioned adipose tissue
   frozen sectioning
      IHC, fluorochrome-conjugated antibodies, 70–72
      slide preparation, 68–72
   paraffin-sectioned adipose tissue, 53–68
      adipocyte size analysis, cell profiler, 59–63
      H&E staining, 58–59
      IHC, horseradish peroxidase substrates, 70–72
      slide preparation, 68–70
Side scatter (SSC), 288f, 292–293

Stromal vascular cells (SVC)
  cell surface marker, staining
    fluorochrome-labeled antibody, 300, 300t
    methods, 301t, 304–306
    PFA, 301
    viability dye, 301
  isolation and preparation
    adipose tissue, 302–303
    centrifuge, 300
    collagenase digestion, 303–304
    digestion buffer, 300
    FACS buffer, 300
    hemocytometer, 300
    incubator set, 300
    lysis buffer, 300
    sterile/ethanol-cleaned surgical instruments, 299
    test tube rocker/mixer, 300
  magnetic beads and positive selection
    CD11b microbeads, 302
    MACS buffer, 302
    methods, 309–310, 309f
    polystyrene round-bottom test tubes, 302
Stromal vascular fraction (SVF), 32
  cell populations, 33
  WAT digestion, 33–36
SVC. See Stromal vascular cells (SVC)
SX-8G IPstar®, 267–268

T
Tyramide signal amplification (TSA), 63, 66–68

U
Uncoupling protein 1 (UCP1)
  brown adipose tissue, 143
  positron emission tomography, 143

V
Volumes of interest (VOI), 134–135, 136f

W
WAT. See White adipose tissue (WAT)
Western blot analysis
  adiponectin
    immunoblot analysis, 252

SDS-PAGE, 252
and immunoprecipitation
  endogenous murine seipin, 163
  HEK293 cells, 163–164, 164f
  procedure, 165–166
  SDS-PAGE and analysis, 164–165, 164f
White adipose tissue (WAT), 48
  adipocyte precursor cells (see Adipocytes)
  C57BL6/J mice, 116f, 118f
  cellular and noncellular components, 18
  chemical denervation
    intra-fat injections, 207–208
    6-OHDA, 207–208
    requirements, 208–209
  confocal microscopy
    adipocyte morphology, 19–20
    advantages, 23–24
    ATM and CLSs, 20–21
    blocking step, 26
    capillaries and blood vessels, 22–23
    3D reconstructions, 28
    fat-associated lymphoid clusters, 21–22
    materials, 24–25
    milky spots, 21–22
    nuclei and lipids staining, 27
    perfusion, collection, and fixation methods, 25–26
    primary antibody incubation, 26–27
    secondary antibody incubation, 27
  curved serrated forceps, 204
  data analysis, 111–112, 112f, 113f
  data expression, 220
  epididymal WAT, 113–114, 115f, 116f
  ImageJ image analysis software, 108–111
    adipocyte labelling, 110
    background subtraction, 109
    binary format, image conversion, 110
    border enhancement, 110
    calibration, 108
    duplicate image, 108
    excess noise, removal of, 109
    image threshold, 109
    open base image, 108
  inguinal WAT, 113–114, 115f
  local sensory denervation, 209–211
  MetaMorph image analysis software
    adding/removing objects, 105

White adipose tissue (WAT) (Continued)
    adipocyte labelling, 104
    adipocyte modification, 105
    adipocytes separation, 103
    border enhancement, 102
    calibration, 101
    data logging, 105, 107
    identification, 103
    image threshold, 102
    open base image, 101
    saving regions, 106
    screen shot, 107
    semiautomated custom image analysis, 100
    steps, 103
    unique adipocyte separation, 103
    unwanted regions, removal of, 107
  morphology and function, 94–95
  origin, 200–201
  physiological response, 201
  razor/depilatory cream, 204
  remodeling, 94–95
  results and experimental guidance, 113–114
  sensory denervation, CGRP ELIA
    lyophilizer, 220
    requirements, 218–219
    tissue homogenization, 219
  surgical denervation
    CGRP, 202–203
    EWAT, 205, 206f
    IBAT, 205, 205f
    identification of, 203
    IWAT, 205, 206f
    norepinephrine (NE) content, 202–203
    RWAT, 205
    sympathectomy, 202
    sympathetic activity assessment, 202–203
  sympathetic denervation, NETO
    AMPT method, 212–214
    HPLC-EC, 217–218
    sample preparation, 214–217
  tissue forceps, 204
  tissue preparation
    expression of adiponectin, 96–97, 97f
    sectioning, deparaffinizing, and staining, 98–100
    sFRP1, 96–97, 97f
    standard necropsy procedure, 97–98
    tissue embedding, 98
  vicryl suture, 204
Whole mounted adipose tissue
  advantage and disadvantage, 48–49
  confocal imaging, 51–53
  immunohistochemistry, 48–49
  slide preparation, 49–50

X

^{133}Xenon washout techniques, 229–230

Z

Zeiss Axio Observer Z1 microscope, 83–86
Zeiss microscope, 100

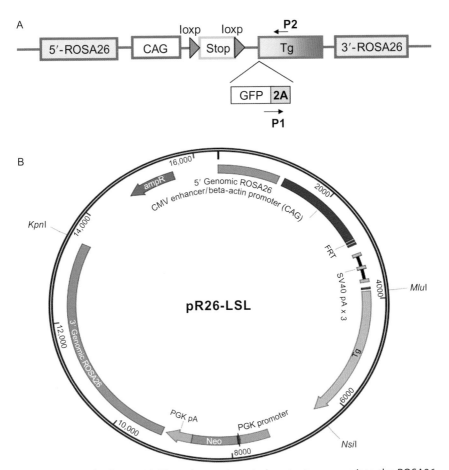

Sona Kang et al., Figure 1.1 Targeting strategy to insert a transgene into the ROSA26 locus. (A) Schematic depicting an example of a targeting construct. ROSA26 targeting arms surround a CAG promoter, a Stop cassette flanked by loxP sites, and a transgene of interest which may be attached to a reporter by a self-cleaving 2A peptide, (B) another view of pR26-LSL showing the entire plasmid.

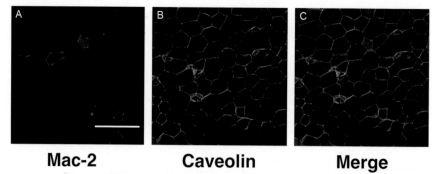

Gabriel Martinez-Santibañez *et al.*, **Figure 2.1** Crown-like structures in white adipose tissue. Gonadal fat pads from a high fat diet fed C57Bl/6 mouse were fixed, isolated, and stained as above. Macrophages stain Mac-2 in red (A), Caveolin-1 plasma membrane in green (B), and images were merged (C). Scale bar = 350 μm.

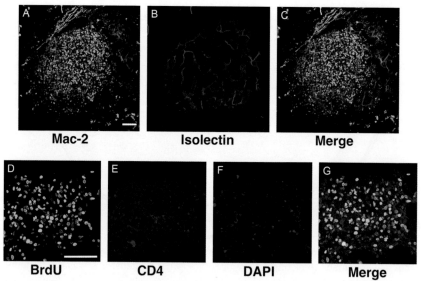

Gabriel Martinez-Santibañez *et al.*, **Figure 2.2** Milky spots and FALCS in adipose tissue. Omental fat pads from a C57Bl/6 mouse was fixed and stained and surface imaged to identify milky spots on the fat pad surface. Macrophages (Mac-2) are shown in green (A) and vasculature (isolectin) shown in red (B) with merged image (C). Gonadal fat pads were stained and imaged for surface collections of monocytes to identify FALCS. Proliferating cells are stained for BrdU in green (D), $CD4^+$ T cells are shown in red (E), nuclei are stained with DAPI in blue (F), and all three channels shown merged (G). Scale bar = 100 μm.

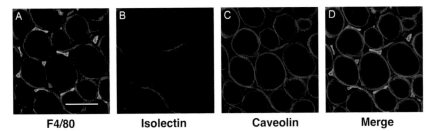

Gabriel Martinez-Santibañez *et al.*, **Figure 2.3** Adipose tissue vasculature. Gonadal fat pads from C57Bl/6 mice fed a chow diet. F4/80$^+$ macrophages are stained in green (A), vasculature (isolectin) in red (B), Caveolin-1 denotes the adipocyte plasma membrane in blue (C), images merged in (D). Scale bar = 100 μm.

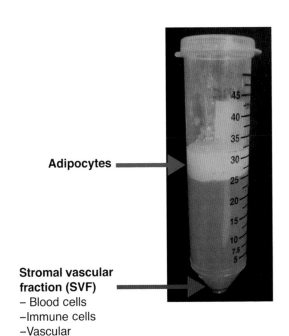

Adipocytes

Stromal vascular fraction (SVF)
- Blood cells
- Immune cells
- Vascular
- Adipocyte precursors

Christopher D. Church *et al.*, **Figure 3.1** Separation of SVF from adipocytes.

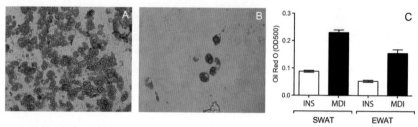

Christopher D. Church et al., **Figure 3.4** Differentiation of primary adipocyte precursor cells. (A) Day 7 Subcutaneous WAT (SWAT), (B) Day 7 Epigonadal WAT (EWAT), (C) Day 7 Oil Red O quantification. INS, insulin-only differentiation; MDI (adipogenic cocktail containing IBMX, dexamethasone, and insulin).

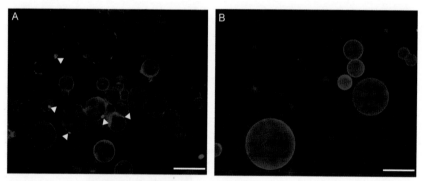

Christopher D. Church et al., **Figure 3.5** Isolation of adipocytes. (A) Intact adipocytes with single DAPI positive nuclei within a cell membrane. (B) Lipid ghosts without DAPI positive nuclei. White scale bar represents 100 μm.

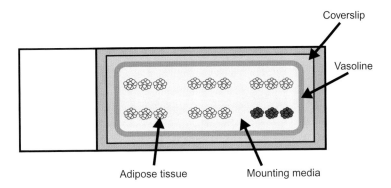
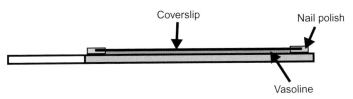

Ryan Berry *et al.*, Figure 4.1 A depiction of a slide prepared for imaging of adipose tissue in whole mount.

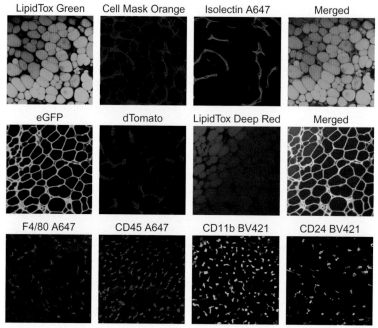

Ryan Berry et al., Figure 4.2 Example images of whole mounted fluorochrome-containing WAT acquired through confocal microscopy. WAT was isolated from fluorescent reporter mice (eGFP and dTomato) and/or incubated in fluorescent stains/antibodies to label lipids (LipidTox), plasma membranes (Cell Mask Orange), endothelial cells (Isolectin GS-IB$_4$), macrophages (F4/80 and CD11b), nonspecific blood lineage cells (CD45), B-cells, and adipocyte progenitor cells (CD24) as described in Table 4.1. Images were acquired with the appropriate laser/filter settings listed in Table 4.1 and laser voltage/gain settings determined by FMO controls.

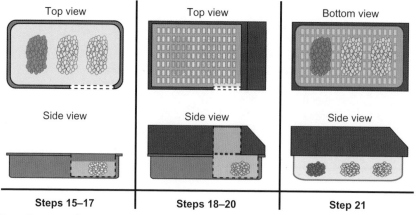

Ryan Berry et al., Figure 4.3 A depiction of paraffin embedding of adipose tissue.

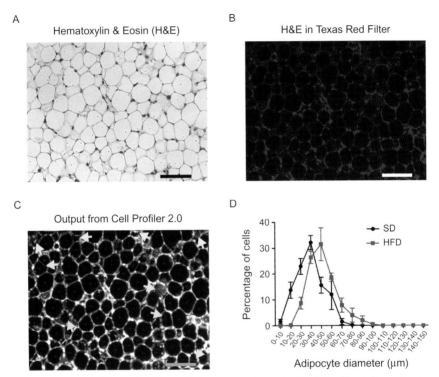

Ryan Berry et al., Figure 4.4 Example images and graph of Cell Profiler Adipocyte size analysis. (A) Bright-field image of H&E-stained paraffin-sectioned WAT. (B) Fluorescent image of H&E-stained paraffin-sectioned WAT using the Texas Red filter cube. (C) Image from (B) outputted from Cell Profiler 2.0 following adipocyte pixel area analysis. Yellow arrows indicate improperly gated cells that must be manually excluded from the dataset. (D) Adipocyte cell size distribution in WAT depots isolated from standard- (SD) and high-fat-diet (HFD) fed mice following Cell Profiler 2.0 analysis and conversion of measured adipocyte pixel area to adipocyte diameter as described in Section 3.1.3.

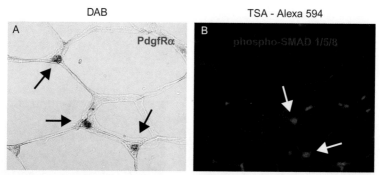

Ryan Berry et al., Figure 4.5 Immunohistochemistry on paraffin-sectioned WAT. (A) HRP-DAB-mediated immunolabeling of PdgfRα in paraffin-sectioned WAT. Black arrows indicate PdgfRα+ cells marked by deposition of oxidized DAB, which is brown in color. (B) HRP-TSA-mediated immunolabeling of phospho-SMADs 1,5,8 in paraffin-sectioned WAT. White arrows indicate phospho-SMAD+ cells marked by fluorescence of oxidized TSA-A594.

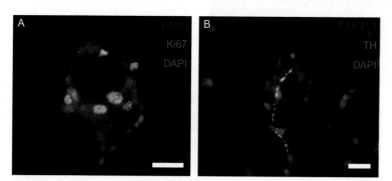

Ryan Berry et al., Figure 4.6 Immunofluorescent staining of frozen-sectioned WAT. (A) A classic crown-like structure stained with antibodies for F4/80 and Ki67 and counterstained with DAPI. (B) Sympathetic innervation of a beige adipocyte as shown through staining with antibodies for UCP-1 and the sympathetic nerve fiber marker tyrosine hydroxylase (TH) and counterstained with DAPI. Scale bar is 20 μm.

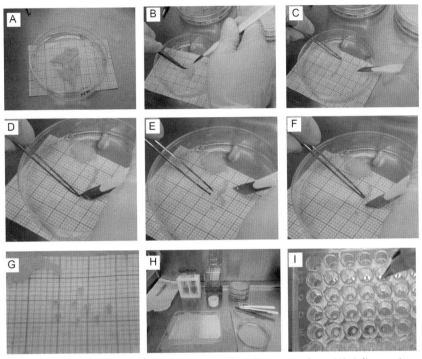

Raziel Rojas-Rodriguez *et al.*, Figure 5.1 Embedding procedure. (A) Adipose tissue samples placed in a 100 cm petri dish containing 25 ml of EGM-2 MV medium. The millimeter (mm) paper placed under the petri dish is used as a size reference. (B) Sample of adipose tissue in plate #2 containing 15 ml of EGM-2 MV medium. The scalpel and forceps are used to hold the fat and cut it into strips. (C) Piece of fat strip cut from the adipose tissue sample. Using the millimeter paper reference, the fat strip is aligned in order to cut the appropriate size of each slice (explant). (D) For the first cut, it is easier to start at one of the ends of the adipose tissue strip. The forceps are used to hold the fat while the scalpel is used to cut the slice. (E) The explant is aligned with one of the quadrants in the millimeter paper to verify adequate size. (F) The rest of the strip is cut into slices. The adipose tissue is held by forceps and the cut is done by the scalpel. While handling the forceps, avoid pulling or stretching the fat, since it may damage the tissue. (G) Individual slices cut to appropriate size and verified with the millimeter paper. (H). Display of workstation in the biocabinet before starting the embedding procedure. Explants were transferred to plate #3, containing 25 ml of EGM-2 MV medium. 96-multiwell plate is kept in a tray filled with ice for the embedding steps. (I) Embedding step. After the Matrigel is dispensed, forceps are used to place the explants, one per well. The explant is positioned at the center of the well.

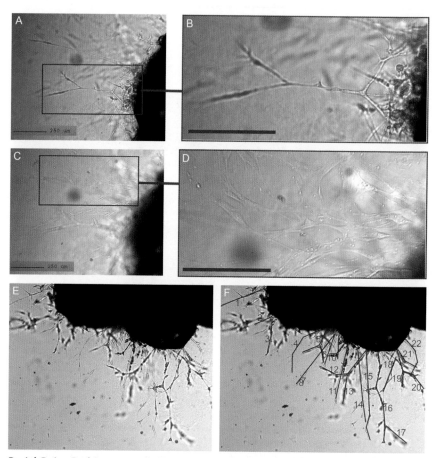

Raziel Rojas-Rodriguez et al., Figure 5.2 Cells emerging from mouse adipose tissue explant. (A, B) Capillary sprout emerging from embedded mouse explant, displaying characteristic linear branching structure. (C, D) Focus set to the surface of the well, where fibroblastic adherent cells can be seen emerging from the explant, observed at a different optical plane of the image. (E) Phase contrast image of the explant and the capillary sprouts 14 days postembedding. (F) Structures shown in red highlight formations that can be considered to be sprouts.

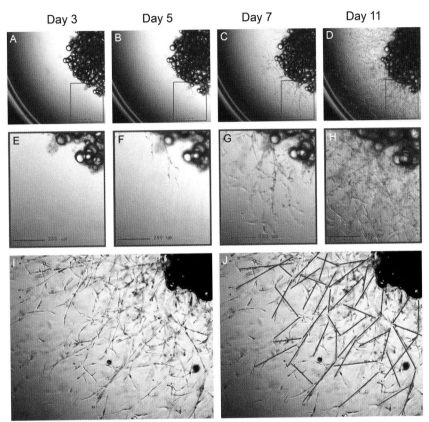

Raziel Rojas-Rodriguez *et al.*, Figure 5.5 Capillary sprouting from human adipose tissue. A human explant from subcutaneous adipose tissue at days 3 (A, B), 5 (C, D), 7 (E, F), and 11 (G, H) postembedding. Capillary sprouting begins to be observed at day 5. After day 11, the growth is highly increased (I, J), making difficult to identify all sprout formation.

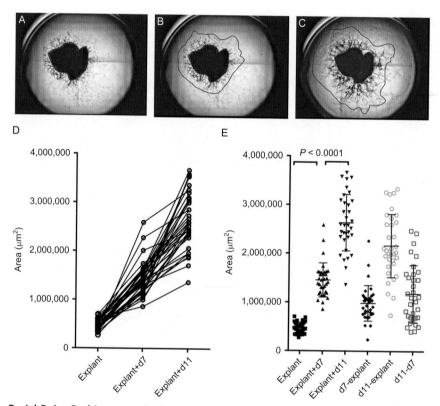

Raziel Rojas-Rodriguez *et al.*, **Figure 5.6** Digital analysis of capillary growth area. An example of montages generated from bright field images of quadrants of a single well from a 96-well-multiwell plate containing an explant from human omental adipose tissue. The region of the explant (A), and of capillary growth at day 7 (B), and day 11 (C) postembedding is delineated. The areas are calculated for the selected regions highlighted in red. (D) Calculated areas of 34 explants from the same tissue sample growing in the same 96-well-multiwell plotted in a before–after format, revealing linear growth in all embedded explants over the culture period. (E) Scatter plot displaying the means and standard deviation of the values obtained for each explant at each time point, and values obtained after subtracting the area of the initial explant. Paired Student's *t*-test between time points reveals highly significant differences, which can be used to compare angiogenic potential among different donors.

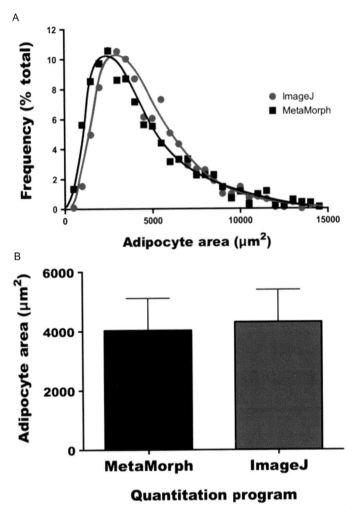

Sebastian D. Parlee et al., Figure 6.3 Adipocytes size and average area are similar when quantified by MetaMorph or ImageJ. Ten-week-old C57BL6/J mice ($N=5$) fed a 60% high-fat diet for 6 weeks were sacrificed and epididymal adipose tissue depots harvested, fixed, sectioned and adipocytes counted using either the MetaMorph or ImageJ method described herein. The frequency distribution of adipocyte sizes is equivalent whether analyzed by MetaMorph or ImageJ. No significant differences were found in either the frequency distribution (A, two-way ANOVA with Bonferroni post-hoc analysis) or average adipocyte area (B, Student's t-test).

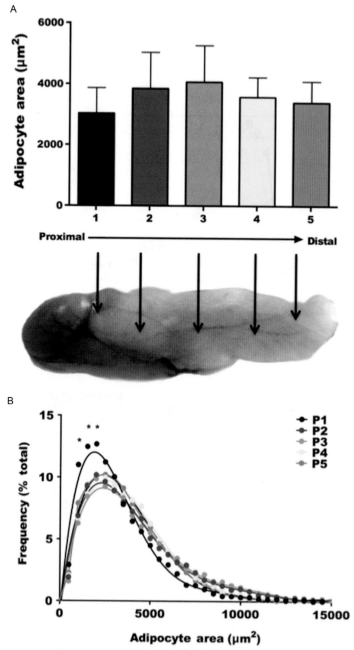

Sebastian D. Parlee et al., Figure 6.5 The frequency of small adipocytes but not the average adipocyte area changes in epididymal adipose tissue located on the testicle. Ten-week-old C57BL6/J mice fed a 60% high-fat diet for 6 weeks were sacrificed and epididymal WAT harvested and divided into five equal sections according to the proximity to the testicle before fixing, sectioning and counting of adipocytes (~1000 per sample) using the MetaMorph method described herein. Whereas the average size of adipocytes did not differ between samples (A), there was a higher frequency of small adipocytes (1000–2000 μm^2) directly adjacent to the testicle (labeled 1). Two-way ANOVA with Bonferroni post-hoc analysis, $^*p<0.05$ compared to samples 2–5.

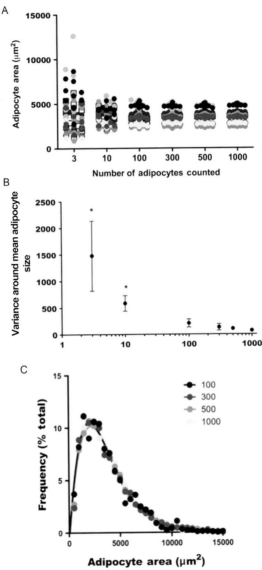

Sebastian D. Parlee et al., Figure 6.6 Counting the area of as few as 100 adipocytes provides an accurate distribution of adipocyte size in WAT. Ten-week-old ($N=10$) C57BL6/J mice fed a 60% high-fat diet for 6 weeks were sacrificed and epididymal WAT harvested, fixed and sectioned. For each individual animal 3, 10, 100, 300, 500 or 1000 adipocytes were randomly counted 10 times and the average adipocyte area calculated. While there was no significant difference in the mean adipocyte area (A) when counting between 3 and 1000 adipocytes, the variance around this mean is significantly greater when counting 3 or 10 adipocytes (B). The distribution of adipocyte areas, however, does not differ whether you count 100, 300, 500 or 1000 adipocytes. Accordingly counting a minimum of 100 adipocytes is sufficient to estimate mean adipocyte size, minimize variance around that mean and provide an accurate estimation of the distribution of adipocytes in WAT. One-way ANOVA with Tukey post hoc analysis, *$p<0.05$ versus 1000 adipocytes counted.

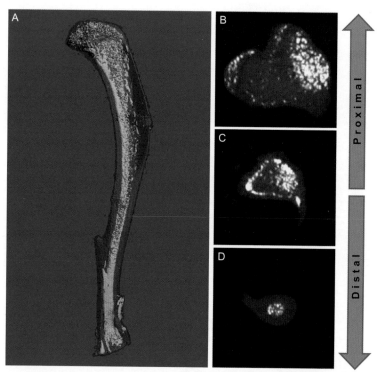

Erica L. Scheller et al., Figure 7.1 Distribution of MAT in the medullary canal. Adipocytes in the BM of the mouse are unevenly distributed throughout the medullary canal. They are most densely clustered in the epiphyses. In the metaphysis and diaphysis, adipocytes are most numerous near the central vascular canal and adjacent to the cortical bone. (A) Three-dimensional reconstruction of a 16-week-old C3H mouse osmium-stained tibia. Light blue, bone; white, MAT. (B–D) Transverse sections of the same bone from more proximal (toward the knee) to distal (toward the ankle) showing defined regional clustering of the marrow adipocytes (white).

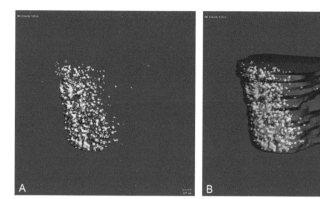

Erica L. Scheller et al., Figure 7.2 Osmium-stained and micro-CT-imaged MAT from Ebf1 null tibia. (A) Osmium-stained and micro-CT image of the proximal tibia from a 4-week-old $Ebf1^{-/-}$ mouse (no bone overlay). $Ebf1^{-/-}$ mice have very high MAT. (B) MAT from (A) with the bone overlaid.

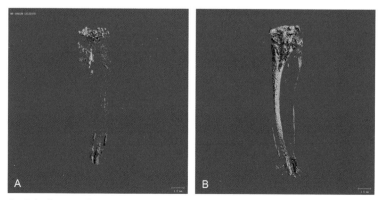

Erica L. Scheller et al., Figure 7.3 Osmium-stained and micro-CT-imaged MAT from C57BL/6 and C3H/HeJ. (A) Osmium-stained and micro-CT image of the tibia from 15-week-old B6 mice (no bone overlay). (B) Osmium-stained and micro-CT image of the tibia from 15-week-old C3H mice (no bone overlay).

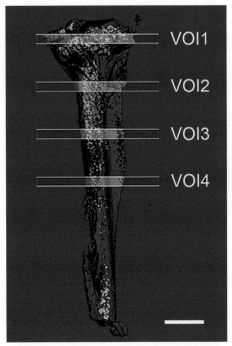

Erica L. Scheller *et al.*, Figure 7.4 Positioning of the volumes of interest. VOIs in mouse femur.

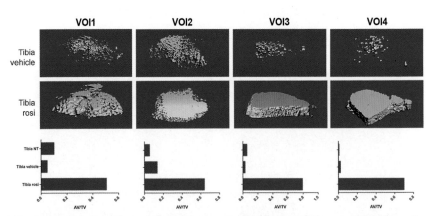

Erica L. Scheller *et al.*, Figure 7.5 Osmium-stained and micro-CT image of MAT from the femur of C57BL/6 mice fed a diet-containing rosiglitazone or control diet. B6 mice were fed a control diet or a diet-containing rosiglitazone for 8 weeks. Rosiglitazone is a PPARγ agonist and as such a potent inducer of MAT. The tibia was collected, stained with osmium tetroxide, and imaged with micro-CT. On control diet (vehicle) MAT is highest in VOI2 and decreases as you move distally down the shaft (VOIs 3 and 4). In contrast, mice on a rosiglitazone diet have very high MAT over the length of the tibial shaft.

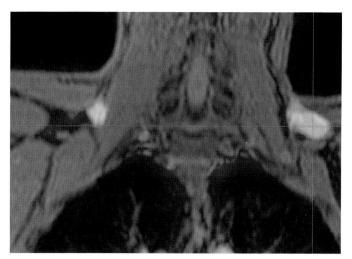

Magnus Borga et al., Figure 8.1 An example PET/MR image of a young male with significant activation of supraclavicular adipose tissue during cold exposure. The light color indicates high uptake of ^{18}FDG.

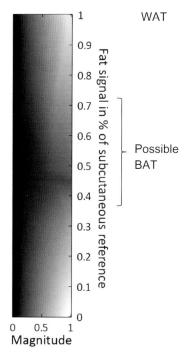

Magnus Borga et al., Figure 8.2 Relative fat content (RFC) color map used on the fat images.

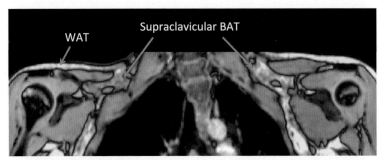

Magnus Borga *et al*., Figure 8.3 Coronal section from an adult male subject with indications of supraclavicular BAT.

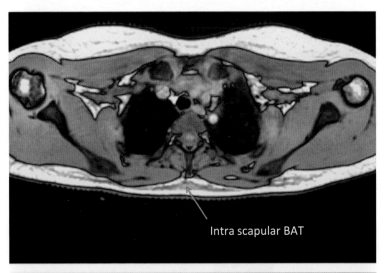

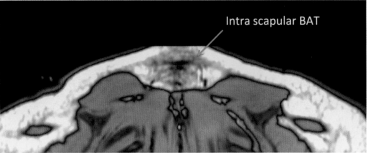

Magnus Borga *et al*., Figure 8.4 Axial and coronal sections from a male subject with indication of intrascapular BAT.

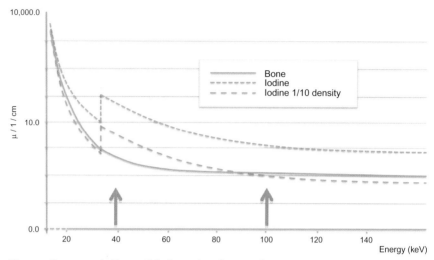

Magnus Borga et al., Figure 8.5 Assuming the use of mono energetic X-rays, at approximately 100 keV, the same linear attenuation coefficients are measured for bone and iodine. Data acquired at approximately 40 keV allows the differentiation of the two materials, regardless of their respective densities.

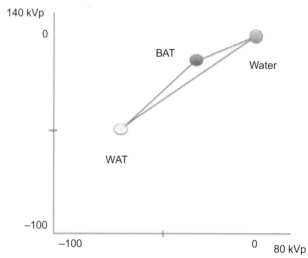

Magnus Borga et al., Figure 8.6 Calibration diagram based on WAT, BAT, and water.

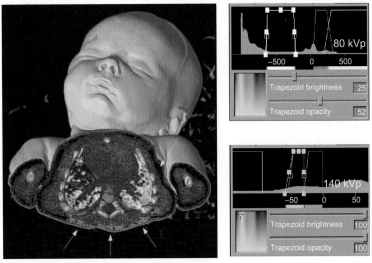

Magnus Borga et al., Figure 8.8 The same child as in Fig. 8.7. Datasets acquired at 80 and 140 kVp were loaded into a Siemens multimodality workstation with volume rendering 3D software (Inspace) in merged mode. For each dataset, voxels were mapped with green color for BAT and black color for WAT. Settings; WAT: 140 kVp, −50HU, 80 kVp, −75HU and for BAT: 140 kVp, −20HU, 80 kVp, −35HU. BAT can be separated from WAT.

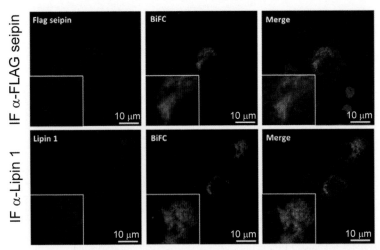

M.F. Michelle Sim et al., Figure 9.2 BiFC analysis of the interaction between seipin and lipin 1 in developing adipocytes. Differentiating 3T3-L1 adipocytes were cotransfected with the Seipin-C-Yn construct with a FLAG tag incorporated at the N-terminus and the Lipin-C-Yc construct. Seipin-C-Yn was detected using antibodies to the FLAG (upper panels), Lipin-C-Yc was detected using an anti-lipin 1 antibody (lower panels). Centre panels show the reconstituted YFP fluorescence, reporting the interaction of seipin and lipin in each case.

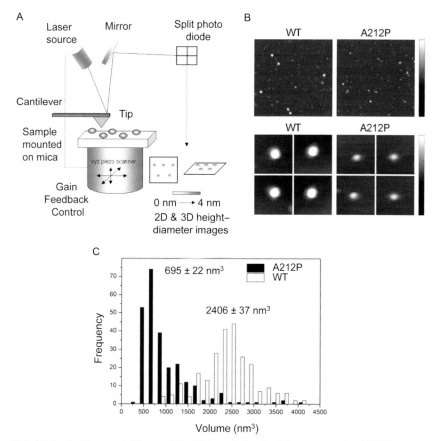

M.F. Michelle Sim et al., **Figure 9.3** AFM analysis of wild-type and the A212P mutant form of seipin. (A) Schematic illustration of AFM analysis. (B) tsA 201 cells were transfected with the long form of wild-type seipin with an N-terminal triple-FLAG tag and a C-terminal Myc tag, or the A212P pathogenic mutant form of this protein that causes lipodystrophy. Representative AFM images of isolated proteins are shown in the upper panels. Scale bar, 200 nm; color-height scale, 0–5 nm. Lower panels show galleries of zoomed images of the wild-type and A212P mutant form of seipin. Scale bar, 20 nm; color-height scale, 0–5 nm. (C) Frequency distributions of molecular volumes of wild-type (white bars) and A212P (black bars) seipin. The means of the distributions ($\pm$SEM) are indicated.

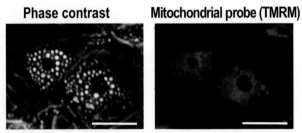

Miwako Nishio and Kumiko Saeki, Figure 10.2 Mitochondrial staining of hESC/hiPSC-derived BA. hESC-derived BAs (left) were stained by tetramethyl rhodamine methyl ester (TMRM) (Life Technologies, Inc.), which detects mitochondria of living cells (right). Similar results were obtained from hiPSC-derived BAs (data not shown). Scale bars indicate 50 μm.

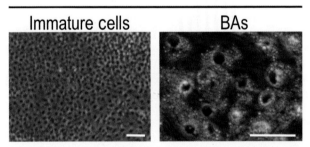

Miwako Nishio and Kumiko Saeki, Figure 10.3 Oil red O staining. Immature hESCs (left) or hESC-derived BAs (right) were subjected to oil red O staining. Similar results were obtained from hiPSC-derived BA (Nishio et al., 2012). Scale bars indicate 50 μm.

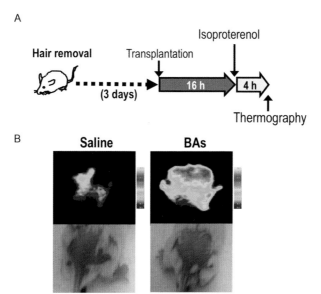

Miwako Nishio and Kumiko Saeki, Figure 10.5 Calorigenic assays *in vivo*. (A) The experimental procedure. (B) Saline (left) or hESC-derived BAs (right) were subcutaneously injected to ICR mice and thermographic assessment was performed. Similar calorinogenic activity was detected in hiPSC-derived BA but not in immature hESC or immature hiPSC (Nishio et al., 2012).

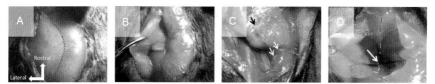

Cheryl H. Vaughan *et al.*, Figure 11.1 IBAT nerve identification. The orientation of all pictures is the dorsal, interscapular surface. (A) Picture of both IBAT lobes. Dotted line delineates left from right IBAT. (B) Forceps revealing the ventral surface of the left IBAT lobe. (C) Picture of right IBAT and associated intercostal nerves supplying sympathetic and sensory innervation (black arrow, 3 nerves; white arrow, 1 nerve each). D. Sulzer's vein draining both BAT lobes (white arrow). Dotted line depicts midline and medial borders of both BAT lobes.

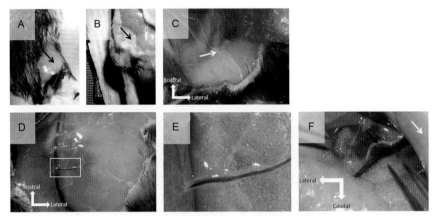

Cheryl H. Vaughan et al., Figure 11.2 IWAT nerve identification. Dorsal (A) and ventral (B) views of animal. Black arrow denotes IWAT pad location. (C) Dorsolateral view of nerves innervating (white arrow) fat and skin. (D) Medial border of the right IWAT (midline to the left of picture). (E) Magnification of white box in D depicting nerves on either side of blood vessel. (F) Ventral surface showing nerve that bifurcates sending one branch to leg and one to IWAT. Do not cut this nerve (black arrow). Peritoneal cavity to the right of this picture (white arrow).

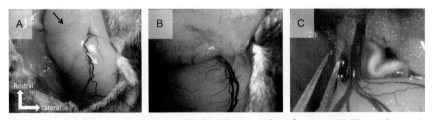

Cheryl H. Vaughan et al., Figure 11.3 EWAT nerve identification. (A) The right testis (white arrow) and right EWAT (black arrow). (B) Toluidine blue dye highlighting nerve innervating EWAT. (C) Forceps holding nerve to EWAT that should not be cut. If cut, loss of testis function occurs.

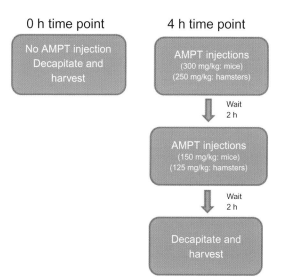

Cheryl H. Vaughan et al., Figure 11.4 Timeline of AMPT injections for NETO measurement.

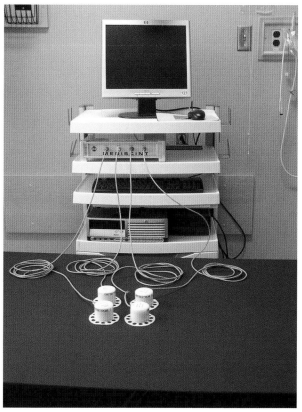

Richard Sotornik and Jean-Luc Ardilouze, Figure 12.1 Mediscint, Oakfield Instruments, Eynsham, UK.

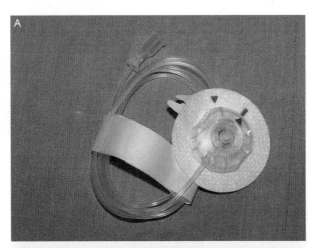
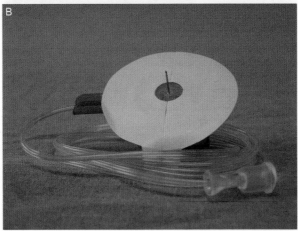

Richard Sotornik and Jean-Luc Ardilouze, Figure 12.2 Quick-set Infusion Set; Minimed, Medtronic of Canada Ltd., Mississauga, Canada.

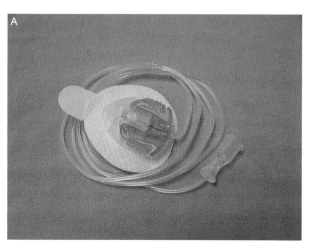

Richard Sotornik and Jean-Luc Ardilouze, Figure 12.3 Inset II Infusion Set, Animas Corporation, LifeScan Canada Ltd., Burnaby, British Columbia, Canada.

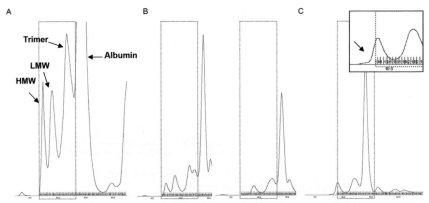

Joseph M. Rutkowski and Philipp E. Scherer, Figure 13.2 Example FPLC gel filtration elution curves assessed at 280 nm detection. Samples are plotted as mAU (*y*-axis) versus volume (*x*-axis) with the red dashed box indicating the first 24 collected eluted fractions. (A) HMW, LMW, and trimeric adiponectin each elute as a distinct protein peak ahead of albumin. (B) Lower sample loading volumes result in more unique peaks (left), but too low of plasma volume coupled with low adiponectin makes discerning the peaks difficult (right). Here, this sample may contain limited HMW adiponectin. (C) Improper elution collection setting can result in adiponectin lost to the waste stream. The first 24 collected contain all of the serum albumin and fail to capture nearly half of the HMW adiponectin (inset, arrow).

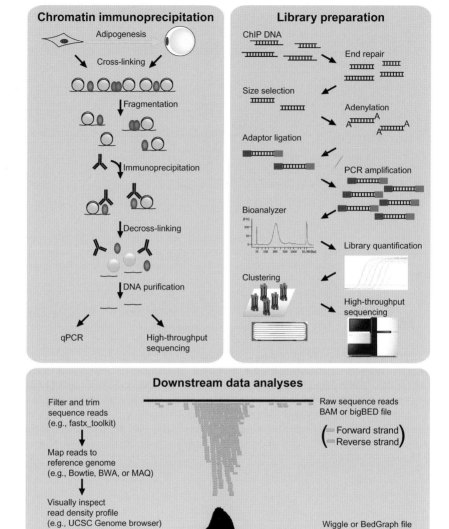

Ronni Nielsen and Susanne Mandrup, Figure 14.1 Schematic representation of the ChIP-seq workflow—ChIP, library preparation, and downstream data analyses. See text for details.

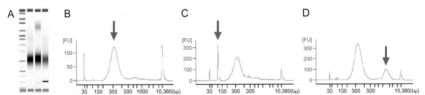

Ronni Nielsen and Susanne Mandrup, Figure 14.2 Bioanalyzer quality control. (A) Bioanalyzer data represented as gel image. (B) Bioanalyzer trace showing the DNA fragments present in the libraries. Red arrow indicates library peak ready for sequencing. (C) Sample containing adaptor dimers (red arrow) indicating the need for further cleanup. (D) Library with extra high molecular band. This will not affect sequencing results but indicates over amplification of the sample (i.e., need to run fewer PCR cycles). FU, fluorescence units.

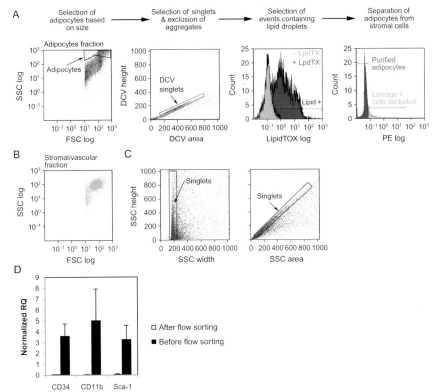

Susan M. Majka et al., Figure 15.3 Multistep flow cytometry/sorting strategy. (A) Gating strategy for adipocyte isolation is diagrammed from left to right. In the first step, adipocytes are identified by their large size and refractile properties in a plot of FSC versus SSC. A gate is placed around the population of cells that are larger and more refractile than those present in the stromal/vascular fractions (B). In the next step, DyeCycle Violet (DCV) fluorescence of the gated adipocytes is evaluated in a plot of peak height versus peak area (note the linear scale). The singlets that form a diagonal distribution are gated, while cell aggregates are ignored. Singlets can also be identified optically by comparing SSC (or FSC) peak height to peak width or peak area as shown in (C). In the third step, LipidTOX fluorescence of the singlets is evaluated on a histogram. We compare the fluorescence signal distribution of LipidTOX-stained cells to a small portion of unstained cells to positively identify events containing a lipid droplet. Finally, any remaining stromal contaminants are excluded based on their labeling with PE-conjugated antibodies to stromal markers. The purified adipocytes can undergo further analysis and/or sorting. (D) QRT-PCR was used to verify the staining and gating strategy. The results show the presence of several stromal cells markers in adipocyte fractions prior to sorting. These markers were virtually undetectable in flow-purified adipocytes.

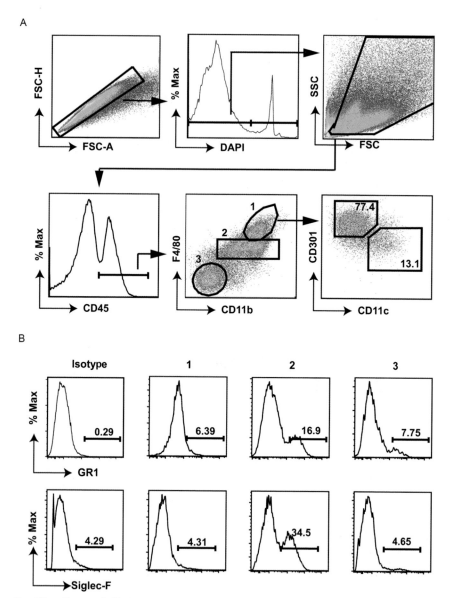

Kae Won Cho et al., Figure 16.1 Gating strategy to identify and characterize ATMs. (A) Gating strategy for the identification of M1 (CD11c$^+$) and M2 (CD301$^+$) ATMs according to the flow cytometry protocol described herein. (B) Histograms of Gr-1 and Siglec-F expression on CD11b^{high}F4/80^{high} (gate 1), CD11b$^{mid/high}$F4/80^{dim} (gate 2), and CD11b$^-$F4/80$^-$ (gate 3) SVCs demonstrates that the CD11b$^{mid/high}$F4/80^{dim} population (gate 2) can be contaminated with neutrophils (Gr-1) and eosinophils (Siglec-F).